Biological Identity

Analytic metaphysics has recently discovered biology as a means of grounding metaphysical theories. This has resulted in long-standing metaphysical puzzles, such as the problems of personal identity and material constitution, being increasingly addressed by appeal to a biological understanding of identity. This development within metaphysics is in significant tension with the growing tendency among philosophers of biology to regard biological identity as a deep puzzle in its own right, especially following recent advances in our understanding of symbiosis, the evolution of multi-cellular organisms and the inherently dynamical character of living systems. Moreover, and building on these biological insights, the broadly substance ontological framework of metaphysical theories of biological identity appears problematic to a growing number of philosophers of biology who invoke process ontology instead.

This volume addresses this tension, exploring to what extent it can be dissolved. For this purpose, the volume presents the first selection of essays exclusively focused on biological identity and written by experts in metaphysics, the philosophy of biology and biology. The resulting cross-disciplinary dialogue paves the way for a convincing account of biological identity that is both metaphysically constructive and scientifically informed, and will be of interest to metaphysicians, philosophers of biology and theoretical biologists.

Anne Sophie Meincke is a Senior Research Fellow at the Department of Philosophy of the University of Vienna. She works on metaphysics, philosophy of biology, philosophy of mind and action and their respective intersections. Her recent publications include the article "Autopoiesis, Biological Autonomy and the Process View of Life" (2019) and the edited volume *Dispositionalism: Perspectives from Metaphysics and the Philosophy of Science* (2020).

John Dupré is Professor of Philosophy of Science and Director of the Centre for the Study of Life Sciences (Egenis) at the University of Exeter. His main field of expertise is the philosophy of biology, but he also has a long-standing interest in metaphysics. His recent publications include *Processes of Life* (2012) and *Everything Flows: Towards a Processual Philosophy of Biology* (2018), co-edited with Daniel Nicholson.

History and Philosophy of Biology
Series Editor: **Rasmus Grønfeldt Winther** *is Associate Professor of Humanities at the University of California, Santa Cruz (UCSC).*

This series explores significant developments in the life sciences from historical and philosophical perspectives. Historical episodes include Aristotelian biology, Greek and Islamic biology and medicine, Renaissance biology, natural history, Darwinian evolution, Nineteenth-century physiology and cell theory, Twentieth-century genetics, ecology, and systematics, and the biological theories and practices of non-Western perspectives. Philosophical topics include individuality, reductionism and holism, fitness, levels of selection, mechanism and teleology, and the nature-nurture debates, as well as explanation, confirmation, inference, experiment, scientific practice, and models and theories vis-à-vis the biological sciences.

Authors are also invited to inquire into the "and" of this series. How has, does, and will the history of biology impact philosophical understandings of life? How can philosophy help us analyze the historical contingency of, and structural constraints on, scientific knowledge about biological processes and systems? In probing the interweaving of history and philosophy of biology, scholarly investigation could usefully turn to values, power, and potential future uses and abuses of biological knowledge.

The scientific scope of the series includes evolutionary theory, environmental sciences, genomics, molecular biology, systems biology, biotechnology, biomedicine, race and ethnicity, and sex and gender. These areas of the biological sciences are not silos, and tracking their impact on other sciences such as psychology, economics, and sociology, and the behavioral and human sciences more generally, is also within the purview of this series.

Biological Identity
Perspectives from Metaphysics and the Philosophy of Biology
Edited by Anne Sophie Meincke and John Dupré

Philosophical Perspectives on the Engineering Approach in Biology
Living Machines?
Edited by Sune Holm and Maria Serban

For more information about this series, please visit: https://www.routledge.com/History-and-Philosophy-of-Biology/book-series/HAPB

Biological Identity
Perspectives from Metaphysics and
the Philosophy of Biology

**Edited by
Anne Sophie Meincke and
John Dupré**

LONDON AND NEW YORK

First published 2021
by Routledge
2 Park Square, Milton Park, Abingdon, Oxon OX14 4RN

and by Routledge
605 Third Avenue, New York, NY 10017

First issued in paperback 2022

Routledge is an imprint of the Taylor & Francis Group, an informa business

© 2021 selection and editorial matter, Anne Sophie Meincke and John Dupré; individual chapters, the contributors

The right of Anne Sophie Meincke and John Dupré to be identified as the authors of the editorial material, and of the authors for their individual chapters, has been asserted in accordance with sections 77 and 78 of the Copyright, Designs and Patents Act 1988.

All rights reserved. No part of this book may be reprinted or reproduced or utilised in any form or by any electronic, mechanical, or other means, now known or hereafter invented, including photocopying and recording, or in any information storage or retrieval system, without permission in writing from the publishers.

Trademark notice: Product or corporate names may be trademarks or registered trademarks, and are used only for identification and explanation without intent to infringe.

Publisher's Note
The publisher has gone to great lengths to ensure the quality of this reprint but points out that some imperfections in the original copies may be apparent.

British Library Cataloguing-in-Publication Data
A catalogue record for this book is available from the British Library

Library of Congress Cataloging-in-Publication Data
A catalog record has been requested for this book

ISBN: 978-0-367-49503-9 (pbk)
ISBN: 978-1-138-47918-0 (hbk)
ISBN: 978-1-351-06638-9 (ebk)

DOI: 10.4324/9781351066389

Typeset in Times New Roman
by codeMantra

To Friedrich, Gabe and Julian

Contents

List of figures ix
List of contributors xi
Series editor's foreword xiii
RASMUS GRØNFELDT WINTHER

Foreword xv
ALAN C. LOVE

Acknowledgements xvii

1 **Biological identity: why metaphysicians and philosophers of biology should talk to one another** 1
ANNE SOPHIE MEINCKE AND JOHN DUPRÉ

2 **Siphonophores: a metaphysical case study** 22
DAVID S. ODERBERG

3 **Biological individuals as "weak individuals" and their identity: exploring a radical hypothesis in the metaphysics of science** 40
PHILIPPE HUNEMAN

4 **What is the problem of biological individuality?** 63
ERIC T. OLSON

5 **The role of individuality in the origin of life** 86
ALVARO MORENO

6 **The being of living beings: foundationalist materialism versus hylomorphism** 107
DENIS M. WALSH AND KAYLA WIEBE

Contents

7 The origins and evolution of animal identity 128
STUART A. NEWMAN

8 Processes within processes: a dynamic account of living beings and its implications for understanding the human individual 149
JOHN DUPRÉ

9 Activity, process, continuant, substance, organism 167
DAVID WIGGINS

10 Diachronic identity in complex life cycles: an organisational perspective 177
JAMES DIFRISCO AND MATTEO MOSSIO

11 Pregnancy and biological identity 200
ELSELIJN KINGMA

12 Processual individuals and moral responsibility 214
ADAM FERNER

13 The nature of persons and the nature of animals 233
PAUL F. SNOWDON

14 Processual animalism: towards a scientifically informed theory of personal identity 251
ANNE SOPHIE MEINCKE

Index 279

Figures

3.1	The space of organismality	46
3.2	Ecological space of individuality	51
3.3	Ecological interactions	53
4.1	[No title]	65
10.1	*Drosophila* metamorphosis	182
10.2	Fission and fusion	188
10.3	Sexual reproduction	193

Contributors

James DiFrisco, Centre for Logic and Philosophy of Science (CLPS), KU Leuven, Belgium

John Dupré, Centre for the Study of Life Sciences (Egenis), University of Exeter, UK

Adam Ferner, Independent researcher and author, London, UK

Philippe Huneman, Centre National de la Recherche Scientifique (CNRS), Paris, and Université Paris 1, Sorbonne, France

Elselijn Kingma, Department of Philosophy, University of Southampton, UK, and University of Eindhoven, Netherlands

Anne Sophie Meincke, Department of Philosophy, University of Vienna, Austria, and Centre for the Study of Life Sciences (Egenis), University of Exeter, UK

Alvaro Moreno, IAS Research Centre for Life, Mind and Society, University of the Basque Country, San Sebastián, Spain

Matteo Mossio, Centre National de la Recherche Scientifique (CNRS), Paris, and Institut d'Histoire et de Philosophie des Sciences et des Techniques (IHPST), Paris, France

Stuart A. Newman, Department of Cell Biology, New York Medical College, Valhalla, New York, USA

David Oderberg, Department of Philosophy, University of Reading, UK

Eric T. Olson, Department of Philosophy

Paul F. Snowdon, Department of Philosophy, University College London, UK

Denis Walsh, Department of Philosophy, Institute for the History and Philosophy of Science and Technology, and Department of Ecology and Evolutionary Biology, University of Toronto, Canada

Kayla Wiebe, Department of Philosophy, University of Toronto, Canada

David Wiggins, Faculty of Philosophy, New College, University of Oxford, UK

Series editor's foreword

Rasmus Grønfeldt Winther

In this volume, co-editors Anne Sophie Meincke and John Dupré gather an impressive array of scholars to address one of the burning themes of contemporary philosophy of biology: *biological identity*. Productively connecting philosophy of biology with analytic metaphysics and ontology, the book traces a complex—and sometimes esoteric—conceptual landscape, while always providing stimulation and inspiration for the journey.

Central concepts covered within include complex systems, self-organisation, persistence, time, process, and inheritance. Where does a living entity begin and end? At what levels of organisation or composition do different kinds of beings exist, and how overlapping, interactive, and easily identifiable are these levels? Could an ecosystem or even Gaia—the earth as a whole—be identified as a biological process or entity? These are some of the questions this volume helps us address.

The series so far expresses diversity and depth: from studies of the fundamental nature of biology to explorations of the relations of biology to philosophical matters such as moral realism, action theory and personal identity, to historical studies of Darwinism and of concrete figures such as William James and Jacob von Uexküll. Meincke and Dupré's volume exemplifies the work showcased in the History and Philosophy of Biology series and opens up avenues for its possible future directions.

Ecology and health are topics we'd like to see addressed more in future volumes. An Amazon rainforest or Australian wildland burning out of control. An Arctic melting at an ever-increasing pace. Oceans acidifying and dying in spots. This is the ecological state of affairs in 2020. Behind it all lies a restless and aggressive species, but one that is also capable of foresight and concerted positive action. What is our role in all of this? What do we owe non-human biological entities, and why might we not be living up to this moral and political responsibility? How does all this affect our health, and our future prospects? How can the history and philosophy of biology help us conceptualise our challenges better and help us work towards concerted, collective positive action?

Foreword

Alan C. Love

Once upon a time (i.e., in the 20th century), professional philosophers were in agreement about a set of fundamental questions and the methodology of formal analysis by which to tackle them. Every day they would apply their methodology to these questions, such as the nature of personal identity and material constitution, and vigorously argue about proposed answers. To the surprise of many, fate shined on the profession and the overall number of philosophers in the academy grew substantially, despite—according to some naysayers—the seeming lack of progress in answering these fundamental questions. However, in part because of their increasing numbers, differences began to emerge among professional philosophers about how to understand their research questions and whether other, distinct methodologies might be needed to answer them. One group found that attending to developments in the natural sciences encouraged a fruitful rethinking of these questions and reconfigured what kinds of answers might be appropriate. Another group thought that philosophy should remain separate from other areas of inquiry and concentrate on *a priori* considerations about what is possible to secure answers to perennial questions. Although both groups described themselves as professional philosophers, they rarely interacted and sometimes intentionally avoided one another by creating affiliations and societal meetings where they would only see those who were likeminded. The situation induced puzzlement in casual observers and fellow professional academics, especially because they appeared to arrive at such discordant conclusions (e.g., realism versus anti-realism about biological identity). Alas, many were convinced that there is no accounting for philosophers. They always had been a peculiar breed.

Fables like this might make good bedtime stories but almost always are bad history. Yet they can play the role of consciousness-raising about our current situation, similar to how Dr Seuss's *The Lorax* prompted reflection on environmental degradation (at least for some). The volume in your hands (or on your screen) is a rigorous and scholarly response to this balkanised situation in the specific context of philosophical questions about identity and individuality. It succeeds in this consciousness-raising far better than my poorly constructed fable. Through an interweaving (literally) of chapters from both analytic metaphysicians and philosophers of biology, it demonstrates the value of bringing both groups together in order to dialogue about these issues. In particular, it shines a light on difficulties with the animalist appeal to biological identity, points to divergent interpretations of complex biological phenomena (e.g., holobionts or colonial invertebrates with complex life cycles), accents controversies about appeals to process ontology or scientific inquiry to help resolve debates about individuality, and exposes fault lines about whether or not neo-Aristotelian metaphysics is consistent with discoveries in contemporary biological science and how it relates to process ontology.

Importantly, this demonstrated value does not imply that all differences have been reconciled or that consensus was achieved. However, the format brings into relief critical features that often remain implicit, including differences in how research questions are understood and differences in the evaluative standards brought to bear on putative answers. For example, some metaphysicians concentrate on the question of what spatiotemporal regions are occupied by an organism, while some philosophers of biology focus on articulating a definition of what it means to be a biological organism. That the questions differ does not mean only one of them is genuinely philosophical. And recognising their distinctness also fosters recognition of distinct standards for evaluating answers. Some metaphysicians demand complete generality for an adequate account of identity, whereas some philosophers of biology require that any adequate account embrace the scientific successes of individuation, however non-general or partial these might be. For the former, individuality is typically an all or nothing matter; for the latter, individuality can exist in degrees as a continuous quantity. The success of dialogue is not measured in agreement on every premise, but rather in identifying explicitly these points of divergence to recognise the true sources of disagreement. Only then is there some hope of moving forward and advancing conceptual understanding.

The editors are to be commended for assembling this volume and bringing these contributors into conversation with one another. It sets a necessary framework for research going forward:

> [...] any metaphysical theory of biological identity, if it is to be convincing, must pay attention to the relevant body of scientific knowledge [...] [and] biological research and its accompanying philosophical reflection in the philosophy of biology are shaped and guided by metaphysical presumptions.
>
> (Meincke and Dupré, this volume, p. 4f.)

More scholarship of this type is desperately needed on a variety of topics beyond personal identity and material constitution. Assuming this book is a harbinger of things to come, then the bedtime story just might end "happily ever after".

Acknowledgements

This book originates from a conference on "Biological Identity" co-organised by the editors of this volume and held in June 2016 at the Institute of Philosophy, London, UK. The conference was the first to bring together metaphysicians and philosophers of biology to discuss questions concerning biological identity. We are most grateful to the Institute for funding the conference through their 2015/16 annual conference grant as well as to the Mind Association and the British Society for the Philosophy of Science (BSPS) for additional funding. We would also like to thank all conference speakers and the audience for making the conference such a memorable, inspiring event.

Our interest in initiating a dialogue between metaphysicians and philosophers of biology on the topic of biological identity grew out of our work carried out under a grant from the European Research Council under the European Union's Seventh Framework Programme (FP7/2007-2-13), Grant Agreement 324186 ("A Process Ontology for Contemporary Biology" (PROBIO)), on which John Dupré was the Principal Investigator, and Anne Sophie Meincke was a Research Fellow. Apart from contributing to the successful realisation of the conference, the grant was essential in allowing us to pursue the plan of making the conference papers available to a wider audience as a book. We are very grateful to the ERC for its support.

The book manuscript developed over several years and its completion was made possible by a Senior Research Fellowship Anne Sophie Meincke held on a research project led by Elselijn Kingma and funded by the European Research Council under the European Union's Horizon 2020 Framework Programme, Grant Agreement 679586 ("Better Understanding the Metaphysics of Pregnancy" (BUMP)), as well as, finally, by an "Elise Richter" research grant which Anne Sophie Meincke received from the Austrian Science Fund, Grant Agreement V-714 ("Bio-Agency and Natural Freedom"). The support of both funding institutions is greatly appreciated.

Most chapters in this volume developed from papers given at the aforementioned conference, and we herewith offer our deepest thanks to the authors for agreeing to contribute their conference papers to this book, and for their dedication and patience throughout the process, from initial submission to the revisions phase to production. We would also like to express our sincere gratitude to those authors who did not speak at the conference for their willingness to add their expertise to this project, even at short notice, and to Cambridge University Press for allowing us to reprint an essay by David Wiggins.

Ambitious projects like the present one draw upon inspirational intellectual environments. In this regard, we foremost would like to thank everyone involved with the PROBIO project, and especially the other Research Fellows, Stephan Guttinger and Daniel Nicholson. Anne Sophie Meincke also wishes to express her gratitude to Elselijn Kingma for stimulating intellectual exchange, including controversial discussions on process ontology, and to the whole BUMP team, Teresa Baron, Suki Finn, Alexander Geddes and Sigmund Schilpzand, for being such inspiring and supportive colleagues. Anne Sophie Meincke has additionally benefitted greatly from discussing questions of biological identity with Antony Galton, Scott Gilbert, Matt Haber, Johannes Jaeger, Alan C. Love, Alvaro Moreno, Eric T. Olson, Thomas Pradeu, Peter Simons, Helen Steward, Peter van Inwagen and David Wiggins. John Dupré has benefitted from discussions of related matters with more people than he can keep track of, including many of those just mentioned as well as a number of colleagues in the Centre for the Study of Life Sciences (Egenis), which he directs, and in the concentric circles of department, college and university in which the Centre is located. He wishes to apologise to all of those not mentioned by name in these acknowledgements.

Last but not least we are indebted to Rasmus Grønfeldt Winther for including this volume in Routledge's *History and Philosophy of Biology* book series, and for his capable and tolerant supervision of the project; to Routledge for providing a congenial environment for the series and this book and to Alan Love for writing a foreword at very short notice.

There is no proper space in jointly written acknowledgments for the editors to thank each other, but we would like to conclude by saying that this has been a very fruitful collaboration and that we wish that the spirit of dialogue may spread further through the often too compartmentalised academic world.

1 Biological identity

Why metaphysicians and philosophers of biology should talk to one another

Anne Sophie Meincke and John Dupré

1.1 Biological identity in metaphysics and in the philosophy of biology

Our world seems populated by individual things that keep their identity over time. The car in which I will soon go home is the same that brought me to the office in the morning. The pebble which lies on the windowsill in my kitchen is the one which I collected at the seaside yesterday, and the rubber plant next to it was some years ago given to me as a birthday present. I also have not the slightest doubt that the dog happily welcoming me now is the same dog that I left alone at home a few hours earlier. And, not least, I am pretty sure that I myself was the very same person on all the various occasions just mentioned.

However, as so often in philosophy, what seems natural and straightforward at first glance, turns out on closer consideration to raise a host of problems. Am I really the same as, say, ten years ago? Have I not changed quite a lot since then? At any rate, I have aged; and the same is sadly true for my car which is at risk of failing the upcoming MOT test. How is it possible to stay the same while changing? What does it take for persons and cars to persist through time? Is persistence tantamount to numerical identity or to some weaker relation, say, relations of continuity? And if the latter, relations between what? As has famously been noted in discussions of identity, self-identity is not a relation between two things, but a necessary truth about one thing. But then, once we move away from strict self-identity, exactly how much continuity is needed for identity through time and how are we to handle hypothetical branching cases? Philosophers are a long way from agreeing on how to answer these questions. In fact, the debate on personal identity in analytic metaphysics is (in)famous for its notoriously aporetic character, and the corresponding debate about the persistence criteria of artefacts fares no better – just think of puzzles such as that of the ship of Theseus which already baffled the ancient Greeks.

Remarkably, an increasing number of metaphysicians currently express their confidence that the difficulties abate, or even vanish, when it comes to biological beings. So-called animalists suggest that key obstacles faced by standard psychological theories of personal identity can be avoided once it is acknowledged that we are animals or organisms, i.e., that the necessary and sufficient conditions of our identity over time are biological – either purely (Olson 1997, Snowdon 2014) or at least in part (Wiggins 2001). Instead of personal identity, the animalists maintain, we should focus on biological identity. This tendency towards a biological approach within the personal identity debate (see also Blatti and Snowdon 2016) is echoed by a parallel development in the metaphysical debate on the constitution of material objects. Peter van Inwagen (1990) has famously argued that inanimate things like pebbles or cars do not exist. In metaphysical terms, he argues, they turn out to be mere collections of particles, which is to say, not things or objects at all. If indeed any composite things exist, so the claim goes, then these are living organisms – such as my rubber plant and myself – due to the strong unity of biological life which makes each organism precisely one rather than many.

The metaphysical appeal to biological identity, which contrasts with the wide-spread scepticism among metaphysicians concerning personal identity and artefact identity, looks less surprising when we consider the broader intellectual context in which it is situated, the current revival of Aristotelian metaphysics. Opposing the empiricist-cum-linguistic orientation of large parts of contemporary metaphysics, a growing number of metaphysicians advocate a return to an a priori approach to metaphysics together with a commitment to Aristotelian concepts, such as substance, essence or potentiality, as real aspects – or "categories" – of being (Wiggins 1980, 2001, Lowe 2001, 2002, 2006, Oderberg 2007, Tahko 2012, Groff and Greco 2013, Novotný and Novák 2014). As is well known, Aristotle argued for the ontological priority of substances, i.e., of composite but individual particular things, over both simple particular things ("atoms") and universal abstract things or concepts ("universals", Platonic "ideas" or "forms").[1] Interestingly, this incorporates the view that living things – organisms – are the paradigmatic cases of substances (Moya 2000, Cohen 2002, chapter 5), whereas Aristotle remained ambivalent with respect to the question of whether artefacts possess the status of substances (see Katayama 1999 for a detailed analysis). These doctrines revolve around Aristotle's theory of hylomorphic unity according to which physical objects are compounds of matter (ὕλη) and form (μορφή), and they continue to influence contemporary metaphysical views of biological identity even where the reference to Aristotle is loose and superficial (as in van Inwagen 1990 and Olson 1997).

Nonetheless, the (re-)discovery of biological identity in contemporary metaphysics, and the optimism associated with it, remains astonishing from the perspective of another philosophical discipline, the philosophy of biology. As it happens, recent years have seen a lively debate in the philosophy of biology on biological identity – biological individuality with respect to both its synchronic and its diachronic dimension[2] – which has brought to light the numerous and intricate problems associated with the interpretation and application of this concept (see, e.g., Clarke 2010, 2013, Pradeu 2012, 2016, Bouchard and Huneman 2013, Ereshefsky and Pedroso 2016, Fagan 2016, Godfrey-Smith 2016, Guay and Pradeu 2016b, Haber 2016, Lidgard and Nyhart 2017, Paternotte 2016, Wilson and Barker 2019).[3] As a result, philosophers of biology are far from considering biological identity as a miracle cure for identity puzzles, typically regarding it rather as a puzzle in its own right. Two recent developments in biology that have been much discussed by philosophers of biology are mainly responsible for this.

First, studies in symbiosis and the evolution of multi-cellular organisms undermine the belief in the organism as a homogeneous, strongly unified unit that is strictly demarcated from its environment. Organisms, it begins to appear, are rather ultimately heterogeneous assemblies of diverse units tied together by varying degrees of cooperation into a more or less stable constellation with fuzzy and fluctuating boundaries (Dupré and O'Malley 2009, Queller and Strassmann 2009, Dupré 2010) – they are so-called holobionts (Margulis 1991, Gilbert, Sapp and Tauber 2012, Bordenstein and Theis 2015, Gilbert and Tauber 2016, Queller and Strassmann 2016, Skillings 2016) and, in the case of eusocial organisms, arguably also superorganisms (Wilson and Sober 1989, Hölldobler and Wilson 2009). At the same time, it is far from clear that individual organisms, if there are any, are the relevant units upon which evolution acts, rather than genes, genomes, cells, groups, species or perhaps all of these (Dawkins 1976, Wilson and Sober 1994, Okasha 2006, Godfrey-Smith 2009; see Lloyd 2017a for an informative overview of the discussion). In particular the hypothesis that natural selection operates at the level of the holobiont (Zilber-Rosenberg and Rosenberg 2008, Doolittle and Booth 2017, Gilbert, Rosenberg, and Zilber-Rosenberg 2017, Lloyd 2017b, Suárez 2018), or even at the level of the superorganism (Wilson and Sober 1989, Folse and Roughgarden 2010, Haber 2013), is complemented and amplified by the recent appreciation of the frequency and evolutionary importance of epigenetic inheritance, developmental plasticity and niche construction (Odling-Smee, Laland and Feldman 2003, Fusco and Minelli 2010, Gilbert 2014, Chiu and Gilbert 2015, Jablonka 2017).[4]

Second, an important strand of systems biology, even while still relying on the concept of an individual organism,[5] stresses the dynamical and environment-dependent character of organisms (Alberghina and Westerhoff 2005, Noble 2006, 2017, Boogerd et al. 2007). Organisms are described as complex hierarchies of biological processes interacting with each other and with environmental processes so as to keep the system as a whole in a far-from-equilibrium state. Accordingly, if there is such a thing as the identity of biological dynamical systems, this reveals itself to be a hard-won achievement, constantly constituted and maintained by the system itself, specifically by maintaining a controlled exchange of matter and energy with the environment. It therefore comes as no surprise that some philosophers of biology, including the editors of this volume, have called for abandoning underlying traditional substance metaphysical conceptions in favour of a new process metaphysical framework regarded as more suitable for the description and understanding of biological phenomena and possibly of reality in general (Bickhard 2011, Dupré 2012, 2017, Jaeger and Monk 2015, Nicholson and Dupré 2018, Meincke 2018, 2019a, 2019b). This process ontological turn in the philosophy of biology is accompanied by a rising awareness of the role scientific practices – qua processes of individuation – play in conceptualising identity and individuality in biology as well as in other sciences (Bueno, Chen and Fagan 2018), followed by calls for a pluralistic stance on individuation and individuality (Dupré 2018, Kaiser 2018, Love 2018 and Waters 2018 (all in Bueno, Chen and Fagan 2018)).

Comparing the perspectives on biological identity in metaphysics and the philosophy of biology, we thus find a striking tension: while Aristotle-inspired metaphysicians tend to be confident that biological identity is a robust part of reality on which we can rely to solve long-standing metaphysical problems such as those of personal identity and material constitution, philosophers of biology, engaging with the latest research in biology, have unveiled the manifold and intricate challenges for a satisfying account of biological identity – challenges that may well make one wonder if there is such a thing as biological identity at all. One may ask, at least, whether there is any unique, objectively given kind of biological identity, or rather various kinds of continuities in the living nexus that may be picked out and employed for particular purposes, an idea that one of us has referred to as "promiscuous individualism" (Dupré 2012, p. 241). At the very least, the existence of competing catalogues of criteria of biological identity, together with the lack of consensus among philosophers of biology as to what the phenomena are to which the respective criteria of biological identity can be applied, indicates the immense difficulty of the task of providing a unified concept of biological identity in the light of recent scientific discoveries. From the point of view of the philosophy of biology, we can no longer appeal to any unproblematic notion of biological identity.

1.2 Why metaphysicians and philosophers of biology should talk to one another

One of the two main aims of the present volume is to investigate whether the tension between the metaphysicians' views of biological identity and those of the philosophers of biology can be resolved. Are we entitled to a realist view of biological identity? And if so, how does such a realist view relate to empirical facts about biological entities revealed by today's biology? We believe that any metaphysical theory of biological identity, if it is to be convincing, must pay attention to the relevant body of scientific knowledge. At the same time, we are strongly aware of the fact that both actual biological research and its accompanying philosophical reflection in the philosophy of biology are shaped and guided by metaphysical presumptions.[6] Thus, exactly the difficulties philosophers of biology have with conceptualising biological identity point, we think, towards fundamental metaphysical questions which need to be explicated and reflected upon. The conclusion we draw from this is that a dialogue is needed across the boundaries of these disciplines. Metaphysicians and philosophers of biology must talk to one another in order to better understand biological identity.

The present volume initiates such a cross-disciplinary dialogue on the subject of biological identity, by bringing together contributions from experts in metaphysics, the philosophy of biology and theoretical biology. As a matter of fact, so far the debates on biological identity in metaphysics and in the philosophy of biology have happened largely in isolation from one another. There has not been much exchange between the two disciplines at all since the philosophy of biology emerged as an independent field of study in the 1960s. Apart from the discussions of essentialism which began in the 1980s, philosophers of biology have only recently started more frequently to contextualise their research within metaphysical debates.[7] Metaphysicians, on the other hand, have only just started to engage with the philosophy of biology, typically with the intention to argue in favour of Aristotelian substance metaphysics (e.g., Oderberg 2007, Simpson et al. 2017, part 2). This includes recent attempts to demonstrate that contemporary biological theory actually supports key principles of Aristotelian substance metaphysics and its scholastic sequel (Boulter 2013) or even to reveal modern natural science, including biology, to be founded upon Aristotelian metaphysics (Feser 2019). This trend is echoed by a cautious but growing interest in (neo-)Aristotelian concepts among philosophers of biology (Ariew 2007, Walsh 2015).

These current movements in metaphysics and the philosophy of biology, indicating some tendency towards a mutual rapprochement, provide a conducive environment for the specific dialogue-initiating project of the present volume. However, they also prompt questions. First, there is the specific question of what particular potential a (neo-)Aristotelian approach may have for resolving the tension between the perspectives of the metaphysicians and the philosophers of biology on biological identity, or whether, rather, it is itself responsible for that tension: may (neo-)Aristotelian concepts help legitimise certain realist intuitions about biological identity or do they facilitate unwarranted claims that fly in the face of empirical findings?

Second, there is a general question about how metaphysics, the philosophy of biology and biology relate to one another. How do metaphysicians on the one hand and philosophers of biology and theoretical biologists on the other methodologically approach the problem of biological identity? What are the respective epistemic and conceptual resources they draw on?

The volume's second aim is to explore the prospects for fruitful interactions between metaphysics and the philosophy of biology with respect to biological identity but also in more general terms. This, as it were, metaphilosophical interest is fuelled by current discussions in particular about the status of metaphysics as an academic discipline: what is metaphysics and what is it good for? Do we need metaphysics at all and if so, in what form?

Recently, some critical voices have accused traditional analytic metaphysics of being a futile armchair exercise, and have called for replacing it with a metaphysics "motivated by currently pursued, specific scientific hypotheses, and having as its sole aim to bring these hypotheses advanced by the various special sciences together into a comprehensive world-view" (Ladyman and Ross 2010, p. vii). According to these scholars, metaphysics has a right to existence only as a "metaphysics of science" as thus defined. Against such moves, traditionally minded – and Aristotle-inspired – metaphysicians emphasise the logical priority of metaphysics to empirical science: "[...] science *presupposes* metaphysics [...]. Scientists inevitably make metaphysical assumptions, whether explicitly or implicitly, in proposing and testing their theories – assumptions which go beyond anything that science itself can legitimate" (Lowe 2001, p. 5). Exactly because metaphysics precedes science in this sense, it cannot do without a priori methods: metaphysics, as understood by the neo-Aristotelians, has "a non-empirical subject-matter" because it establishes what *possibly* exists in reality, while empirical science tells us – at least "tentatively and provisionally" – "which of the many incompatible possibilities for the fundamental structure of reality actually obtains" (Lowe 2002, p. 10f.).[8]

Our view is that there is an intermediate space between these two extreme conceptions, between an understanding of metaphysics as a strictly a priori venture on the one hand and what proponents of such an understanding typically dismiss as a "handmaid of science" view of metaphysics on the other, a view that restricts metaphysics to serving the epistemic interests of (natural) science. A major difficulty with the former is that no exploration of mere possibilities can ever get close enough to what happens to be real: not only are so many possibilities conceivable without being contradictory that it could only be by chance if we were to classify the right possible constellation of things as real, but it also remains unclear how any such theory of what is conceivably real could ever be comprehensive given its abstract nature. Peter Simons (2006) cogently demonstrates this problem with respect to mereological explanations of material constitution and parthood relations. Formal mereological principles, he argues, are not up to the task of answering the questions under what conditions two or more individuals compose another individual and what are the real parts of an individual, because both composition and parthood crucially involve various and specific aspects of causal and functional unity not captured by those principles.[9] As Simons points out:

> One may draw opposing conclusions from this fact. One conclusion is that because the composition principles are indeed so general and abstract, they need take no lessons from the discussion of concrete cases,

which pertain to material detail rather than formal ontology. The contrapositive conclusion, which I uphold, is that because the abstract considerations are so far removed from actual cases, they cannot be taken as reliable guides to a realistic ontology of part and whole.
(Simons 2006, p. 611)

We concur with Simons on this point. More generally put: constraints of metaphysical conceivability underdetermine biological reality, which tends to be much richer and also much more puzzling than we commonly imagine it to be. There is an inevitable element of surprise in the scientific study of biological reality. Frequently we discover, with the help of empirical methods, phenomena the existence of which we lacked the imagination to anticipate. And some of these phenomena are such that their discovery may well lead us to reconsider – and possibly revise – key ontological commitments and metaphysical principles. This emphatically applies to the scientific findings that challenge traditional concepts of biological identity.

If they "wish[...] to remain in accord with the real world" (Simons 2006, p. 597), metaphysicians thus are well advised to take account of, or even to take a lead from, the sciences. However, we do not think this implies that metaphysics collapses into science or into some theoretical extension thereof (perhaps called "metaphysics of science").[10] Metaphysics may be – and actually should be – scientifically informed but it is not thereby identical with the philosophy of the special sciences, nor with philosophy of science in general. Why? Because it asks different questions. These questions are different not, or at least not primarily, insofar as they concern different entities – often they concern the same entities as those that are studied by science. Metaphysical questions are rather located at a different level of abstraction; they have a different scope and pursue different epistemic interests.

The metaphysical questions about composition and parthood are good examples of this. Biologists and philosophers of biology do not normally ask under what conditions two or more individuals of *whatever* type would compose another individual of *whatever* type, nor what the real parts are of an individual of *whatever* type. Biologists and philosophers of biology are concerned with individuals of a *specific* type – biological individuals (if there are any), so they may ask questions about composition and parthood with respect to biological individuals; and if they do, this usually is meant to help answer other questions concerning biological entities, for instance, questions about how to count biological entities, how multi-cellular organisms have evolved, what different forms of living organisation there are, etc. In contrast, metaphysicians, when reflecting on material constitution and parthood, typically intend to understand what these are in the most general terms. Even when focusing on the material constitution and part-whole relations of biological individuals, this is usually motivated by the desire to understand these biological phenomena in the context of material constitution and parthood in reality as such – as we would expect from a discipline that investigates "the fundamental structure of reality as a whole" (Lowe 2002, p. 3).

Having argued that there is intermediate space between viewing metaphysics as an a priori exercise on the one hand and as a "metaphysics of science" that merely elaborates a posteriori on scientific discoveries on the other, we note that different positions are possible within that intermediate space. Interestingly enough, there is considerable disagreement among proponents of the newly emerging discipline called "metaphysics of science", or also "scientific metaphysics" (Ross, Ladyman and Kincaid 2015), as to what exactly is its agenda (see, e.g., Slater and Yudell 2017 and the helpful discussion in Guay and Pradeu 2017). According to Mumford and Tugby (2013), for instance, the metaphysics of science is not – as Ladyman and Ross (2010) argue – concerned with the construction of a scientific worldview, but rather investigates the metaphysical preconditions of science, by clarifying the most general and most fundamental concepts on which science relies.[11] And again unlike Ladyman and Ross (2010), Mumford and Tugby (2013) are content with promoting the metaphysics of science thus understood as a subfield of metaphysics rather than as the only legitimate form of metaphysics.[12] In the same way, those broadly sympathetic with the Ladyman-Ross approach may consider proposing that metaphysics, qua metaphysics of science, should look at science as a source for more general views about the world, while acknowledging that there may be other resources that it can call on. The view we advocate is that metaphysics as a whole may have both concerns and methodological resources that transcend those of the sciences, but that this does not release metaphysics from according with science.[13] If metaphysical theses prove inadequate to encompass the empirical phenomena, they must be revised. Metaphysics – however exactly understood – and science need to be mutually informed.

It remains to be seen how the discipline-specific agendas and strategies of metaphysics and the philosophy of biology, and the metatheoretical questions they give rise to, bear upon the study of biological identity. Our hope is that the chapters published here demonstrate the usefulness of a cross-disciplinary approach to the problem of biological identity even where they disagree on both the subject matter and how to approach it methodologically. Indeed, one of our aims is to bring these differences and disagreements into the open. The attentive reader will find in this book a rich and robust variety of positions which we hope will stimulate further discussions about how we can account for biological identity as well as about what specific contributions can be made by metaphysicians on the one hand and philosophers of biology and theoretical biologists on the other. We would be gratified if the dialogue inaugurated by the present volume were to pave the way for a convincing account of biological identity that is both metaphysically constructive and scientifically informed.

1.3 Overview of the dialogue to follow

This volume contains 12 original contributions written by scholars working on metaphysics, the philosophy of biology, biology and their intersections. We have also included a previously published essay by David Wiggins because it presents a significant contribution to the debate on biological identity from the perspective of metaphysics and a point of reference for several of the original contributions in this volume. We have deliberately refrained from grouping the chapters in sections according to themes or disciplines. Instead, we have aimed to implement a dialogical principle by ordering the chapters as a sequence of contrasting perspectives – a sequence that moves from metaphysical and evolutionary reflections of biological individuality, raising questions about when a set of constituents should be said to constitute an individual whole, towards questions about the persistence of such wholes over time. The latter question, in turn, is reflected in the debate about whether organisms are better understood as substances or as processes and how this bears on our understanding of diachronic biological identity, including implications for personal identity and moral responsibility. This organisation of the chapters, we hope, will help make visible both the areas of shared interest and the tensions between discipline-specific approaches to the problem of biological identity that call for future efforts of reconciliation.

The dialogue opens with **David S. Oderberg**'s chapter **"Siphonophores: A Metaphysical Case Study"** (Chapter 2). Oderberg takes up the challenge posed by siphonophores to the neo-Aristotelian thesis that all concrete biological particulars are either organisms or parts of organisms or collectives of organisms, and do not belong to more than one of these categories (thesis T). The tripartite distinction between parts, organisms and collectives of organisms corresponds with Aristotle's strict distinction between parts of substances, substances and pluralities of substances. But philosophers of biology commonly categorise siphonophores as belonging to both the class of organisms and the class of collectives of organisms, or as being located on the borderline between these two classes. In contrast, Oderberg argues that siphonophores should be regarded as individual organisms, on the basis of three considerations: (i) the zooids that constitute siphonophores qualify as specialised parts of a whole in terms of their structure, function and overall morphology; (ii) the zooids' colonial origin does not entail their colonial status; (iii) zooid budding is a developmental process of growth, not to be confused with reproduction. Oderberg draws the conclusion that thesis T, and Aristotelian substance metaphysics in general, is in full conformity with natural science.

Philippe Huneman, in **"Biological Individuals as 'Weak Individuals' and Their Identity: Exploring a Radical Hypothesis in the Metaphysics of Science"** (Chapter 3), proposes to apply the kind of metaphysics of science advocated by Ladyman and Ross to the question of biological individuality.

Beginning with recent insights into the prevalence of symbiosis, and remarking recent research strategies that treat organisms as ecosystems, he proposes that we should think of biological individuals as ecosystems. A familiar problem with this idea is that given the range of kinds of relation between organisms in the traditional sense (microbes or genomically homogeneous macrobes) some criterion is required for membership of the organism as ecosystem. Huneman advocates a concept of "weak individuality" that can address this problem: roughly speaking, if you are part of an individual, then the chances of something else interacting with you are greater if it is part of the same individual than if it is not. As Huneman explains, this criterion implies that individuality is not an all-or-nothing matter, but a matter of degree. While all this will seem quite plausible to a metaphysician of science of the kind to which Huneman affiliates himself, it is in stark contrast with views held by neo-Aristotelian metaphysicians, such as Oderberg.

Having seen how different metaphysical frameworks impact on the understanding of biological identity, the time has come for some foundational reflections. **Eric Olson**'s chapter addresses, from a metaphysical perspective, the question **"What Is the Problem of Biological Individuality?"** (Chapter 4) with the aim of better understanding what could count as a solution of this problem. Olson argues that philosophers of biology tend to misstate the problem of biological individuality as a result of certain metaphysical presumptions that are inadequately reflected upon: instead of asking what regions are occupied by an organism, philosophers of biology are searching for a definition of "biological individual" or, more specifically, of "organism". Using the genetic theory and the functional-integration theory of biological individuality as examples, Olson identifies what he calls the principle of material plenitude as being mainly responsible for this definition-orientated approach. This principle, following from the conjunction of temporal-parts ontology and the doctrine of unrestricted composition, holds that every matter-filled spacetime region is occupied by a material thing. Olson concludes by proposing an existential rather than definitional statement of the problem of biological identity, which does not presuppose the principle of material plenitude or any other controversial metaphysical thesis.

In contrast with Olson's analytical approach, **Alvaro Moreno**'s chapter **"The Role of Individuality in the Origin of Life"** (Chapter 5) draws our attention to the evolutionary history of biological individuality, the study of which, he believes, is required for an adequate understanding not only of biological individuality but also of life as such: the successful proliferation of life on Earth would not have been possible without the emergence of forms of biological individuality. This emergence, Moreno explains, is a long process with three stages. Biogenesis begins with the evolution of self-maintaining (autopoietic) networks of chemical processes which become organisationally integrated as a result of a process of encapsulation. Second, as these minimal forms of individuality acquire primitive capacities of inheritance they undergo a process of complexification which triggers the appearance of proto-ecosystems, i.e., synchronic networks of interactions between different groups of proto-species. This process is accompanied by the emergence of semi-autonomous genetic quasi-individuals which replicate themselves using the organisation of autopoietic protocells. Only when, finally, the latter develop an early form of immune system to protect themselves against invasion do we see the advent of true individuals in the sense of hierarchically and cohesively organised systems, which through cooperation may in turn generate new cohesive associated entities.

Denis Walsh and Kayla Wiebe's chapter **"The Being of Living Beings: Foundationalist Materialism versus Hylomorphism"** (Chapter 6) takes another look at the relationship between metaphysics and science. It proceeds from the observation that the study of organisms in the biology of the 20th and early 21st centuries has predominantly been framed by a theory of being that stresses the commonalities between living and non-living matter. Walsh and Wiebe aim to show that this theory of being – they call it Foundationalist Materialism – unhelpfully constrains the scientific study of organisms by preventing us from appreciating the distinctive features of organisms, namely their self-building, self-maintaining, processual and emergent capacities. In response, they propose Aristotelian hylomorphism as a more suitable ontological framework. According to hylomorphism, organisms, qua substances, are interactions between matter and form such that while the properties and capacities of an organism as a whole are sustained by the properties and capacities of its parts, the former cannot be reduced to the latter. The hylomorphist assumption of a dynamical reciprocity between matter and form, Walsh and Wiebe argue, does justice to the distinctive status of organisms as processual emergents – a thesis neo-Aristotelian metaphysicians are likely to welcome.

Questions of the identity of an individual are widely recognised as being intimately connected with the question of a kind to which the individual belongs. While metaphysicians have traditionally appealed to Aristotle's hylomorphist idea of an essential form shared by different individuals, philosophers of biology have generally assumed that the relevant kinds should be the species (or perhaps higher taxa) provided by broadly Darwinian evolutionary biology. In his chapter, **"The Origins and Evolution of Animal Identity"** (Chapter 7), theoretical biologist **Stuart A. Newman** argues that this is a mistake. Because of their variability, both synchronic and diachronic, species provide poor candidates for natural kinds, a fact which has contributed to scepticism about the uniqueness and objectivity of biological individuality. However, focusing on the underlying processes by which liquid tissues diversify into a relatively small number of basic forms offers a more promising approach. What Newman calls physicalist evolutionary developmental biology can provide categories prior to phylogenetic kinds, and suited to the application of concepts of natural kind, essence and even, in a broadly Kantian sense, natural purpose. Newman offers a sketch of the evolution of individuality that emphasises the interaction of genetic factors with the physics of biological materials, and thus the gradual emancipation of biological form from these genetic factors through the emergence of new, genuinely processual biomaterials.

John Dupré's chapter, **"Processes within Processes: A Dynamic Account of Living Beings"** (Chapter 8), aims to show that we ought to take seriously in metaphysical terms the emerging processual picture of biological life, and this not only with respect to the evolution of biological individuality, as sketched by Moreno and Newman. However, unlike Walsh and Wiebe, Dupré does not think that the concept of substance can play any constructive role in this. Instead, Dupré argues that an organism is not a thing, or a substance, but a process. Here the principal grounds for this claim are laid out explicitly: organisms are in thermodynamic disequilibrium with their environments; organisms have life cycles (discussed also in DiFrisco and Mossio's chapter); and organisms (typically) are dependent on many symbionts. Of course, since humans are organisms, it follows on this view that they are also a kind of process, and this has consequences for a number of central philosophical issues, most notably for the present volume, personal identity as discussed in Snowdon's and Meincke's chapter. Dupré's chapter concludes with discussions of two less standard issues, the possibility of immortality, which is argued to be even less plausible than from traditional substance perspectives, and the ontological nature of pregnancy, which is described as the gradual bifurcation of a process. The latter topic is also discussed in the chapters by Kingma and Meincke.

David Wiggins, in his chapter **"Activity, Process, Continuant, Substance, Organism"** (Chapter 9), argues against the view put forward by Dupré, Meincke and others that organisms are processes and that the appropriate ontological framework for biological science is provided by process ontology. Wiggins's rejection of this view is mainly motivated by considerations about persistence: insofar as organisms persist through time they are, Wiggins claims, continuants, i.e., material things or substances. Thus, Wiggins criticises attempts to explain biological identity through time with the help of the concept of genidentity: organisms are not concatenations of states. Assuming that any plausible scientific explanation of biological reality should comprise a plausible account of the persistence of organisms, Wiggins concludes that an ontology that does not allow for material things or substances in addition to processes fails. He therefore proposes a plural ontology which assumes process, activity, event and substance, or continuant, as fundamental categories of being. Such an ontology attributes a characteristic principle of activity for each kind of organism and, Wiggins claims, is also able to handle difficult questions convincingly, such as whether siphonophores and slime moulds are individuals and how to count blackberry plants.

James DiFrisco and Matteo Mossio's chapter, **"Diachronic Identity in Complex Life Cycles: An Organisational Perspective"** (Chapter 10), is concerned, too, with the diachronic dimension of biological identity, and hence with the problem of variability within the life cycle. However, they argue that conditions of diachronic identity cannot be found in any constant properties of the organism nor in a Wiggins-style principle of activity, but must be sought in relations of causal continuity (also known as genidentity) between temporal parts of the organism understood as a four-dimensional causal process. More specifically, and in contrast to common genidentity accounts, DiFrisco and Mossio describe a sufficient condition for diachronic identity, organisational continuity, which besides spatiotemporal continuity has as a second component a causal structure they call closure of constraints. Whereas a constraint, in their terminology, is a feature that acts to limit possible transformations without itself being affected by the interaction, closure of constraints refers to a number of such individual constraints which together determine the thermodynamic flow within a system. It is this that allows a properly organised system to persist through time. An obvious problem that needs to be addressed within this approach is that it does not immediately distinguish development from reproduction, and a large part of the chapter is devoted to exploring ways of dealing with fission, fusion and sexual reproduction within this general framework.

Elselijn Kingma's chapter **"Pregnancy and Biological Identity"** (Chapter 11) explores a hitherto largely neglected facet of the problem of biological identity, mammalian pregnancy. What is the metaphysical relationship between the fetus ("foster") and the gestating organism ("gravida")? And what are the respective implications of this relationship for biological identity over time? As regards the first question, the traditionally predominant view is the "containment view" according to which the foster is merely contained in the gravida but is not a part of the latter. But Kingma points out that the very few arguments that have been provided in favour of the containment view in fact support the parthood view. The parthood view is substantiated by the satisfaction of four criteria of organismality currently discussed in the philosophy of biology: (i) homeostasis and physiological autonomy, (ii) metabolic unity and functional integration, (iii) topological continuity, (iv) immunological tolerance. As to the second question, Kingma argues that assuming a traditional substance ontological framework for arguments in favour of the parthood view carries with it the implication that biological individuals cannot begin to exist before birth, but this implication may be prevented by switching to a revised version of substance ontology which allows for parts of substances to be substances themselves.

Adam Ferner's chapter **"Processual Individuals and Moral Responsibility"** (Chapter 12) critically examines the processual view of biological individuals which Ferner, like Wiggins, interprets as maintaining that organisms, qua processes, are composed of numerically different temporal parts. Ferner's particular concern is with the resulting shift in our self-conception. He argues that Wiggins's arguments against the (four-dimensionalist type of) process view of organisms are compelling not only ontologically but also with respect to moral responsibility; it seems hard to conceive of one temporal part being responsible for the actions committed by another temporal part. Our ordinary concept of moral responsibility presupposes the idea that we are substances, i.e., continuants. However, Ferner acknowledges that what is held to be a piece of descriptive metaphysics – the neo-Aristotelian substance view – may actually contain normative elements. Looking at the criticism directed by post-modern philosophers against the (neo-)liberal concept of the unified, boundaried, atomic, sovereign human subject, Ferner concludes that the processual understanding of the human subject may prove useful in counteracting the marginalisation of collective responsibility inherent in substance ontology.

Paul F. Snowdon's contribution **"The Nature of Persons and the Nature of Animals"** (Chapter 13) undertakes to elucidate the notion of an animal that is pivotal to animalism, i.e., to the theory that we are identical with animals. More specifically, Snowdon's aim is to discern where exactly the biologists' empirical knowledge about animals can help advance philosophical controversies about animalism. Snowdon argues that when it comes to anti-animalist dissociation arguments (arguments that allege that the person and the animal which initially are co-located can come apart), constructive contributions from biologists will be confined to actual cases, such as cases of conjoined twinning. Pro-animalist arguments like the "too many minds problem", however, are not in need of biological certification as their presumptions about animals are uncontentious. As to the general nature of animals, Snowdon, like Wiggins and Ferner, criticises a fourdimensionalist process view, distinguishing two versions thereof: a moderate version which explains animals in terms of processes and an eliminativist version which replaces the concept of an animal with the concept of process. According to Snowdon, both involve questionable assumptions about persistence, contradict ordinary thinking about animals and are guilty of wrongly inferring from the necessity of processes for the existence of organisms to the latter's identity with processes.

Anne Sophie Meincke concludes the dialogue by arguing for **"Processual Animalism: Towards a Scientifically Informed Theory of Personal Identity "** (Chapter 14): a version of animalism that is based on the view that organisms are (continuant) processes. This proposal is motivated by the critical diagnosis that the animalist understanding of the key notion of biological identity is in conflict with our best contemporary science. Meincke shows this with respect to two criteria of biological identity through time proposed by the animalists Eric T. Olson and Peter van Inwagen, the Biological Continuity Criterion and the Life Criterion. The former proves unable to handle branching cases, such as the case of monozygotic twinning, prompting empirically questionable ad hoc claims, while the latter invokes an empirically implausible view of biological life as a well-individuated event. Meincke further argues that these difficulties have their ultimate roots in the thing ontological framework presupposed by animalism, and suggests that animalism adopt a (non-four-dimensionalist) process ontological framework instead. Meincke then explains how such a processual and scientifically informed notion of biological identity, complemented by a processual theory of mammalian pregnancy, resolves the branching problem for animalism and lays the foundations for a convincing comprehensive account of our identity through time that integrates biological and personal aspects.

Acknowledgements

Anne Sophie Meincke acknowledges funding from the Austrian Science Fund (FWF), Grant Agreement V-714 ("Elise Richter" research grant), which allowed her to write this chapter.

Notes

1 Aristotle develops his theory of substance most prominently in the *Categories* (Aristotle 2002) and in books Z, H and Θ of his *Metaphysics* (Aristotle 2004).
2 The investigation of 'biological individuality' is typically restricted to the question of synchronic identity, i.e., the question of what criteria need to be fulfilled in order for something to belong to the class of biological individuals, if there is such a class (it may turn out that no coherent set of criteria can be specified). The concept of 'biological identity' as used in the present volume is wider so as to encompass also questions related to the diachronic dimension.
3 Only recently has the debate in the philosophy of biology on biological identity or individuality been linked to reflections on the concept of an individual in other sciences (see Guay and Pradeu (2016a) (with the programmatic title 'Individuals Across the Sciences'), where some of the articles listed above have appeared).
4 These latest developments in biology gave rise to entirely new research fields, such as evolutionary developmental biology (evo-devo) and ecological evolutionary developmental biology (eco-evo-devo), together with a new overall research programme (the so-called Extended Evolutionary Synthesis) (Gilbert 2013, Gilbert and Epel 2015, Laland et al. 2015, Walsh and Huneman 2017).
5 This strand of systems biology is programmatically non-reductionist, denying the existence of a privileged level of causation (such as the level of genetics) and emphasising the emergence of novel system properties at higher levels of organisation (O'Malley and Dupré 2005, Cornish-Bowden 2006, Noble 2006, 2017).
6 See Andersen, Anjum and Rocca (2019) for an interesting discussion of what they call the philosophical bias of scientific research, regarding it as both unavoidable and beneficial.
7 This holds true also for some recent contributions on biological individuality though the reference to metaphysical debates here remains mostly very general (see, e.g., Ereshefsky and Pedroso 2016, Godfrey-Smith 2016, Guay and Pradeu 2016b (chapters 5, 6 and 16 in Guay and Pradeu 2016a)). An exception is Haber (2016) (chapter 15 in Guay and Pradeu 2016a) who offers a detailed analysis of the metaphysical implications of the thesis that species are individuals (Hull 1989, Ghiselin 1997).
8 See also Lowe (2001), p. 5:

> Empirical science at most tells us what *is* the case, not what *must* or *may be* (or happens not to be) the case. Metaphysics deals in *possibilities*. And only if

we can delimit the scope of the *possible* can we hope to determine empirically what is *actual*. This is why empirical science is dependent upon metaphysics and cannot usurp the latter's proper role.

Note the discrepancy between the neo-Aristotelians' and the naturalists' perception of analytic metaphysics: whereas the latter criticise analytic metaphysics for ignoring the sciences in favour of *a priori* armchair thinking, the former complain that analytic metaphysics misunderstands the very business of metaphysics by becoming too friendly with science. This makes us wonder whether analytic metaphysics really constitutes a unified philosophical venture or whether there are rather many different versions of it exhibiting no more than Wittgensteinian family resemblance.

9 For an instructive parallel argument against a mereological interpretation of biological composition and biological parthood with a special focus on the supposed individuality of species, see Haber (2016).
10 As we will explain in a moment, there are different interpretations of this term.
11 More precisely, they define 'metaphysics of science' as the

[m]etaphysical study of the aspects of reality, such as kindhood, lawhood, causal power, and causation, which impose order on the world and make our scientific disciplines possible [...], and also the study of the metaphysical relationship between the various scientific disciplines.

(Mumford and Tugby 2013, p. 14)

It is debatable whether this conception of 'metaphysics of science' cares enough about actual science, or rather covertly continues the traditional armchair approach.
12 Schrenk (2017) follows this moderate approach; see also his discussion of different 'styles' of metaphysics of science in Schrenk (2017), chapter 7.
13 See Guay and Pradeu (2017) for a detailed proposal broadly congenial to ours.

References

Alberghina, L. and Westerhoff, H. V. (Eds.) (2005) *Systems Biology: Definitions and Perspectives*, Berlin and Heidelberg: Springer.
Andersen, F., Anjum, R. L. and Rocca, E. (2019) 'Philosophy of Biology: Philosophy is the One Bias that Science Cannot Avoid', *eLife* 8: e44929, doi:10.7554/eLife.44929.
Ariew, A. (2007) 'Teleology', in: Hull, D. and Ruse, M. (Eds.), *Cambridge Companion to Philosophy of Biology*, Cambridge: Cambridge University Press, pp. 160–181.
Aristotle (2002) *Categories and De Interpretatione*, transl. J. M. Ackrill, Oxford: Clarendon Press.
Aristotle (2004) *Metaphysics*, transl. H. Lawson-Tancred, London: Penguin Books.
Bickhard, M. (2011) 'Systems and Process Metaphysics', in: Hooker, C. (Ed.), *Handbook of Philosophy of Science. Philosophy of Complex Systems*, Vol. 10, Amsterdam: Elsevier, pp. 91–104.
Blatti, S. and Snowdon, P. F. (Eds.) (2016) *Animalism. New Essays on Persons, Animals, and Identity*, Oxford: Oxford University Press.
Boogerd, F. C., Bruggeman, F. J., Hofmeyr, J.-H. S. and Westerhoff, H. V. (Eds.) (2007) *Systems Biology: Philosophical Foundations*, Amsterdam: Elsevier.
Bordenstein, S. R. and Theis, K. R. (2015) 'Host Biology in the Light of the Microbiome: Ten Principles of Holobionts and Hologenomes', *PLOS Biology* 13(8): e1002226, doi:10.1371/journal.pbio.1002226.

Bouchard, F. and Huneman, P. (Eds.) (2013) *From Groups to Individuals: Evolution and Emerging Individuality*, (*The Vienna Series in Theoretical Biology*), Cambridge, MA: MIT Press.
Boulter, S. (2013) *Metaphysics from a Biological Point of View*, London: Palgrave Macmillan.
Bueno, O., Chen, R.-L. and Fagan, M. B. (Eds.) (2018) *Individuation, Process and Scientific Practices*, Oxford: Oxford University Press.
Chiu, L. and Gilbert, S. F. (2015) 'The Birth of the Holobiont: Multi-species Birthing through Mutual Scaffolding and Niche Construction', *Biosemiotics* 8(2), pp. 191–210.
Clarke, E. (2010) 'The Problem of Biological Individuality', *Biological Theory* 5, pp. 312–325.
Clarke, E. (2013) 'The Multiple Realizability of Biological Individuals', *The Journal of Philosophy* 110(8), pp. 413–435.
Cohen, S. M. (2002) *Aristotle on Nature and Incomplete Substance*, Cambridge: Cambridge University Press.
Cornish-Bowden, A. (2006) 'Putting the Systems Back into Systems Biology', *Perspectives in Biology and Medicine* 49, pp. 475–489.
Dawkins, R. (1976) *The Selfish Gene*, Oxford: Oxford University Press.
Doolittle, W. F. and Booth, A. (2017) 'It's the Song, Not the Singer: An Exploration of Holobiosis and Evolutionary Theory', *Biology and Philosophy* 32, pp. 5–24.
Dupré, J. (2010) 'The Polygenomic Organism', in: Parry, S. and Dupré, J. (Eds.), *Nature After the Genome*, (*Sociological Review Monograph Series)*, Oxford: Blackwell, pp. 19–31 (Reprinted as Dupré 2012, ch. 5.).
Dupré, J. (2012) *Processes of Life: Essays in the Philosophy of Biology*, Oxford: Oxford University Press.
Dupré, J. (2017) 'The Metaphysics of Evolution', *Interface Focus*, published online, August 18, 2017, http://rsfs.royalsocietypublishing.org/content/7/5/20160148\.
Dupré, J. (2018) 'Processes, Organisms, Kinds, and the Inevitability of Pluralism', in: Bueno, O., Chen, R.-L. and Fagan, M. B. (Eds.), *Individuation, Process and Scientific Practices*, Oxford: Oxford University Press, pp. 21–28.
Dupré, J. and O'Malley, M. A. (2009) 'Varieties of Living Things: Life at the Intersection of Lineage and Metabolism', *Philosophy, Theory and Practice in Biology* 1(3), http://hdl.handle.net/2027/spo.6959004.0001.003 (reprinted as Dupré 2012, ch. 12.).
Ereshefsky, M. and Pedroso, M. (2016) 'What Biofilms Can Teach Us about Individuality', in: Guay, A. and Pradeu, T. (Eds.), *Individuals Across the Sciences*, Oxford: Oxford University Press, pp. 103–121.
Fagan, M. B. (2016) 'Cell and Body. Individuals in Stem Cell Biology', in: Guay, A. and Pradeu, T. (Eds.), *Individuals Across the Sciences*, Oxford: Oxford University Press, pp. 122–143.
Feser, E. (2019) *Aristotle's Revenge. The Metaphysical Foundations of Physical and Biological Science*, Neunkirchen-Seelscheid: editions scholasticae.
Folse, H. J. and Roughgarden, J. (2010) 'What Is an Individual Organism? A Multilevel Selection Perspective', *The Quarterly Review of Biology* 85(4), pp. 447–472.
Fusco, G. and Minelli, A. (2010) 'Phenotypic Plasticity in Development and Evolution: Facts and Concepts', *Philosophical Transactions of the Royal Society: Biological Sciences* 365, pp. 547–556, doi:10.1098/rstb.2009.0267.

Ghiselin, M. T. (1997) *Metaphysics and the Origin of Species*, Albany: State University of New York Press.
Gilbert, S. F., Sapp, J. and Tauber, A. I. (2012) 'A Symbiotic View of Life: We Have Never Been Individuals', *The Quarterly Review of Biology* 87(4), pp. 325–341.
Gilbert, S. F. (2013) *Developmental Biology*, 10th ed., Sunderland, MA: Sinauer Associates.
Gilbert, S. F. (2014) 'A Holobiont Birth Narrative: The Epigenetic Transmission of the Human Microbiome', *Frontiers in Genetics* 5: 282, doi:10.3389/fgene.2014.00282.
Gilbert, S. F. and Epel, D. (2015) *Ecological Developmental Biology. The Environmental Regulation of Development, Health and Evolution*, 2nd ed., Oxford: Sinauer Associates.
Gilbert, S. F. and Tauber, A. I. (2016) 'Rethinking Individuality. The Dialectics of the Holobiont', *Biology and Philosophy* 31(6), pp. 839–853.
Gilbert, S. F., Rosenberg, E. and Zilber-Rosenberg, I. (2017) 'The Holobiont with its Hologenome is a Level of Selection in Evolution', in: Gissis, S. B., Lamm, E. and Shavit, A. (Eds.), *Landscapes of Collectivity in the Life Sciences*, (*The Vienna Series in Theoretical Biology*), Cambridge, MA: MIT Press, pp. 305–324.
Godfrey-Smith, P. (2009) *Darwinian Populations and Natural Selection*, Oxford: Oxford University Press.
Godfrey-Smith, P. (2016) 'Individuality and Life Cycles', in: Guay, A. and Pradeu, T. (Eds.), *Individuals Across the Sciences*, Oxford: Oxford University Press, pp. 85–102.
Groff, R. and Greco, J. (Eds.) (2013) *Powers and Capacities in Philosophy: The New Aristotelianism*, New York and London: Routledge.
Guay, A. and Pradeu, T. (Eds.) (2016a) *Individuals Across the Sciences*, Oxford: Oxford University Press.
Guay, A. and Pradeu, T. (2016b) 'To Be Continued. The Genidentity of Physical and Biological Processes', in: Guay, A. and Pradeu, T. (Eds.), *Individuals Across the Sciences*, Oxford: Oxford University Press, pp. 317–347.
Guay, A. and Pradeu, T. (2017) 'Right Out of the Box: How to Situate Metaphysics of Science in relation to Other Metaphysical Approaches', *Synthese*, doi:10.1007/s11229-017-1576-8.
Haber, M. H. (2013) 'Colonies are Individuals: Revisiting the Superorganism Revival', in: Bouchard, F. and Huneman, P. (Eds.), *From Groups to Individuals: Evolution and Emerging Individuality*, (*The Vienna Series in Theoretical Biology*), Cambridge, MA: MIT Press, pp. 195–218.
Haber, M. H. (2016) 'The Biological and the Mereological. Metaphysical Implications of the Individuality Thesis', in: Guay, A. and Pradeu, T. (Eds.), *Individuals Across the Sciences*, Oxford: Oxford University Press, pp. 295–316.
Hölldobler, B. and Wilson, E. O. (2009) *The Superorganism: The Beauty, Elegance, and Strangeness of Insect Societies*, London: W. W. Norton.
Hull, D. L. (1989) *The Metaphysics of Evolution*, Albany, NY: SUNY Press.
Jablonka, E. (2017) 'The Evolutionary Implications of Epigenetic Inheritance', *Interface Focus* 7: 20160135, doi:10.1098/rsfs.2016.0135.
Jaeger, J. and Monk, N. (2015) 'Everything Flows: A Process Perspective on Life', *EMBO Reports* 16(9), pp. 1064–1067.
Kaiser, M. (2018) 'Individuating Part-Whole Relations in the Biological World', in: Bueno, O., Chen, R.-L. and Fagan, M. B. (Eds.), *Individuation, Process and Scientific Practices*, Oxford: Oxford University Press, pp. 63–89.

Katayama, E. G. (1999) *Aristotle on Artifacts: A Metaphysical Puzzle*, Albany, NY: State University of New York Press.
Ladyman, J. and Ross, D. (2010) *Every Thing Must Go: Metaphysics Naturalized*, Oxford: Oxford University Press.
Laland, K. N., Uller, T., Feldman, M. W., Sterelny, K., Müller, G. B., Moczek, A., Jablonka, E., and Odling-Smee, J. (2015) 'The Extended Evolutionary Synthesis: Its Structure, Presumptions, and Predictions', *Proceedings of the Royal Society B* 282, doi:10.1098/rspb.2015.1019.
Lidgard, S. and Nyhart, L. K. (Eds.) (2017) *Biological Individuality. Integrating Scientific, Philosophical, and Historical Perspectives*, Chicago, IL and London: University of Chicago Press.
Lloyd, E. (2017a) 'Units and Levels of Selection', in: Zalta, E. N. (Ed.), *The Stanford Encyclopedia of Philosophy* (Summer 2017 Edition), https://plato.stanford.edu/archives/sum2017/entries/selection-units/.
Lloyd, E. (2017b) 'Holobionts as Units of Selection: Holobionts as Interactors, Reproducers, and Manifestors of Adaptation', in: Gissis, S. B., Lamm, E. and Shavit, A. (Eds.), *Landscapes of Collectivity in the Life Sciences*, (*The Vienna Series in Theoretical Biology)*, Cambridge, MA: MIT Press, pp. 351–368.
Love, A. C. (2018) 'Individuation, Individuality, and Experimental Practice in Developmental Biology', in: Bueno, O., Chen, R.-L. and Fagan, M. B. (Eds.), *Individuation, Process and Scientific Practices*, Oxford: Oxford University Press, pp. 165–191.
Lowe, E. J. (2001) *The Possibility of Metaphysics*, Oxford: Clarendon Press.
Lowe, E. J. (2002) *A Survey of Metaphysics*, Oxford: Oxford University Press.
Lowe, E. J. (2006) *The Four-Category Ontology*, Oxford: Oxford University Press.
Margulis, L. (1991) 'Symbiogenesis and Symbionticism', in: Margulis, L. and Fester, R. (Eds.), *Symbiosis as a Source of Evolutionary Innovation: Speciation and Morphogenesis*, Cambridge, MA: The MIT Press, pp. 1–14.
Meincke, A. S. (2018) 'Persons as Biological Processes. A Bio-Processual Way Out of the Personal Identity Dilemma', in: Nicholson, D. J. and Dupré, J. (Eds.), *Everything Flows. Towards a Processual Philosophy of Biology*, Oxford: Oxford University Press, pp. 357–378.
Meincke, A. S. (2019a) 'Autopoiesis, Biological Autonomy and the Process View of Life', *European Journal for Philosophy of Science* 9: 5, doi:10.1007/s13194-018-0228-2.
Meincke, A. S. (2019b) 'The Disappearance of Change. Towards a Process Account of Persistence', *International Journal of Philosophical Studies* 27(1), pp. 12–30, doi: 10.1080/09672559.2018.1548634.
Moya, F. (2000) 'Epistemology of Living Organisms in Aristotle's Philosophy', *Theory in Biosciences* 119(3–4), pp. 318–333.
Mumford, S. and Tugby, M. (2013) 'What Is the Metaphysics of Science?', in: Mumford, S. and Tugby, M. (Eds.), *Metaphysics and Science*, Oxford: Oxford University Press, pp. 3–28.
Nicholson, D. J. and Dupré, J. (Eds.) (2018) *Everything Flows: Towards a Processual Philosophy of Biology*, Oxford: Oxford University Press.
Noble, D. (2006) *The Music of Life. Biology Beyond the Genome*, Oxford: Oxford University Press.
Noble, D. (2017) *Dance to the Tune of Life: Biological Relativity*, Cambridge: Cambridge University Press.

Novotný, D. D. and Novák, L. (Eds.) (2014) *Neo-Aristotelian Perspectives in Metaphysics*, New York and London: Routledge.
Oderberg, D. S. (2007) *Real Essentialism*, London: Routledge.
Odling-Smee, F. J., Laland K. N. and Feldman, M. W. (2003) *Niche Construction: The Neglected Process in Evolution*, Princeton, NJ: Princeton University Press.
Okasha, S. (2006) *Evolution and the Levels of Selection*, Oxford: Clarendon Press.
Olson, E. T. (1997) *The Human Animal. Personal Identity without Psychology*, Oxford: Oxford University Press.
O'Malley, M. A. and Dupré, J. (2005) 'Fundamental Issues in Systems Biology', *BioEssays* 27, pp. 1270–1276.
Paternotte, C. (2016) 'Collective Individuals. Parallels between Joint Action and Biological Individuality', in: Guay, A. and Pradeu, T. (Eds.), *Individuals Across the Sciences*, Oxford: Oxford University Press, pp. 144–164.
Pradeu, T. (2012) *The Limits of the Self: Immunology and Biological Identity*, Oxford: Oxford University Press.
Pradeu, T. (2016) 'The Many Faces of Biological Individuality', *Biology and Philosophy* 31, pp. 761–773.
Queller, D. C. and Strassmann, J. E. (2009) 'Beyond Society: The Evolution of Organismality', *Philosophical Transactions of the Royal Society B: Biological Sciences* 364(1533), doi:10.1098/rstb.2009.0095.
Queller, D. C. and Strassmann, J. E. (2016) 'Problems of Multi-species Organism: Endosymbionts to Holobionts', *Biology and Philosophy* 31(6), pp. 855–873.
Ross, D., Ladyman, J. and Kincaid, H. (Eds.) (2015) *Scientific Metaphysics*, Oxford: Oxford University Press.
Schrenk, M. (2017) *Metaphysics of Science: A Systematic and Historical Introduction*, London and New York: Routledge.
Simons, P. (2006) 'Real Wholes, Real Parts: Mereology without Algebra', *The Journal of Philosophy* 103(12), pp. 597–613.
Simpson, W. M. R., Koons, R. C. and Teh, N. R. (Eds.) (2017) *Neo-Aristotelian Perspectives on Contemporary Science*, (*Routledge Studies in the Philosophy of Science*), London and New York: Routledge.
Skillings, D. (2016) 'Holobionts and the Ecology of Organisms: Multi-species Communities or Integrated Individuals?', *Biology and Philosophy* 31, pp. 875–892.
Slater, M. and Yudell, Z. (Eds.) (2017) *Metaphysics and the Philosophy of Science*, Oxford: Oxford University Press.
Snowdon, P. F. (2014) *Persons, Animals, Ourselves*, Oxford: Oxford University Press.
Suárez, J. (2018) 'The Importance of Symbiosis in Philosophy of Biology: An Analysis of the Current Debate on Biological Individuality and Its Historical Roots', *Symbiosis* 76(2), pp. 77–96.
Tahko, T. E. (Ed.) (2012) *Contemporary Aristotelian Metaphysics*, Cambridge: Cambridge University Press.
Van Inwagen, P. (1990) *Material Beings*, Ithaca, NY and London: Cornell University Press.
Walsh, D. (2015) *Organisms, Agency, and Evolution*, Cambridge: Cambridge University Press.
Walsh, D. and Huneman, P. (2017) *Challenging the Modern Synthesis. Adaptation, Development, and Inheritance*, New York: Oxford University Press.

Waters, C. K. (2018) 'Ask not 'What *Is* an Individual?'', in: Bueno, O., Chen, R.-L. and Fagan, M. B. (Eds.), *Individuation, Process and Scientific Practices*, Oxford: Oxford University Press, pp. 91–113.
Wiggins, D. (1980) *Sameness and Substance*, Oxford: Blackwell.
Wiggins, D. (2001) *Sameness and Substance Renewed*, Cambridge: Cambridge University Press.
Wilson, D. S. and Sober, E. (1989) 'Reviving the Superorganism', *Journal of Theoretical Biology* 136, pp. 337–356.
Wilson, D. S. and Sober, E. (1994) 'Reintroducing Group Selection to the Human Behavioral Sciences', *Behavioral and Brain Sciences* 17(4), pp. 585–608.
Wilson, R. A. and Barker, M. J. (2019) 'Biological Individuals', in: Zalta, E. N. (Ed.), *The Stanford Encyclopedia of Philosophy* (Fall 2019 Edition), https://plato.stanford.edu/archives/fall2019/entries/biology-individual/.
Zilber-Rosenberg, I. and Rosenberg, E. (2008) 'Role of Microorganisms in the Evolution of Animals and Plants: The Hologenome Theory of Evolution', *FEMS Microbiology Reviews* 32, pp. 723–735.

2 Siphonophores
A metaphysical case study

David S. Oderberg

2.1 Introduction

The problem of biological individuality, as it is called, is as important as it is unfortunately named. Its importance lies in its role as a point around which we can organise a good deal of metaphysical thinking about biology. The poor name lies in the use of the term "individuality". If we do not allow ourselves to be misled by terminology, we can get a clearer view of what the problem is and why it matters. Once we have that, we can *test* our preferred solution against the biological data. For metaphysics must be answerable to science even as, from my neo-Aristotelian vantage point, it reigns over science as an a priori discipline supplying the conceptual framework without which science itself makes no sense.

The unwise reliance on the term "individuality" or "individual" is that it is too general; therefore, it is capable of interpretations that put biologists and philosophers at cross purposes among themselves and with each other. An individual is, if nothing else, a *countable unit* of some kind. (This is all there is to individuality for Lowe 1998, p. 160, 2009, p. 55.) Oak trees and salamanders are individuals but water and gold are not (though pools of water and lumps of gold *are*). Bees are countable; so are their colonies. The legs of kangaroos are countable units: all normal kangaroos have two of them. The species of kangaroo are also countable units, of which there are four. Kangaroo DNA, by contrast, is not countable, though we can count kangaroo chromosomes (16 for both species of grey kangaroo).

We might insist on more than mere countability for individuality, but we will still have a very generous and not highly discriminatory bag of biological particulars.[1] Adding a spatial criterion, for example – a particular location, a set of boundaries (however fuzzy), contiguous matter – will still let in entities that, for important metaphysical reasons, we should keep separate. Colonies of organisms and parts of organisms have locations and boundaries (however rough and ready for the former), and the skeletons of calcium carbonate connecting the polyps of many coral species turn a colony of organisms into a contiguous unit.

Metaphysically, we should want to distinguish *parts* of organisms from *organisms* themselves, and both from *groups* of organisms (of which colonies are the ones of special biological interest). As Ellen Clarke puts it:

> The problem of biological individuality can be distinguished from the problem of defining living systems by focusing on what properties separate living individuals from living *parts* and from living *groups*, while taking the property of life itself for granted.
>
> (Clarke 2010, p. 316; emphases in original)

Here, by "individual" she clearly means "individual organism". It is the status of organism rather than the status of individual that should be our preoccupation. Stephen Boulter frames the question as being whether "the working biologist can readily identify the individuals of a given species, and distinguish these individuals from their biological parts and from the groups they may join" (Boulter 2013, p. 608). Again, the tripartite distinction is arguably correct, but the question should *not* be thought of as solely – perhaps even primarily – one for the working biologist. Nor, evidently, does Boulter himself think so. Rather, the metaphysician should *inform* biological practice by giving biologists the conceptual tools for placing biological entities into one of these three ontological categories.

Is the tripartite distinction – between (i) parts, (ii) organisms, and (iii) collectives, colonies, or groups of organisms – correct? By this I mean: is the distinction exhaustive and mutually exclusive? For a neo-Aristotelian, the answer must be yes, given Aristotle's own strict distinction between substances, parts of substances, and pluralities of substances. As a neo-Aristotelian, I subscribe to thesis **T**: all concrete biological particulars are either organisms or parts of organisms or collectives of organisms, and do not belong to more than one of these categories.

We all accept the existence of members of each category, but the interesting question is whether any biological particular can belong to more than one of the categories or to none. If the answer is affirmative, it is likely to be because of the truth of the second disjunct – that it can belong to none of the categories – rather than the first. For, at least on the traditional understanding of these things, to be a part of an organism is precisely *not* to be an organism, to be an organism is *not* to be a collective, and so on. My gut bacteria and I live in a symbiotic relationship, but it would be plain bad science to say that the bacteria were literally *part* of me – organs like my kidneys or liver, or parts like my skin or blood cells. A lion might belong to a pride but is not *itself* a pride, nor a member of a school of fish itself a school, nor is any organism literally *composed* of other organisms, and so on.

It would be more plausible to wonder whether there were a biological particular that fit into *none* of the three categories – perhaps because it had enough features of more than one of the categories for there to be no sufficient reason to place it in any particular category rather than another. In any case, whether one allowed that such a particular might demonstrate the categories not to be mutually exclusive or whether one thought that it showed them merely to be non-exhaustive, it would refute the tripartite distinction as usually understood by Aristotelians. In other words, *thesis* **T** *is falsifiable*. Does this make it an empirical hypothesis? Or does it mean that metaphysical theses can sometimes be falsified by empirical data? Probably the best way to understand **T** is as a hybrid – an empirical hypothesis essentially informed by metaphysical considerations. If a falsifier exists, this will be an empirical matter, but its existence will show the underlying metaphysics behind **T** to be wrong. Whatever the correct interpretation of the methodology, thesis **T** opens up an excellent opportunity to test a broadly Aristotelian understanding of life.

2.2 Parts, organisms, colonies

The specific test case on which I focus as a possible falsifier of thesis **T** is that of the siphonophores, in many ways the most curious of the marine animals. Before examining them, however, we need to understand the content of **T**. (For more detail, see Oderberg 2018.) What is it that differentiates parts, organisms, and colonies? (Rather than looking at groups in general I limit the analysis to colonies since this is where the controversy over siphonophores is most acute.) I do not propose to give a formal definition of these three kinds of biological particular, as this would require an article of its own. In general, I adhere to an "ontological independence" view of substance as understood in the Aristotelian tradition. (The view broadly agrees with that of Lowe 1998, chapter 6.) An organism is an individual substance of the biological kind. Its independence consists in its not being *essentially* dependent on any other particular except for its own parts. The concept of essential dependence here is quite strict. Loosely, organisms essentially depend on food, water, oxygen, an ecosystem, and so on, but this loose dependence is *causal* in kind. Food, water, and the like are necessary for keeping an organism in existence, enabling it to develop, or flourish, or adapt to its environment. Organisms are the *kinds* of things that need food, water, and so on. Moreover, taking in some kinds of nutrition and not others (carnivore versus herbivore, for instance) can be part of *what it is to be* a certain kind of organism, but only in the sense that organisms have certain *properties* – Aristotelian *propria* in the strict sense – that are necessary for belonging to the essential kinds to which they belong. (See Oderberg 2011.)

Strictly, however, by which I mean the strictness of *essential constitution*, an organism – as a living substance – is not *defined* in terms of things other than itself, by which I mean other than its own body plan, structure, arrangement of parts, suite of functions, and so on. All organisms will have at least one part, and one function, that relates essentially to things distinct from the organism and its parts – other organisms, organic entities such as types of food, inorganic entities such as air and water, and so on. But that does not mean the *organism* is defined in a way that depends necessarily on the *existence* of any of these other things. A colonial organism might not live long without a colony to inhabit, but its *definition* – its classification in the ontological scheme of organisms – does not require that any colony actually *exist*.

Contrast organisms with parts and colonies. A part of an organism *cannot* be defined except in terms of things other than itself – that is, things other than its own structure, functions, arrangement of its own parts, and the like. The part must be defined in terms of the whole organism it subserves – an entity beyond the boundaries of the part itself. For the part to exist – for it to *be* a part of an organism – there must also exist an organism of the right kind for the part to subserve. Aristotle was right to hold that a part without a whole to subserve is a part in name only – his famous "homonymy" principle (*Metaphysics* Z:10, 1035b23 and elsewhere, Ross 1928). We can speak of organs grown in labs, organs kept on ice after the organism has died, and other "edge" cases, but strictly such entities are better thought of as *proto-organs*, or *former* organs that may (if transplanted) become literal organs once again. (For more on edge cases, see Oderberg 2018.)

A colony looks as though it can, like an organism, be defined wholly in terms of its own structure, functions, arrangements of parts, and so on. This, however, masks the asymmetry between colonies and organisms. Colonies do not, strictly speaking, have parts: they have *members*. Or perhaps more precisely, they cannot *only* have parts but must have members as well, whereas organisms do not have *members* except in the archaic biblical sense synonymous with *organs*. If, for example, we take a beehive (the artificial housing structure) to be part of a bee colony – though I find this a more dubious way of talking than speaking of the hive as a structure used by the colony though not strictly part of it – then we can see that bee colonies cannot be defined solely in terms of parts such as these. They must be defined – not merely additionally but *primarily* – in terms of *members*. And these members – the individual bees – are themselves organisms that are *not* defined in terms of the colonies to which they belong, even if the colonies are obligate. As I suggested earlier, a colonial organism might need a colony to survive, just as it needs food, but neither its food nor its colony has to *exist*, definitionally speaking, in order for the organism to exist.

Still, it might be objected, the asymmetry between organisms and colonies is not yet clear. If the organism is a substance because it is defined wholly in terms of its own structure, body plan, functions, and arrangement of parts, without the existence of anything else being metaphysically *entailed* by this definition, why isn't the colony also a substance? When defining colonies we still do not "stray outside" the colony itself: we can define it solely in terms of its functions, the identity and function of its parts (if it has any parts), the identity and function of its members, the structural and organisational relations between them, and so on. The existence of nothing else is entailed by defining the colony in this way. In reply, the objection does not show that the cases are symmetrical. When defining the colony in terms of its members we *do* stray outside the colony proper, so to speak. This is because the members, as organisms, do not have a *complete definition* in terms of their functions in respect of the colony; unlike a part of an organism – an *organ*[2] – which *is* completely defined by its subservience to the whole of which it is a part. So while it is true that an organism's existence entails, metaphysically, the existence of its parts,[3] and a colony's existence entails, metaphysically, the existence of members,[4] the former entailment does not take us, metaphysically, beyond the organism, but the latter entailment does take us beyond the colony.

It is crucial to appreciate that by "metaphysically" in this context the neo-Aristotelian means "definitionally, as constitutive of the kind of thing an entity is". When we "zoom in" on the organ, ontologically speaking, our purview never goes beyond the organism – since the organ can only be completely understood in terms of its function in respect of the whole. When we "zoom in" on a colonial member, by contrast, our purview can and does go beyond the colony. This is because a complete definition of, say, the individual bee, or ant, or coral polyp, or bacterium includes non-colonial elements – individual structure, internal function, parts, and so on. In other words, the colonial members must be understood as organisms in their own right before their colonial functions are taken into account. Even if the colonial behaviour is obligate, and even if this behaviour enters into the essential definition of the kind (about which we need not have any settled view), it cannot be completely defined by such behaviour. Hence the earlier metaphor of "straying outside" the entity in question cannot be cashed out in purely spatial terms. We have to grasp what it is we must look at when we seek to define the entity. When we see things in this way, the asymmetry between colonies and organisms is as manifest as that between organisms and organs.

2.3 Siphonophores

With this stage setting in place, let us now examine perhaps the hardest case for Aristotelians and their adherence to thesis **T** – the siphonophores. These are an order of marine animal[5] belonging to the phylum *Cnidaria*, the latter defined by the possession of cnidocytes or explosive cells containing nematocysts or organelles that fire toxic structures into the cnidarians' prey. The phylum includes sea anemones, corals, jellyfish, and multifarious hydrozoa or "sea serpent animals" that have either or both of a polypoid form (vase-shaped structure with a base at one end and a circular mouth surrounded by small tentacles at the other) or a medusoid form (floating, bell-shaped structure with trailing tentacles).

The siphonophores themselves come in three sub-orders – physonects, cystonects, and calycophorans (Totton 1965). These are distinguished by the combination of primary structures present, and there are three of these as well. The first, moving from top to bottom, is the pneumatophore, which is a gas-filled float enabling flotation and vertical motion through the water column. The second is the nectosome: a stem-like region containing nectophores, individual bell-like structures or medusae enabling the siphonophore to swim through the water forward, backwards, and in turns. The third is the siphosome: a stem-like region containing a large number of diverse structures performing specialised functions of the type we would expect from any organism – feeding, predation, reproduction, protection, and excretion being the main functions. The physonects have all three primary structures; the cystonects have a pneumatophore and siphosome only; the calycophorans have only a nectosome and siphosome. *Physalia physalis*, known popularly as the Portuguese man-of-war, is a cystonect and is the best known of the siphonophores. Although I will speak generally of the siphonophores and refer to other species, I will refer more frequently to *Physalia physalis* as an exemplar of the order: certainly it exemplifies the metaphysical problems that make all of the siphonophores such a threat to thesis **T**. It is impossible even to scratch the surface of the entire order of siphonophores in a single, relatively short discussion. Many species are poorly understood, some barely at all. If, though, the Aristotelian can defend thesis **T** against the claims of *P. physalis* and a few other specific examples, it is highly probable, albeit not certain, that it can be defended against siphonophores as a whole.

2.4 "Colonial organism" and "poly-person"

Ever since Thomas Huxley, Ernst Haeckel, and Louis Agassiz did their heroic work in the 19th century expanding our knowledge of the siphonophores, including their taxonomy, biologists have wrestled with how they are to be understood at the basic level: are they individual organisms or colonies of organisms? The standard contemporary answer is given by Stephen Jay Gould: "Are siphonophores organisms or colonies? Both and neither; they lie in the middle of a continuum where one grades into the other" (1984, p. 29). This idea is embedded in the view expressed by Jack Wilson. He first states: "the same intuitions that allow us to count puppies and tomato plants with confidence leave us perplexed when we try to count colonial siphonophores like the Portuguese man-of-war" (1999, p. 1). He goes on to propose a kind of ontological pluralism, whereby a siphonophore counts as a "sufficiently functionally integrated living entity" to be what Wilson calls a "functional individual", lying on a spectrum of greater and lesser integration (1999, p. 63f.). This indifference to the strictures of thesis **T** is found also in E. O. Wilson, who first writes that although siphonophores "resemble organisms", each one "is a colony". This looks like adherence to **T**, but Wilson immediately goes on to say: "The resolution of the paradox is that siphonophores are both organisms and colonies" (Wilson 1975, pp. 383, 385).

Indifference to thesis **T** is seen in the common nomenclature, whereby siphonophores are routinely described as "colonial organisms" or "colonial animals" (Wilson 1975, p. 386, Dunn 2009, p. 233, Siebert et al. 2015).[6] If **T** is true, nothing can be *both* a colony *and* an organism or animal. It might be a colony with some organism-like properties, or an organism with some colony-like properties, but this does not seem to be the way the description is applied. The implicit rejection of **T** goes back to Ernst Haeckel, who coined or at least brought to prominence the terms "poly-organ" and "poly-person" to describe the theories that siphonophores were, respectively, single organisms with organs or else colonies of organisms. He considered both the poly-person and the poly-organ theories to be right in some ways, but that each exaggerated the phenomena supporting its side. Rather, he claimed, "the truth lies midway between the two interpretations" (Haeckel 1888, p. 2f.).

Biologists typically do appeal to the sorts of phenomena that metaphysicians are also concerned with when it comes to overarching metaphysical classifications of biological particulars. To attribute to them any self-conscious, let alone systematic, *metaphysical* thinking about what they are investigating is usually, however, hasty. For instance, George Mackie, probably the leading contemporary expert on siphonophores, in one place describes them as "complex, highly polymorphic creatures, whose 'colonies' are composed of many polypoid and medusoid 'individuals', and yet they function physiologically as single individuals" (Mackie, Pugh, and Purcell 1987, p. 98). Yet in another, published in the previous year, he characterises them unambiguously as colonies (Mackie 1986). In neither place, though, does he recognise the apparent inconsistency of his claims. What is required is a biologically informed, metaphysical analysis of siphonophores, rather than relatively breezy or relaxed assertions whether from biologists or philosophers.

2.5 Structure and function: the zooid

It is useful to follow Gould by dividing into three the kinds of phenomena that give rise to so much classificatory difficulty: structure and function; phylogeny; and reproduction and growth. (Gould speaks of "growth and form" for the third, but form belongs to structure and function, and the points he raises for the third category touch reproduction and growth more than form, despite some overlap (Gould 1984, p. 28).) The first concerns the very structure and function of the siphonophore. Everything ever written about siphonophores, at least since the end of the 19th century, will tell you that they are constituted by *zooids* – literally, "animal-like" beings. The term "zooid" itself is as slippery as the metaphysical classification of siphonophores is thought to be. It is used to denote locomotive biological particulars produced by and living within an organism, in particular spermatozoa. Is a spermatozoon an organism? If thesis **T** is correct, then given that it is neither a colony nor in any obvious sense a *part* of the organism that produces it, it must itself be an organism. That said, it has to be admitted that spermatozoa are rather strange organisms, given that they do not themselves fall within the usual biological taxonomy – they do not belong to species although they belong to the producing organisms that *themselves* belong to species. They are highly specialised; indeed, they have a single function: find an ovum and fertilise it. They do not themselves reproduce; on the other hand, they possess locomotion, which is at least necessary for animality (Oderberg 2007, pp. 183–193). Whether, however, they are sentient – necessary and sufficient for animality (ibid.) – is barely explored, though they may well have a form of memory (Brugger, Macas, and Ihlemann 2002). The difficulty of assessing the metaphysical status of spermatozoa is, so many believe, found equally if not more so in the zooids of the siphonophores.

Taking *Physalia* as our exemplar, its zooids divide into medusoid and polypoid: the medusoid zooids are bell-like structures usually with tentacles or tentacle-like attachments near the mouth and hanging down, and the polypoid zooids are polyp-like, that is, roughly tubular in shape with mouth and tentacles pointing up. Its main zooids are: gastrozooids for feeding; gonozooids for reproduction, containing further zooids called gonophores – sacs containing gametes; various protective and defensive zooids such as palpons and dactylozooids attached to long tentacles descending from the pneumatophore, which is the large, gas-filled float that keeps *Physalia* floating and drifting on the ocean surface. The pneumatophore is part of the protozooid – a polyp that develops from the embryo and gives rise, through budding, to all the other zooids. The tentacles contain the nematocysts, or stinging cells that deliver the well-known, painful venom to many a hapless swimmer. An important zooid lacking in the Portuguese man-of-war is the nectophore, a medusoid structure used for swimming. This places *Physalia* in the sub-order of cystonects, siphonophores having only a pneumatophore and a siphosome – the part of the stem with the feeding, reproductive, and protective zooids. *Physalia* might, according to some observation, have nectophores and vestigial nectophores in the siphosome, but it is doubtful that they are for locomotion and they are poorly understood (Bardi and Marques 2007, p. 432; see Totton and Mackie 1960 for the most detailed study of *Physalia physalis*).

There is an immense amount of polymorphism among the siphonophores and their constituent zooids, but virtually all the zooids are for the purpose of feeding, locomotion, reproduction, or defence – as one would expect from any animal. Why, then, should they not be categorised as *parts* of the organism – as *organs*? Critics of thesis **T**, including those whose criticism is implicit in the relaxed use of the term "colonial animal", claim that siphonophores "have a life cycle wherein multiple asexually produced zooids, each of which is homologous to a free-living solitary animal, remain attached and physiologically integrated throughout their lives" (Dunn and Wagner

2006, p. 743). Homology is the sharing of traits derived from a common ancestor, the implication being that the zooids are so similar to their free-living homologues, both derived from ancestral animals, that they are "persons" as much as "organs" – impoverished organisms whose function is to serve their colonial master.

The response to this point is twofold. First, whether the zooids are homologous to free-living animals is irrelevant, as I will argue when discussing the phylogenetic criterion of metaphysical status in Section 2.6. Second, the bare assertion that they are homologous begs the question whether the implication of the assertion is that the zooids are *as good as* organisms. Consider Julian Huxley's view (1912, p. 120):

> In the majority of Siphonophora, the persons of the colony have mostly only a historical individuality: some of them are sometimes so much modified and reduced that it has baffled all the zoologists to decide whether they are homologous with individuals or with mere appendages of individuals: and in function each is devoted so little to itself, so wholly to serving some particular need of the whole, that if one were separated from the rest, it would appear a perfectly useless and meaningless body to an investigator who did not know the whole to which it belonged.

Huxley's use of the term "person" notwithstanding, the thrust of his remarks is that the zooids are *parts* of the whole rather than *members* in the colonial sense. On this, at least as far as structure and function go, I submit that he is correct. The zooids do not function like organisms, do not look like organisms, and have no free-living phase (Gould 1984, p. 24). Each zooid is highly specialised, performing a single task – whether feeding, protection, locomotion, reproduction, among others.

Note that the free-living eudoxid stages of calycophorans are *not* independent zooids that have detached from the polygastric siphonophore. The siphosome on all siphonophores is a repeating column of sections called cormidia, consisting of zooids in similar arrangements. In *P. physalis*, each cormidium contains a gastrozooid, gonodendron, and dactylozooid with tentacle (Totton and Mackie 1960, Bardi and Marques 2007). Most calycophorans release a *cormidium* that develops into a free-living sexual unit – the eudoxid stage – that releases gametes via its gonophores, with the gametes fertilising to form the larval stage of a new polygastric calycophoran. This gives us no reason to think the cormidium that is shed was an individual organism while belonging to the polygastric siphonophore. In its free-living state, it is plausible to consider it an organism whose sole purpose, like that of a locomotive gamete such as a spermatozoon, is reproductive. (The same can be said of the released sexual medusoids of other siphonophores such as the physonect agalmids.) It has a gastrozooid and tentacle for feeding, a bract for protection, and the gonophore also serves for locomotion due to the lack of nectophores (Carré and Carré 1991, p. 30). As part of the polygastric calycophoran, however, it is no different to any of the undetached cormidia, and as such should be regarded as a part.

As for the transition from part to free-living organism, we can simply invoke the Aristotelian homonymy principle broadly understood. Just as a dead, severed part of an organism is a part in name only, and a corpse of an organism is an organism in name only, so a *living* part produced through autotomy, as in the case of calycophorans, is no longer truly a part but an organism in its own right, having undergone a substantial change on detachment. This is a form of asexual reproduction, accompanied by the sexual reproduction resulting from gametes released by the eudoxid. It is obviously not alternation of generations, which is both haploid and diploid, but the alternating of sexual and asexual reproduction in the calycophoran parallels alternation of generations and so is hardly a source of mystery on this score.

The status of the zooids as organs is clear from their extreme specialisation. The nectophores are purely for locomotion (Mackie 1964): they rely on other zooids for capture and distribution of food and for protection. In *Physalia*, the gastrozooids and gonozooids rely on the tentacles and dactylozooids for protection, the gastrozooids rely on the tentacles for food capture, the gastrozooids supply nutrition to the other zooids, they all rely on the pneumatophore for flotation and motion, and so on. Contrast this with, say, a bee or ant colony. We should generally, I submit, treat their colonial behaviour as a property of the individual organisms – in the strict Aristotelian sense of a characteristic metaphysically "downstream" from their constitutive essence and so not part of their *definition*. But even if, in some cases, the colonial behaviour is part of the very definition of the kind, it will not exhaust our understanding of the function and behaviour of the colonial organism, which must always be an organism *first* and colonial in nature *second*. With siphonophore zooids, however, their constitutive essence – what they are in their very nature – is to be specialised parts of a whole.

Again, although siphonophore zooids have "nerve nets", loose distributions of neurons considered primitive precursors of central nervous systems in animals with bilateral symmetry; this does not mean that they are independent organisms working together as members of a colony. In physonects, for example, there is a "giant axon" running along the stem and a connected ectodermal nerve net linking the giant axon to all the zooids, making possible a quick contractile reaction to stimuli (Grimmelikhuijzen, Spencer and Carré 1986, p. 474). Indeed, according to Mackie, the interactions between the stem and the other zooids give siphonophores "behavioural capabilities equalling or surpassing those of other cnidarians", including by implication true jellyfish – which are agreed on all sides to be individual organisms and yet with nervous systems of similar primitiveness (or sophistication!) to siphonophores (Satterlie 2011). In physonects and calycophorans, moreover, although there are nerve nets and nerve rings in each nectophore, they are all connected by nerve tracts to the stem, enabling their typically graceful and totally co-ordinated swimming movement (Gould 1984, p. 24, echoing Mackie 1973). Contrast this with corals, which form a true colony: each has a nerve net, but these are not connected (though corals do sit on a shared exoskeleton).

In short, there is nothing in the structure and function of siphonophore zooids themselves that points to their being individual organisms within a colony. Their extreme specialisation, co-ordination, nervous connections, and inability to survive as free-living units indicate clearly, albeit not infallibly, their metaphysical status as true *parts* of a whole organism, defined entirely in terms of their function in respect of the whole. This is not to say, given our relatively thin understanding of siphonophores, that we might not discover something about the zooids that overturns this judgement. But if we are after absolute certainty, biology is not always the best place to look.

2.6 Phylogeny: the evolution of coloniality

According to Gould (1984, p. 24), by the criterion of evolutionary history the siphonophore zooids are "individual polyp or medusa organisms" since siphonophores are colonies that "evolved from simpler aggregations of discrete organisms, each reasonably complete and able to perform a nearly full set of functions (as in modern coral colonies)". He goes on to assert: "But the colony has become so integrated, and the different persons so specialized in form and subordinate to the whole, that the entire aggregation now functions as a single individual, or superorganism". Note first that the term "superorganism" is not merely useless but positively invites confusion. Strictly, it is applied to colonies of conspecific organisms such as ants and termites (Hölldobler and Wilson 2009); loosely, it is abused to the point where even human beings have been described as superorganisms (Kramer and Bressan 2015), with the implication that very few metazoa end up as individual organisms at all. In neither the loose nor strict sense does the "super" prefix yield ontological illumination. Gould's usage here is perhaps a *ne plus ultra* of obfuscation, since "single individual" is disjoined with "superorganism" – the former is qualified by "functions as", and the disjunction is qualified by "entire aggregation". Not that we should have expected a philosophical treatise from a biologist writing a popular article, but rather the cloudiness of expression is symbolic of the overall confusion and (perhaps understandable) lack of philosophical care taken by biologists in general.

That said, the thrust of Gould's claims is clear enough: since siphonophores evolved from single organisms coming together into increasingly integrated colonies, the zooids of siphonophores are persons and the whole is a colony. Mackie (1986, p. 176) quotes Dendy (1924) as stating an "important biological truth", namely that "evolution consists, to a very large extent, if not mainly, in the progressive merging of individualities of a lower order in others of a higher order". Lest one think Mackie concludes that the "individualities of a higher order" include siphonophores as individual organisms, note his insistence that "while the general concept (individuality) is useful, we run into problems when it comes to calling specific objects individuals" (1986, p. 176). This does not mean he regards them as true colonies either; rather, the question of metaphysical status is, as it is for Gould, not deep enough to be important or precise enough to be answerable.

Yet why should we even think that the conclusion of Gould's argument follows from its premises? As background, we know very little about how the cnidarians evolved, let alone the siphonophores in particular. As Mackie, Pugh, and Purcell (1987, p. 121) put it, citing Scrutton (1979): "The numerous gaps in the palaeontological record hamper any attempt to establish a phylogeny and in the case of siphonophores it is debatable whether any fossil record has been found". Moreover, they add, citing Werner (1973), recourse to inferring ancestral structure from investigation of current morphology and life history is perilous, since "such information can be interpreted in different ways and any evidence can be taken to support totally different theories".

There does seem to be general agreement that so-called "colonial organisms" such as the siphonophores represent a stage in the evolutionary road to truly multicellular organisms, if not chronologically – since the zooids are already themselves multicellular – but conceptually, inasmuch as they are at the highest level of colonial integration short of full individuality as organisms. Whereas, some speculate, the triploblastic animals (most animals on earth) were able to develop true organs due to the presence of a mesoderm, the diploblastic siphonophores escaped their limitations by developing functional specialisation of the zooids that were once, presumably, free-living organisms that found an adaptive advantage in attaching to each other and acting in a highly co-ordinated manner (see further Wilson 1975, pp. 384–386, Mackie, Pugh, and Purcell 1987, p. 110, Niklas and Newman 2013).

Suppose all of this is true. How is it relevant to the metaphysical status of actual siphonophores? It does not follow, from the assumption that siphonophores evolved from free-living, unspecialised organisms, that their specialised zooids are themselves organisms. Compare the speculation about siphonophores to the endosymbiotic theory of Margulis (1970) concerning the origin of eukaryotic cells. According to the theory, these originated from the colonial integration, via symbiosis, of various prokaryotes – most notably the bacteria that became the mitochondria of the eukaryotic cell. If the theory is true, this does not imply that actual eukaryotic cells are colonies. If it did, then it would turn out that there were far fewer individual organisms in existence than anyone ever thought:[7] only the bacteria and archaea would make the cut! The implication is absurd, as is the thought that if multicellular organisms evolved colonially from free-living single-celled organisms, the former must themselves *be* colonies. In other words, a colonial *origin* does not entail a colonial *status*. More generally, the thesis that *ontology* recapitulates phylogeny is as untenable as the discredited biogenetic law, associated with Haeckel but going back euphonically to Meckel, that *ontogeny* recapitulates phylogeny. There is simply no good reason to think that we can read off, from the phylogenetic origin of a species (where I use "species" in the broad, metaphysical sense[8]) its metaphysical status as colony or organism. What it *is* and where it *comes from* are distinct questions. To be sure, homologies between zooids on the one hand, and free-living polyps and medusae on the other might give us good indicators of the evolutionary origin of siphonophores. From that, however, we cannot infer that zooids, metaphysically speaking, are just like their free-living homologues except for serving a colonial master. On the contrary, what we know of siphonophore zooids, in terms of their structure, function, and overall morphology, tells us that they are organs – parts of organisms – even if it is true that they arose from a radical transformation of prior, free-living polyps and medusae.

Indicative of the confusion between ontology and phylogeny is Beklemishev's account of the evolution of colonial animals, as recounted by Wilson (1975, p. 387). Beklemishev, says Wilson, was influenced by "two venerable ideas: the concept of the superorganism and the view that biological complexity evolves by the dual processes of the differentiation and integration of individuals". He identified

> three complementary trends as the basis of increasing coloniality: (1) the weakening of the individuality of the zooids, by physical continuity, sharing of organs, and decrease in size and life span, as well as by specialization into simplified, highly dependent heterozooids; (2) the intensification of the individuality of the colony, by means of more elaborate, stereotyped body form and closer physiological and behavioral integration of the zooids; and (3) the development of cormidia, or "colonies within colonies".
>
> (Wilson 1975, p. 387)

It does not, however, take much reflection on (1) and (2) to see that these are not criteria of increasing *coloniality* but of increasing *individuality* – the individuality of the organism, of which each part is defined wholly in terms of its subservience of the whole. As to the third criterion, it cannot be used as a criterion of coloniality without begging the question. That aside, we have no more reason for thinking of cormidia as colonies within colonies than we do for thinking of them as organisms in their own right.

2.7 Growth and development

Gould believes that growth and development provide an *"embarras de richesses* by presenting evidence for and against both theories", namely the poly-person and poly-organ theories (1984, p. 28). In support of the latter

Gould observes, correctly, that "a siphonophore begins life as an unambiguous person", whose later development is but the "elaboration of this one individual". Why? Because all siphonophores develop from a *single embryo* (leaving aside eudoxids, which, if they are organisms at all, are, I tentatively suggest, no more siphonophores than the gametes they produce). The embryo develops into a protozooid, the original polypoid zooid that then buds all the other zooids of the siphonophore. At its most general, metaphysically speaking, the budding of zooids from an embryonic siphonophore is no different to the budding of limbs and organs in a human embryo, each individually, or in groups, specialised for a given suite of tasks serving the organism.

Given this pattern of embryological development, a problem for the colonial view immediately arises: if the mature siphonophore is a genuine colony, is the larva also a colony? What about the planula, a free-swimming cylindrical entity with no zooids yet budded? Why would one call this anything other than a unitary organism, except because of retrospective bias in favour of the view that siphonophores *must* be colonies because of the zooids? Yet if one weakens the view to one of indifference as to whether the larva is a colony or an organism, how is this a more reasonable position? There is no reason for indifference about the status of the larva other than the magic – as opposed to the non-existent virtue – of 20/20 metaphysical hindsight.

For Gould, however, ontogeny is not as decisive as I claim it to be. He argues:

> Admittedly, each colony begins life as a single ovum, but it then develops a series of entities – full persons in this view – by budding from a common stem. This is a familiar mode of growth for many aggregations conventionally regarded as colonies. A stand of bamboo or a field of dandelions may trace its origin to a single seed, yet we usually view each budded stem or flower as an individual.
>
> (Gould 1984, p. 28)

We must immediately put an irrelevant consideration to one side, namely that the genetic identity of the zooids of a single siphonophore entails their belonging to a single organism. Clonal colonies, such as bamboo, are also genetically identical (copying errors and mutations aside). What differentiates the bamboo from the siphonophore for present purposes is the nature of the processes in each case. Gould is guilty of conflating the way in which siphonophores *grow and develop* with the way in which bamboo *reproduce.* When siphonophores bud zooids they are not reproducing themselves; reproduction is a wholly separate process involving the gonozooids and their gametes, resulting in a new protozooid from which bud new zooids, giving rise to a new siphonophore. By contrast, when a bamboo grows rhizomes that bud new culms, what results are precisely that – new culms, that is, new mature bamboo stems *reproduced* (asexually) by the parent, which itself was reproduced via rhizomes from another culm, or from a seed produced by another culm, or from a cutting. What results each time are new bamboo trees, completely the same in morphology and function to their conspecific asexual parent. To analogise this *reproductive* process to the *developmental* process of zooid budding is either to miss the point or to beg the question in favour of prior indifference to thesis **T**.

The same point can be made against Gould's analogy between zooid budding and dandelion reproduction. The *Taraxacum* genus, containing the many species of dandelions, reproduces new plants by sexual or asexual methods, including sexual self-pollination. Each method results in a new plant morphologically and functionally identical to the parent and, like bamboo, sometimes physically separated and sometimes not. Again, the disanalogy with siphonophores should be apparent. Gould's phrase (1984, p. 28), "trace its origin to a single seed", is therefore highly ambiguous since there are many ways in which this can apply to an organism or a colony. Moreover, to regard the siphonophore as tracing its *origin* to a single embryo is, at least on the interpretation implying numerical distinctness between siphonophore and original entity, to make a metaphysical mistake – since the embryo *is* the siphonophore at a juvenile phase. If it is not, then what is it? No biologist suggests anything other than what I have just stated to be a metaphysical truth, albeit such a truth is low level enough to be a staple of biological literature. If the siphonophore is a true colony, then so is the embryo, the planula, and the later immature larval stages before zooid differentiation has taken place. Yet this seems absurd on its face – for what is colonial about a planula?

2.8 Conclusion

There is no question but that the siphonophores are a fascinating class of marine creatures, about which our ignorance is still great despite over a century of investigation. That there is a real debate to be had about their metaphysical status is evident from the historic disagreements among biologists over how to understand them, a disagreement that biology itself, without being philosophically informed, is unable to resolve. Moreover, the more recent indifference to thesis **T** – a thesis that implies a binary way of answering the question whether the siphonophore is an organism or a colony, in other words whether the zooids are themselves organs or organisms – itself belies the metaphysical position such indifference implies. For to take a pluralistic approach to status whereby several incompatible understandings apply, or an indifferentist approach whereby no particular status truly applies, is itself to take a metaphysical stand – to endorse the proposition that, as Gould (1984, p. 29) puts it: "since nature has built a continuum from colony to organism, we must encounter some ambiguity at the center. Some cases will be impossible to call as a property of nature, not an imperfection of knowledge".

If my analysis is correct, then siphonophores provide no evidence for the truth of this proposition. If my analysis is merely plausible but not determinative, then at the very least both biologists and philosophers should be less hasty than they often are when pronouncing in favour of the claim that Gould, and others such as Mackie, put forward. What should be most welcome on all sides is a proper debate about the metaphysical status of the siphonophores (and other magnificently troublesome creatures) so as to determine finally whether – as I have argued – thesis **T**, with the Aristotelian metaphysic behind it, withstands attack on this front and retains its place as being, so Aristotelians believe, in full conformity with natural science.

Notes

1 I speak of "biological particulars" here rather than "biological entities" so as to make clear that my concerns in this chapter are not about processes, events, properties, species, other taxa, and so on, all of which are biological entities in a broad sense. A biological particular, so understood, is not merely a unitary entity, a countable thing, but a *functioning* entity – a biological thing that operates, behaves, exercises powers, is a participant in events and processes, and possesses properties. It should therefore be evident that I do not subscribe to the "process ontology" account of biological particulars found most notably in Dupré (2012), though it would take a separate article to evaluate processual theories in the detail they deserve.
2 Note my somewhat loose use of the term "organ". I do not mean merely something recognised as an organ in an anatomy textbook, but any part of an organism – such as a patch of skin, a lump of tissue, a fragment of bone – that has a function in respect of the whole, whether as part of an organ proper or of some other discrete structure or functional entity within the organism.
3 Again, not any particular parts, just parts of a certain kind; and not in *abnormal* cases, where an organism can exist without a part of some kind (such as a leg or tail).
4 Not any particular members, just members of a certain kind (queen, drones, workers, etc.).
5 They belong to the kingdom *Animalia*, but by calling them animals I do not mean to prejudge the very question of whether they are organisms or colonies!
6 The meaning here being quite distinct from the usual one of "organism/animal that *lives* in a colony".
7 Unless one entertained the idea that an individual metazoan organism could literally be composed of billions of colonies; not an idea I recommend anyone take seriously.
8 Meaning, in other words, the sense of the term "species" – as in Aristotelian genus and species – from which the biological sense historically derives. In biology, the species is what an Aristotelian metaphysician regards as the *infima* or *lowest* species in the taxonomic hierarchy. Metaphysically, the infima species is one of many levels of species, which for the biologist are all taxa with their own technical names (such as *order*, *kingdom*, or *family*). Note that the metaphysical sense

of "species" carries across all of reality rather than being confined to biology. Even non-Aristotelian metaphysicians happily speak of "species of artefact" or "species of chemical compound".

References

Bardi, J. and Marques A. C. (2007) 'Taxonomic Redescription of the Portuguese Man-of-War, *Physalia physalis* (Cnidaria, Hydrozoa, Siphonophorae, Cystonectae) from Brazil', *Iheringia, Série Zoologia, Porto Alegre* 97, pp. 425–433.

Boulter, S. (2013) 'Aquinas on Biological Individuals: An Essay in Analytical Thomism', *Philosophia* 41, pp. 603–616.

Brugger, P., Macas, E. and Ihlemann, J. (2002) 'Do Sperm Cells Remember?', *Behavioural Brain Research* 136, pp. 325–328.

Carré, C. and Carré, D. (1991) 'A Complete Life Cycle of the Calycophoran Siphonophore *Muggiaea kochi* (Will) in the Laboratory, under Different Temperature Conditions: Ecological Implications', *Philosophical Transactions of the Royal Society London B* 334, pp. 27–32.

Clarke, E. (2010) 'The Problem of Biological Individuality', *Biological Theory* 5, pp. 312–325.

Dendy, A. (1924) *Outlines of Evolutionary Biology*, New York: Appleton & Co.

Dunn, C. (2009) 'Quick Guide: Siphonophores', *Current Biology* 19, pp. 233–234.

Dunn, C. W. and Wagner, G. P. (2006) 'The Evolution of Colony-Level Development in the Siphonophora (Cnidaria: Hydrozoa)', *Developmental Genes and Evolution* 216, pp. 743–754.

Dupré, J. (2012) *Processes of Life: Essays in the Philosophy of Biology*, Oxford: Oxford University Press.

Gould, S. J. (1984) 'A Most Ingenious Paradox', *Natural History* 93(12), pp. 20–29.

Grimmelikhuijzen, C. J. P., Spencer, A. N. and Carré, D. (1986) 'Organization of the Nervous System of Physonectid Siphonophores', *Cell and Tissue Research* 246, pp. 463–479.

Haeckel, E. (1888) *Report on the Scientific Results of the Voyage of H.M.S. Challenger During the Years 1873–76: Zoology, Vol. XXVIII, Report on the Siphonophorae*, London: Eyre and Spottiswoode.

Hölldobler, B. and Wilson, E. O. (2009) *The Superorganism: The Beauty, Elegance, and Strangeness of Insect Societies*, New York: W. W. Norton & Co.

Huxley, J. S. (1912) *The Individual in the Animal Kingdom*, Cambridge: Cambridge University Press.

Kramer, P. and Bressan, P. (2015) 'Humans as Superorganisms: How Microbes, Viruses, Imprinted Genes, and Other Selfish Entities Shape Our Behavior', *Perspectives on Psychological Science* 10, pp. 464–481.

Lowe, E. J. (1998) *The Possibility of Metaphysics: Substance, Identity, and Time*, Oxford: Clarendon Press.

Lowe, E. J. (2009) *More Kinds of Being: A Further Study of Individuation, Identity, and the Logic of Sortal Terms*, Oxford: Wiley-Blackwell.

Mackie, G. O. (1964) 'Analysis of Locomotion in a Siphonophore Colony', *Proceedings of the Royal Society London B* 159, pp. 366–391.

Mackie, G. O. (1973) 'Report on Giant Nerve Fibres in *Nanomia*', *Publications of the Seto Marine Biological Laboratory* 20, pp. 745–756.

Mackie, G. O. (1986) 'From Aggregates to Integrates: Physiological Aspects of Modularity in Colonial Animals', *Philosophical Transactions of the Royal Society London B* 313, pp. 175–196.
Mackie, G. O., Pugh, P. R. and Purcell, J. E. (1987) 'Siphonophore Biology', in: Blaxter, J. H. S. and Southward, A. J. (Eds.), *Advances in Marine Biology*, Vol. 24, London: Academic Press, pp. 97–262.
Margulis, L. (1970) *Origin of Eukaryotic Cells*, New Haven, CT: Yale University Press.
Niklas, K. J. and Newman, S. A. (2013) 'The Origins of Multicellular Organisms', *Evolution and Development* 15, pp. 41–52.
Oderberg, D. S. (2007) *Real Essentialism*, London: Routledge.
Oderberg, D. S. (2011) 'Essence and Properties', *Erkenntnis* 75, pp. 85–111.
Oderberg, D. S. (2018) 'The Great Unifier: Form and the Unity of the Organism', in: Simpson, W. M. R., Koons, R. C. and Teh, N. J. (Eds.), *Neo-Aristotelian Perspectives on Contemporary Science*, London: Routledge, pp. 210–234.
Ross, W. D. (1928) *Aristotle: Metaphysics*, 2nd ed, Vol. VIII of *The Works of Aristotle*, Oxford: Clarendon Press.
Satterlie, R. A. (2011) 'Do Jellyfish Have Central Nervous Systems?', *The Journal of Experimental Biology* 214, pp. 1215–1223.
Scrutton, C. T. (1979) 'Early Fossil Cnidarians', in: House, M. R. (Ed.), *The Origin of Major Invertebrate Groups*, Systematics Association Special Volume 12, London: Academic Press, pp. 161–207.
Siebert, S., Goetz, F. E., Church, S. H., Bhattacharyya, P., Zapata, F., Haddock, S. H. D. and Dunn, C. W. (2015) 'Stem Cells in *Nanomia bijuga* (Siphonophora), A Colonial Animal with Localized Growth Zones', *EvoDevo* 6, p. 22.
Totton, A. K. (1965) *A Synopsis of the Siphonophora*, London: Trustees of the British Museum of Natural History.
Totton, A. K. and Mackie, G. O. (1960) 'Studies on Physalia Physalis', *Discovery Reports*, Vol. XXX, Cambridge: Cambridge University Press, pp. 301–408.
Werner, B. (1973) 'New Investigations on Systematics and Evolution of the Class Scyphozoa and the Phylum Cnidaria', *Publications of the Seto Marine Biological Laboratory* 20, pp. 35–61.
Wilson, E. O. (1975) *Sociobiology: The New Synthesis*, Cambridge, MA: Harvard University Press.
Wilson, J. (1999) *Biological Individuality: The Identity and Persistence of Living Entities*, Cambridge: Cambridge University Press.

3 Biological individuals as "weak individuals" and their identity

Exploring a radical hypothesis in the metaphysics of science

Philippe Huneman

3.1 Introduction

At first glance, individuals are what we divide the world into through our common language: living beings, nations, cars, cities, etc. We constantly name, compare, describe, count and identify individuals – at least in a manner relative to a general field of enquiry or discussion (cities and nations in geography and politics, etc.). We identify individuals, and question their proper identity, notwithstanding what this latter concept means. Strawson's book on *Individuals* (1959) precisely intended to explore the nature of this notion in our conceptual scheme, along with the criteria according to which we judge that things fall under the concept of individual, and its relationship with metaphysical notions like universals, particular properties and identity. This he called "descriptive metaphysics", in the sense that the goal is to discover the conceptual resources according to which humans in general experience the world and talk about their experience.

However, ascriptions of individuality are nowadays often challenged by scientific facts. Leaving physics aside – where for instance the individuality of fundamental particles such as quarks is a difficult issue, because of entanglement and other quantum phenomena – biology provides us with many cases where what individuals are conflicts with our daily ascriptions of individuality. Seeing many dandelions in a field, as Janzen (1977) famously described, would lead a layperson to identify dozens of individuals there, individualised by the body of each visible plant: however, their genetic sameness entitles some biologists to say that there is a single individual here. Clearly, the identity conditions for each of those dandelions would differ according to those views.

Biological individuality and identity are, therefore, disputed issues. Many controversies concern the notion itself, its criteria (e.g., Folse III and Roughgarden 2010, Clarke 2014), the concept of "organism" and its relation to the notion of individuality (e.g., whether "organism" should be replaced by the notion of individual in general; Pepper and Herron 2008, Bouchard 2010, Haber 2013). What follows is that standard descriptive metaphysics may not be adequate to address these issues, because the models used to distinguish individuals don't accept the usual individualising linguistic procedures. Actually, many metaphysicians are less interested in descriptive metaphysics and its method of conceptual analysis oriented towards our inherited linguistic habits, than in grasping the most general features of what can exist, and the correct metaphysical concepts needed to capture these features. Metaphysics here would "provide, at least for some relatively modest purposes, a useful conceptual framework for the theoretical sciences" (Lowe 2016, 59f.), and could therefore not conflict with these.

However, instead of thinking of metaphysics as capturing this "conceptual framework" in a somehow a priori way – since, as Lowe says, "all other forms of inquiry rest on metaphysical presuppositions" (Lowe 2001, p. v) – another approach consists in assuming our best science first, and then construing the metaphysical concepts in conformity with the teachings of this science. This aligns with Quine's idea that the ontology should be sought in the equations of the theories of current science, and not elsewhere (Quine 1948). Under the name "metaphysics of science", such a programme has been recently advocated by many philosophers of science (Bird 2007, Ladyman and Ross 2009, Slater and Yudell 2017). The credo is that we should look for a naturalised metaphysics, i.e. a metaphysical inquiry that relies on what our best sciences tell us regarding the furniture of the world: "Naturalized metaphysics is metaphysics that is inspired by and constrained by the output of our best science" (Chakravarty 2013, p. 33). Yet, as attested by the edited books just cited, most of the work done centres on physics, which may present the most obvious challenges to our daily conceptual schemes.

Here, I'll apply such an approach to the notion of individuality in biology, and its consequences onto biological identity. I start by explicating some issues raised by individuality in biological sciences, and why recent philosophical attempts to characterise biological individuality are problematic (Section 3.2). Then I consider a hypothesis emerging from current evolutionary research, which states that individuals – including organisms, but not only them – should best be conceived of as ecosystems (Section 3.3). I subsequently introduce a conception of individuality that metaphysically addresses such a hypothesis, defend it and explore some of its consequences (Section 3.4).

3.2 Individuality in biology: the issues

3.2.1 Individuality in our conceptual scheme

If an individual is, according to one etymology, something that cannot be divided, then what are the criteria that tell us what is not divisible? Since "individual" is a basic piece of ordinary language, possibly of our most common "conceptual scheme" (Davidson 1974, who criticises the notion), we clearly have within this scheme some resources to answer these questions pragmatically: we would for instance say that a chair or Nemo the fish is an individual, but half a chair or half of Nemo is not (or at least not a Nemo, not the same individual). Yet, this specific kind of spatial divisibility, as a main feature of the ordinary concept of individual, does not capture all the aspects of the concept.

According to the Aristotelean etymology of the word, individual means the final result of a logical division – genera subdivide into species, which subdivide in varieties, etc., and finally we're left with in-divisible items that are the individuals. One individual is what Aristotle labelled "*todè ti*", meaning "this something" – as a correlate of an ostension; an ostension does not designate one of those more general or abstract items like genera or species that would be, in turn, logically divisible. Here, an individual is indeed in principle distinct from another individual: think of two chairs, two fishes, but also two battles, two thoughts. Even if they are the same species of teleost fish, or belong to the same war (for instance, the Battle of Britain and the Battle of Iwo Jima), they are distinct (otherwise they would not be two). Therefore, the concept of an "individual" X includes the reasons why X will be distinct from Y, that is, the criteria for distinguishing individuals X and Y, which provides some of their identity conditions. As Lowe (2001, p. 33) writes,

> a principle of individuation, we might say, is not so much a criterion of *identity* as a principle of *unity*: countable items are singled out from others of their kind in a distinctive way that is determined by the sortal concept under which they fall [...].

Strawson (1959) construed the concept of individual as a kind of particular, mainly tied to the notion of recognisability, or *re-identifiability*. Not only can individuals be recognised when compared to the rest, but they should also be able to be recognised as the same through time, and hence as being self-identical through time. This involves a dependence of empirical individuals on our cognitive capacities of re-identification. It also entails that temporal persistence is part of what individuality should mean. This feature essentially connects the concept of individuality to that of identity, to the extent that the identity of X is what accounts for X remaining X across time, since "identity", in our ordinary discourse, includes identity through time. Note that this is not the merely logical identity as substitution *salva veritate*, which only concerns the truths about an individual but does not have any temporal dimension. We are concerned with a metaphysical and not a logical concept of individuality, as the latter would be too abstract to be relevant to what the sciences have to tell us. This Strawsonian view correctly captures many of the uses of "individuality" in our linguistic and social practices. Yet what about the sciences, and especially biology?

3.2.2 Biological challenges to metaphysical notions of individuality

Biology *prima facie* presents us with clear cases of individuality, in the sense of particulars likely to be re-identifiable across time, and satisfying clear conditions of discrimination from the rest of things. Actually, according to descriptive metaphysics large metazoans such as vertebrates or forest trees such as oaks (but not organisms in general) are choice examples for our daily concept of individuality. Horses persist in time, one can re-identify them through time, horses are not divisible (in the sense that no division divides them into horses).

Nonetheless, even in familiar nature around us we find organisms challenging the two conditions of temporal persistence and indivisibility. Regarding *indivisibility in space*, 18th-century biologists were amazed when they heard about Trembley's discovery of the polypus, a small shallow water animal that, once cut into two halves, gives rise to two similar organisms. Several analogous organisms, in the family of planarians, have been studied more recently, such as *Nereis* (Umesono and Agata 2009); others like starfish may regenerate from a discarded leg.

With respect to *temporal persistence*, it seems that butterflies raise problems for this concept, as does any organism that displays distinct life forms at larval and adult stages. "Stages" have been called *semaphoronts* by Hennig (1966), and are connected by ontogenetic relationships to form the individual organism. As such, the semaphoront, rather than the organism, is the character bearer in phylogenetics and, according to some, therefore the basic unit of phylogenetic systematics (Mishler and Theriot 2000), which challenges the metaphysical intuition that organisms should be the biological individuals.

Moreover, biological individuality is generally nested, as Leibniz already emphasised in the *Monadology* (with respect to what he called monads, exemplified by metazoan organisms). There exists a hierarchy of individuals, each one presupposing the former one and being constituted by an assembly of those individuals of lower levels: genes assemble within cells, cells assemble into multicellular organisms, organisms into demes (if one considers the relations of interactions) or into species (if one considers their relations of genealogy; Eldredge 1985). Hence, individuals of a given level can be nested into individuals of a higher level: cells exist within organisms, organisms live within demes; and organisms exist within species, while they include genes in themselves.[1] Low levels of individuality, such as genes or cells, while currently often nested, have lived for themselves in a remote past, as giant macromolecules possibly replicating (Maynard Smith and Szathmáry 1995) or still live like this today, like bacteria or archaea.

Here, even a metaphysics such as Lowe's, which is not tied to an analysis of our ordinary language, may fall short of providing us with a criterion to identify individuals, just like a metaphysics based on the analysis of ordinary language. In Lowe's view, "Something, x, is an *individual* if and only if (1) x determinately counts as *one* entity and (2) *x* has a determinate *identity*" (Lowe 2016, p. 50; emphases in original). Yet, "determinately counts" is impossible to apply as a criterion, if we have no additional theoretical tools to tell us what to count.[2]

3.2.3 Hullian views of individuals and transitions in individuality

In biology, in some contexts cells are individuals – for instance we might count cancer cells in an organism, or lymphocytes, when some tests have to be done. In other contexts, the organism that the cells make up is the individual. In the context of evolutionary biology, some authors have attempted to provide principled criteria that decide which are the most relevant contexts for ascribing individuality in an ontologically robust sense – and more precisely, to decide when sets of cells should be counted ontologically as an organism. The whole approach subscribes to the general conception put forth by Hull (1980), according to which, since individuality concepts must be based on our best theories – namely, here, evolutionary biology – individuality is defined by selective processes: individuals are the targets of selection. Subsequently, when it comes to distinguishing among various collections of entities – sets of cells in an organism, sets of bacteria, etc. – the ones that make up genuine individuals, the Hullian answer is given by the consideration of those collected individuals that are targeted as a whole by natural selection. For instance, organisms, and not their cells, have offspring likely to directly interact with other, genetically distinct, organisms – and this allows one to define the fitness of the organism (as a probabilistic expectation of number of offspring) and then the intensity of selection; hence organisms, and not their cells, are under selection, so they are individuals.

Many concepts of individuals have been proposed that elaborate on this general view that units of selection and individuality are intrinsically related. Drawing on Hull's essential connection between individuality and selection, Goodnight (2013) distinguishes three notions of individuality, defined respectively at the level at which fitness is ascribed, the lowest level at which selection acts and the lowest level at which a response to selection occurs. Each of them is more encompassing than the following, and, Goodnight argues, each of them may be relevant to some purposes. Clarke (2014), considering the mechanisms of policing and of demarcating that objectively warrant the univocal action of selection on a set of entities, proposes that the possession of those mechanisms is definitional of the biological individuals made up by this set of entities (e.g., cells, genes). Thus, while Goodnight's answer to the question of what is an individual depends upon the choice of an explanatory project, this is not the case with Clarke's mechanisms-based conception.

Those notions of individuality directly accord with Hull's insight that natural selection is the main ground of biological individuality; an interesting feature of this family of accounts is that they acknowledge that individuality is an evolved character: multicellular organisms are individuals that evolved from unicellular eukaryotes, which were a kind of individuals existing prior to them (Maynard Smith and Szathmáry 1995).

From this viewpoint, multicellular organisms seem to share essential features with any colony of hymenopteran insects: sterile workers, like ant or bee workers, do not reproduce, and – according at least to one perspective on the beehive – mostly work for the queen, which is solely allowed to reproduce. Evolutionary biologists have for quite some time understood that altruism here evolves because it provides benefits to individuals that are most genetically related to the altruist individuals, a process known as *kin selection* (Hamilton 1964, Gardner et al. 2011).[3] The subtleties of various organisations in particular hymenopteran species have been traced back to the genealogical systems of these various species, and the proper sex ratio, conflicts of interest and peculiar behaviour (such as killing many would-be queens, etc.) have been explained in detail on this basis (e.g., West and Gardner 2013).

Concerning the "major evolutionary transitions in individuality" (Maynard Smith and Szathmáry 1995), namely how evolution leads to more and more complex collectives behaving and reproducing as "one" (genes, chromosomes, cells, multicellular organisms, colonies), a crucial explanatory scheme has been *multilevel selection* (Damuth and Heisler 1988, Sober and Wilson 1998) – namely, considering two selective processes, one taking place at the level of competing groups of individuals and in which altruism is favoured because it's good for the group, and the other taking place at the level of individuals themselves, which have an evolutionary interest in reproducing for themselves, that is, to be "selfish". Many controversies are raging within evolutionary biology regarding the relations between kin selection and multilevel selection (for instance West et al. 2008, Nowak et al. 2010, etc.). There are surely formal identities between both (Kerr and Godfrey-Smith 2002, West et al. 2008) but what is especially striking here is that both the evolution of multicellular organisms and the organisation of these hymenopteran colonies are explained via forms of natural selection. The use of multilevel selection thus provides a clear field of application for Hull style views of biological individuality.

3.2.4 Evolutionary transitions as processes: the space of biological individuality

Regarding evolutionary transitions, one could view the appearance of individuals such as metazoan organisms as a *complete* transition, but beehives, where bees retain some autonomy from the colony, as a transition that did not reach a state of total, cell-like, subordination of the parts to the whole. We would shift therefore from a question of what individuals are, to a question of understanding the various transition processes and their differences, especially the difference between *complete transitions* and what have been termed *"component transitions"* (Huneman 2013) such as bee or ant colonies.

This theoretical question of the degree of completion of a transition therefore translates into the metaphysical question of the degree of individuality. Thus, one may doubt that the question "is this an individual or not?" and the underlying issue regarding the criteria according to which X falls under the concept "individual" are accurate: the right question to ask is rather "is A more or less an individual than B?".

But this continuum of individuality is still too simple to account for the complexity of the picture of individuality as it emerges from the research on transitions and in current evolutionary biology. Possibly individuality exists in more than one dimension (see e.g., Godfrey-Smith 2009). Strassmann and Queller (2007, 2009) elaborated upon such a view, to provide an even more subtle notion of individuality: in biology, depending upon how much cooperation and also how much conflict-decrease occurs between individuals, there will be a genuine biological individual made of these low-level individuals. Because those two properties determine two axes (Figure 3.1), individuality is not even a question of degree: to be individual is to be somewhere within the space of individuality. Note that what Strassmann and Queller designed as a hyperspace of "organismality" should be actually renamed hyperspace of individuality, since some of the positions in this space (such as "human bands") are surely not organisms.

In this perspective, contrary to a descriptive metaphysics centred on ordinary language analysis, the concept of individual is not a concept under which items fall or don't fall ("to be an individual", or not), not even a continuum, but a position in a two- or three-dimensional conceptual space. And with respect to Lowe's metaphysics, in which x is an individual iff "(1) x determinately counts as *one* entity and (2) x has a determinate *identity*"

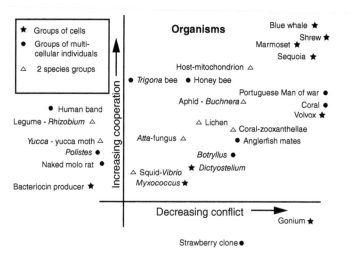

Figure 3.1 The space of organismality (figure reprinted from Strassmann and Queller 2010, p. 607, with permission from Wiley).

(Lowe 2016, p. 50, emphases in original), one sees that his major criterion ("determinately counts") is too coarse-grained to allow for picking out a set of entities as an individual in this space of individuality. This criterion would require identifying the appropriate sortal terms under which a putative individual falls, but the possibility of doing this presupposes that one has already distinguished the individual so as to be able to recognise it as falling under one particular sortal. Does a colony of *Trigonium* bees count as one entity because it has lots of cooperation, while the *Dictyostelium*, those "slime moulds" that are only cohesive entities when resources get lacking (Bonner 2009), does not? But then, those bees may have many conflicts because some new queens can hatch, while *Dictyostelium* has much less conflict. So shouldn't we count as "one entity" the *Dictyostelium* instead?

Yet a second teaching of this space of biological individuality is the following. Strassmann and Queller's table includes the mutualistic couple formed by aphids and the bacteria *Buchnera*. In mutualisms like these, of course, no genetic homogeneity can be assumed, unlike the cases of multicellular organisms or bee colonies where much homogeneity is the basis of the evolutionary transition towards some individuality. Queller (1997) calls such cases "egalitarian transitions", for the cooperators are not brothers (since there is no genetic relationship between them, and, hence, it's not a "fraternal transition"), but they are rather on a par. If one considers such egalitarian transitions to individuality then, in contrast to the discussions presented in Section 3.2.3, multicellular organisms and ant colonies should not be taken as the essential paradigm of biological individuality.[4] The models of transitions to individuality centred on altruism, multilevel selection or kin selection, describe individuals that are more or less genetically related; but we have just seen that this is absolutely not all there is to biological individuality. In the next section, I'll consider the consequences of this state of affairs for the Hullian-inspired views of individuality and introduce an alternative view.

3.3 Organisms as ecosystems and the individuality issue

3.3.1 Egalitarian transition and symbioses

Evolutionary biology presents us with transitions towards individuality, and this research raises key questions for concepts of individuality. As explained above, these transitions can be either *complete* or *component transitions*; and they can be *fraternal* transitions, with respect to which we have many powerful models that comply with the Hullian approaches to individuality, or *egalitarian* transitions. The latter occur each time individuality is gained by lasting assemblages that involve some degree of mutualism – i.e. cooperation between individuals of unrelated species (Noe and Hammerstein 2001). But it is now obvious that egalitarian transitions are all over the place in biology, as indicated by the pervasiveness of symbiosis (Gilbert et al. 2012).

It has indeed turned out that many kinds of living beings arose from symbioses, in such a way that the symbionts finally can't live alone. Eukaryote cells, for instance, did not arise from the mere grouping of related cells that finally cooperated, somehow constraining the fact that selfish mutants can, in the short term, have an evolutionary advantage: they were rather made up by the symbiosis between an archaeon and a bacterium (Alvarez-Ponce et al. 2013, Meheust et al. 2015). In the same way, mitochondria appeared within the cell as an ancient symbiosis between two bacteria (Margulis 1970); and the chloroplasts in the cells of green algae and land plants that allow them to photosynthesise are plastids that originated through endosymbiosis between a primitive eukaryote and a cyanobacterium that it engulfed. Thus here genetically unrelated individuals make up new individuals.

The role of interspecies associations in the evolution of individuality is widespread: most strikingly, placental mammals acquired their major characterising trait (as compared to marsupia), i.e. the placenta, with the help of which the embryo gestates, through having integrated genes from endogenous retroviruses[5] (Dupressoir et al. 2012). Mammals also digest through the symbiotic relationship with thousands of kinds of commensal bacteria, which constitute the gut microbiota. We are now acquainted with some surprising facts about this microbiota: the amount of non-human genetic material in us exceeds that in our own proper genome, and the weight of bacteria within mammals constitutes a significant part of a mammal's own weight (Pradeu 2010). Standard mammalian functions such as epithelial scarification require a functional microbiota.

3.3.2 *The metaphysical issue raised by multispecies individuals – a challenge to Hullian scientific metaphysics*

These egalitarian transitions and the symbiotic individuals they yield raise issues for the concepts of individuality that I have considered in the context of Hull's scientific metaphysics. First, it is not obvious how to model natural selection to capture the principle of individuality since there is no intuitive answer about who are the individuals in the sense of units of selection (van Baalen 2014). Granted, regarding these multispecies assemblages, one could think of the interaction between two processes of evolution, one in which symbionts have been selected against rival symbiont lineages, and one in which hosts have been selected between competing host lineages. This would reintroduce a classical understanding of evolution by natural selection, as in cases of fraternal transitions described above; yet this raises mathematical problems and still has to be legitimated as a correct simplification of the real process (van Baalen 2014).

Second, mutualism affects even the supposed paradigms of individuality that are multicellular metazoan organisms, which evolved through the complete fraternal transitions that the Hullian view properly addressed. These individuals comprise parts that are not cells or parts of the cellular machinery, and that may live within or outside them, but that are needed to perform their functions: some viruses, as I indicated; some plasmids; even prions may be involved in important biological functions such as memory (Shorter and Lindquist 2005) and beyond.[6] The "Roscoff flatworms" (*Symsagifera roscoffensi*), for instance, eat some green algae but don't digest them – yet this seeming lack of functionality allows them to rely on these algae's photosynthesis in order to get resources, and that's how they survive and reproduce (and, incidentally, look like algae). Dupré and O'Malley (2009) give many examples of this sort. In all of them the question is: A relies on B's genes to perform a proper function of A, therefore what exactly is the individual? The functional criterion of A's individuality won't meet another criterion, gene-based *à la* Janzen[7] or selection-based *à la* Hull. This is all the more challenging for the idea that well-defined units of selection, with a proper biological characterisation, can be identified to constitute the individuals of the biological world.

Finally, a major challenge for the Hullian view of biological individuality appears when we consider the crucial role played by *reproduction* in the process of natural selection. The considerations here sketched, which insist on the collective and heterogeneous character of living things, make salient that reproduction itself – the prerequisite of any natural selection mechanism – is not a process likely to involve only entities of one kind. In many species successful reproduction involves several agents of different lineages, as exemplified by the well-studied mutualism or parasitism between the bacteria *Wolbachia* and many arthropods: a wide-ranging phenomenon since 25–88% (according to various estimates) of insect species are estimated to be infected by *Wolbachia*. The parasitic wasp *Asobara tabida* has indeed been shown to be unable to reproduce when *Wolbachia* are killed by antibiotics (Dedeine et al. 2001). Those cases prove problematic because, even though the measure of selection can be made by counting the wasp's offspring, the process of reproduction itself involves more than the wasp, and low fitness may come due to the *Wolbachia* and not due to the wasp.

Consequently, the Hullian views of individuality that rely on natural selection face a major issue when it comes to the "symbiotic view of life" (Gilbert et al. 2012) that is currently emerging.

3.3.3 A radical hypothesis: individuals as ecosystems

Let's focus on cases like slime moulds, siphonophores, *Botryllus schlosseri* and chimera-like organisms, that behave in quite different ways (Wilson 1999), and are scattered across the space of individuality. When one acknowledges their main features, especially the role of symbionts as well as the division of labour between mutualistic entities, etc., a controversial claim arises: a biological individual *is* an ecosystem; many species hang together in the same space, having many interactions, and somehow collaborate to foster a form of stability, e.g., constancy of some physiological variables, as

Tilman (1994) defined stability of an ecosystem by the constancy of its biomass (van Baalen and Huneman 2014).

Proposing that biological individuals are ecosystems raises an obvious objection: ecosystems, or ecological communities, are too ontologically weak, too explanation-dependent, to be considered as individuals. Thus, the issue one will face in order to support the emerging claim about individuals as ecosystems is to make sense of some communities and, moreover, ecosystems as possible *bona fide* individuals.

But for now, what does it mean to equate biological individuals and ecosystems?

It first means a promising research avenue, at least from a heuristic point of view: physiological processes in metazoans could be seen as ecological interactions, with the focus on mutualisms that allow the fulfilment of some functions, or on the niches that are built for stem cells (Costello et al. 2012; see van Baalen and Huneman 2014 for some examples), or even on the competition between cell lineages that can account for both cancer and immunological defences against cancer (Featherston and Durand 2012). Ecological approaches to cell biology at the molecular level (Nathan 2014) have been developed for almost a decade.

This research perspective embodies a specific ontological view of biological individuality. Classically, there is an epistemic privilege conferred to metazoan multicellular organisms arising through a one-cell bottleneck and genomic homogeneity, for they have always been conceived of as paradigmatic individuals; this supports the view that individuals' identity is fixed by their genes. The research perspective advocated here denies such a privilege. Here metazoans are rather individuals in the same sense as *Dictyostelids*, or *Botryllus schlosseri* and other creatures that include in their life cycles stages of fusion of genomically distinct individuals: all of them are individuals in the sense that they are ecosystems. This is essentially the case because their parts with (genomically, epigenetically and phylogenetically) heterogeneous natures are related through all kinds of ecological interactions: mutualism, but also predation, trophic interactions of many kinds, parasitism, competition and "niche construction" (some biologists talk of the niches of stem cells (Scadden 2006)) or "ecosystem engineering",[8] together help to build these ecosystems that we see and label as "biological individuals", and even often as "organisms" (the difference between these concepts will be addressed below).

It is possible that the rate, range or input of each of these interactions differs according to which kind of individual is considered. Metazoans surely display lots of mutualism, which has sometimes evolved into endosymbiosis: the microbiome is a crucial example. And if, following Turner (2000), we believe that termite mounds are biological individuals, then the interaction involved in ecological engineering, which is crucial in the making of the mound, will obviously be a major kind of interaction.

In such a perspective, the axes of the conceptual space of individuality would be the distinct ecological interactions, and one could then specify each "individual", i.e. ecosystem, on this map of ecological interactions, these being for instance classified by degree of proximity (e.g., intimate vs. distant: ant-plant vs. fig-wasp mutualisms) and the number of entities involved. Interactions can be antagonistic (competition, predation) or positive (mutualisms such as pollination, while obligate pollinisation interaction may ground possible individuals) in the sense that the existence of a species increases the abundance of one or other species. This latter interaction has been recently called "facilitation" (Bruno et al. 2003). The idea is that whereas antagonistic interactions such as competition reduce the "fundamental niche"[9] of the least successful competitor, facilitation increases its fundamental niche (because for instance some new resources become available). The way an ecosystem embodies those interactions characterises therefore its functional cohesion, its capacity for persistence and its consistency. But the list is not exhaustive and other interactions may be added as axes (Figure 3.2).

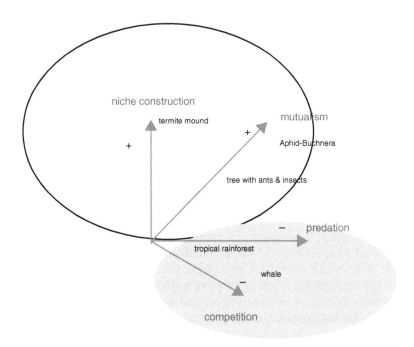

Figure 3.2 Ecological space of individuality. Possible ecosystems are mapped onto this four-dimensional space. Facilitation interactions are oriented towards increasing strength; antagonistic interactions are oriented inversely. The positions of the target systems are only indicative (in grey shading: antagonistic interactions; in no shading: facilitation interactions).

The perspective shift I propose here moves from the space of individuality sketched by Strassmann and Queller (Figure 3.1) to a space of ecological individuality (Figure 3.2), where axes are ecological interactions. The links between the two spaces are complex. Natural selection indeed supervenes on selective pressures, which are the ecological demands on a biological individual; those demands are defined by the ecological interactions it undergoes. Hence, the evolutionary space of individuality (as designed by Strassmann and Queller (2009)), which is based on effects of natural selection (cooperation, conflict, etc.), supervenes in some sense on the space of ecological individuality whose axes are the ecological interactions, which together make up the selective pressures (see Figure 3.3). The whole issue of their relation relies on the question of the articulation between ecological interactions and natural selection, and controversies are raging between those who see natural selection as a statistical result of the ecological interactions, and those who grant natural selection a proper causal efficacy (see e.g., Walsh 2010). A discussion of those relations would be too long here, and somehow remains independent of our question.

What unifies the various things that inhabit the individuality space sketched above and that seem to be individuals is that they all are ecosystems and characterised as specifically unified ecosystems – this specificity determining their position in the ecological space just described. To sum up, if something appears as a biological individual, and if one wants to be as encompassing as one can be regarding the space of individuality – therefore granting some individuality to assemblages like aphid-*Buchnera* that Strassmann and Queller plotted in their diagram – then the nature of those things should consist in being an ecosystem. Yet this assumes that there is a general metaphysical sense of "individuality" that is fulfilled by ecosystems in general, or at least by some ecosystems, a claim that has been denied since Clement's defence of superorganisms (1916) within debates in ecology (e.g., Cooper 2003), and then by many philosophers till now.

Thus, in order to support this conception of biological individuality understood as ecosystem, a view embedded in current programmes of research, one should provide a metaphysical concept of individuality that suits some ecosystems, while descriptive metaphysics and some revisionary metaphysics as well as some ecologists and philosophers of biology concur in denying that ecosystems in general could be individuals. It is not argued here that such an ecosystem-based conception is the only true one, but as I have just suggested, it's an interesting, possible, and in some ways very attractive view; and I will now show that a convincing metaphysical basis can be provided for it.

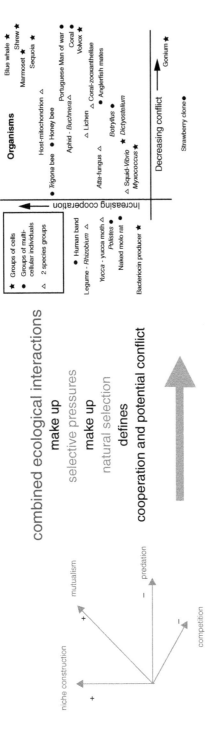

Figure 3.3 The ecological interactions defining the space of ecological individuality and ultimately the cooperation/conflict relations that define the evolutionary individuality space (partial reprint from Strassmann and Queller 2010, p. 607, with permission from Wiley).

3.4 Ecosystem individuality as weak individuality

A major reason for contesting that ecosystems in principle could be individuals precisely relies upon the Hull-inspired views of individuality. In this perspective, ecosystems would be individuals to the extent that selection indeed targets ecosystems or at least some of them; but many biologists are still unconvinced that natural selection can target ecosystems, and in this case, ecosystems would not be individuals; hence, the nature of organisms and biological individuals could not be ecosystem-like. This scepticism relies first on a conceptual issue: how could an ecosystem give rise to *response to selection*? In quantitative genetics for instance, we measure a response to selection as the change in frequency of traits of alleles that is triggered intergenerationally by selective pressures – this change being determined by *heritability*. Some have tried to define a response to selection in the case of *community selection* (Goodnight 2011), but even this remains complicated because in a community different timescales may pertain to different genomes. Thus, even though community selection is plausible, it should be in practice very hard to find. For ecosystems themselves it seems therefore to be even more puzzling, because these by definition include many *abiotic* elements, which in principle do not participate in a response to selection since they instantiate no heritability. Then we would be left with no ontologically robust ecosystems that are *bona fide* individuals.

In this context it has been argued (Huneman 2014b) that ecosystems or communities can still be individuated on the basis of a scheme of decoupling between kinds of interactions, even though natural selection can't provide us with an individuating principle. This relies on the view of modularity advanced by Simon (1980). It gives rise to a very general "weak concept of individuality".

Weak individuality can be summed up in the following terms: assuming that individuals are made up of some entities, so that the individuality issue is about which assemblages of entities count as individuals and which ones don't, then among a set of entities whose interactions are known and modelled with respect to a range of parameters (defined in one of our best theories of these interactions), individuals are the subsets of entities that interact most between themselves rather than with the rest.

Of course, mentioning assemblages of *entities* in the definition raises the spectrum of a regression ad infinitum; here it suffices to say that the concept is intended to be a scientific metaphysical concept, and therefore its use is intrinsic to a theory, which assumes the validity of other theories; those assembled entities under focus are given by such other theories – for instance in ecology organisms are given by biological theories assumed by ecologists, and the individuality question is about which ones, once assembled, constitute ontologically grounded communities and then ecosystems.

Translated in probabilistic terms the basic intuition of weak individuality here is that if you are part of an individual, the chances that something that is strongly relevantly interacting with you is also part of this same individual are higher than the chances that it is something external. For instance, in

Exploring a radical hypothesis   55

any metazoan organism the chances that some cell in strong interactions with a given organism's cell is part of this organism are higher than the chances that it is a cell from another organism. Having such a property would thereby imply the following: considering a set S, if A is in S, then "S is an individual" means that the chance of A having one of its strongest interactions with something within S is higher than the chances that it is in interaction with something outside S.

A general definition of an individual can therefore be written as (following the theory presented in Huneman 2014a):

Consider S a set of entities; i is a given entity in S.
$x_{i,j}$ is a link between i and j (namely, any kind of interaction; j is in S)
H, n are constant values defined in advance (n between 0 and 1; the variable h is the value of the strength of interaction, n is a significance threshold)
$P(x_{i,j}, h)$: probability that a link $x_{i,j}$ has strength h
We define the set of i-centred strong interactions: $H_i = \{x_{i,j}$ such that $P(x_{i,j}, H) > n\}$
We define the set of i-strongly interacting entities: $J_i = \{j$ such that $x_{i,j}$ belongs to $H_i\}$
If S is an individual system then for all i in S, J_i is included in S.

n is an arbitrary fixed value, but it is very likely that some value between 0.7 and 0.9 – or in general the value of significance in a traditional statistics framework, like 0.95 – in many cases will yield a result close to our intuitive ascriptions of individuality.

As to h, the strength of a given interaction, several parameters proper to the science under focus have to be estimated within its definitional computation. For instance, if we consider that all the entities are in a network of entities like in the case of networks in community ecology (Dunne et al. 2002, Montoya and Solé 2002, Montoya et al. 2006), and can be represented in a graph (each interaction being a link), then the main parameters would include at least the number of the various pathways between entities (the more the pathways, the stronger the interactions), and the number of steps between i and j (the fewer the steps, the stronger the interaction).

H is the strength of "strong" interactions; it is reasonable to choose a quite high value, determined for example by the threshold defining the first quintile of interactions when all interaction strengths are measured. Yet, the fact that we take n as a constant, plausibly reflecting high chances of strong interaction, can be alleviated by another option: considering the *average* value of $P(x_{i,j}, H)$, written $\underline{P}$. Then the sets J_i so defined will be sets where the probability of interactions between entities is clearly higher than average, and therefore could be considered as individuals. Hence, the clause about defining the sets H_i can be rewritten as:

$$H_i = \{x_{i,j} / P(x_{i,j}, H) > \underline{P}\}.$$

In this case one has a statistical threshold of significance defined by the system itself, namely the general set of interactions within which one begins to look for individuals. Applying this scheme onto our models of interactions allows one to discriminate subsystems that are indeed more cohesive, self-contained and distinct than the set consisting of the remainder of the entities; it is reasonable to say that these subsystems or subsets are individuated through their interaction patterns, these interactions being of the types considered by the scientific theory under focus.

Of course, the size of the initial set of entities, the is, impacts upon the number of individuals likely to be found. But here it's methodologically reasonable to start with quite large sets; this is indeed what always happens in ecology where the number of interactions in a given setting is usually very large.

Note also that the notion of individuality here appears as a relative one: in the case where all interactions are of the same strength, even if it's a high strength, there would be no individuals, exactly as all molecules in a mass of water may have comparable strength of interactions and don't result in distinct watery individuals. But this suits well the idea that "individuality" exists in a space of individuality so that "individual" in fact means that some collection is more an individual than others, so that to be an individual only means to be "more or less an individual" (even though we saw there is no total order on the set of possible individuals). A feature of the weak concept of individuality is therefore this intrinsic relativity. This might be counterintuitive, but, as recalled by David Lewis (1986), metaphysics may assert some counterintuitive claims (he thought here of his own modal realism), though it's the price to pay for reaching one's epistemic goals (e.g., intellectual clarity, systematicity, accounting for more features than the natural image of the world, etc.).

Importantly, the characterisation of weak individuality given here only defines a "formal" concept of individuality, in the sense that what makes an interaction "strong", measured by the h variable in the above scheme, is only determined by a specific theory, since the concept of the strength of an interaction is only defined within a theory by a weighted combination of factors that cannot be the same in a different theory (for instance, measuring the strength of interactions in particle physics involves considerations that make no sense in community ecology; or, frequency of interaction would make sense for strong interactions in community ecology, but not in immunology).

Even though scientists don't formally use this procedure to pinpoint individual ecosystems, this is a rational reconstruction of the individuation procedures as they are allowed by the knowledge of interactions parallel to the case of causality according to Woodward (2003). According to the formal concept of weak individuality, "individuality" is not a concept in the usual sense (e.g., a predicate whose ascription is defined by necessary and sufficient conditions), but *rather a scheme for extracting individuals on the basis of the theories that elaborate our best models of interactions* in a given ontological domain.

To sum up, in biology applying concepts that describe ecological interactions allows theoreticians to use this formal decoupling scheme of weak individuality, and thus to single out those sets of interactions that define ontologically individual ecosystems (in the weak sense of individuals). Weak individuality provides thereby a way to answer the ontological objection stated above against ecosystem individuality and can differentiate genuinely ontological ecosystems or communities from merely indexical communities or ecosystems. Since there is a sense in which ecosystems are individuals, i.e. some ecosystems studied are really and objectively individuals (in the sense of being grounded in our best theories of objects, and not purely subjectively), then the controversial claim that biological individuality should be conceived of in terms of ecosystems can be supported. The nature of biological individuality in its highest generality consists in being an ecosystem; and the metaphysical concept of individuality required to think of such individuality is the weak concept of individuality just forged. Given that we now have a metaphysical science-based concept of individuality according to which some ecosystems are genuine individuals, one is allowed to adopt the hypothesis that biological individuals are ecosystems in order to make sense of research programmes dealing with individuals such as organisms as if they were ecosystems.

The weak concept of individuality is intrinsically pluralist, since various instantiations of the variable h representing the force of interactions make for distinct "material" (as opposed to "formal") concepts of individuality. Such a pluralism concurs with a critique often addressed to Hullian views of biological individuality: by favouring an evolutionary viewpoint, they forget another aspect of individuality, especially the functional metabolic coherence dealt with by physiologists, and often captured by the concept of organism. Godfrey-Smith (2013) distinguishes to this extent "Darwinian individuals", defined by their ability to undergo natural selection, from "physiological individuals" such as biological organisms (ordinary metazoans are indeed comprised within the overlap of the two classes); and Dupré and O'Malley (2009) try to integrate lineages (hence genes) with metabolism to capture where life occurs. This "evolutionary individuals/organismality" pluralism is attractive but is still unable to make sense of some of the very weak individuals that inhabit the individuality space, and more generally of the fact that individuality seems to happen in gradual ways. Inversely, the organism exhibits a kind of individuality that emerges within specific theories, especially developmental theory and physiology: to this extent, a "material concept" of weak individuality (in which the variables are implemented on the basis of the theories of interactions that those sciences propose) may be likely to capture the organism concept, and therefore the duality "organism/evolutionary individual" would be integrated in the more general pluralism that I suggested, where various local material concepts of weak individuality coexist with the specific evolutionary perspective on individuality offered by the Hullian concepts.

Obviously, this pluralism regarding biological individuality entails a pluralism regarding the concept of biological identity – both as distinctiveness, and as persistence of the individual – which could in turn lead to various sets of identity conditions depending on the respective perspective through which a biological item X is considered as individual.

3.5 Conclusion

A plausible view of organisms, and of biological individuals in general, as ecosystems emerges from various explorations of individuality reviewed here, which emphasised the insufficiencies of a Hullian view in this regard, and is supported by a metaphysical concept of individuality conceived of as weak individuality. Such a concept captures the general decoupling scheme according to which in various sciences the models of interactions allow one to identify, parse and distinguish individuals. This scheme is very general but is differently instantiated in each theory, because the theoretical terms likely to describe interactions and therefore to measure and compare them are proper to each science. It pertains to a science-based metaphysics, and notions of biological identity should therefore be conceived of along the same lines.

Acknowledgements

I am grateful to various audiences in conferences where this paper was presented, especially the summerschool "Superorganisms, Organisms and Suborganisms as Biological Individuals" in Gut Siggen, Germany, in July 2015. Many thanks for great comments to Sébastien Dutreuil, Marie Kaiser, Thomas Pradeu, Thomas Reydon, and Christian Sachse, and to Andrew McFarland for a language check. Criticisms and comments, as well as a thorough editing, by the editors of this volume, Anne Sophie Meincke and John Dupré, have been crucial for improving this chapter. This work is supported by the Laboratoire International Associé CNRS Paris-Montréal "Epistemic and Conceptual Issues in Evolutionary Biology" and the GDR CNRS 3770 Sapienv.

Notes

1 Notice that hierarchies encompass individuals, but not all levels of the ecological hierarchy are individuals: genes, demes or clades are not obviously individuals.
2 For instance, how, with respect to a given Portuguese man-of-war, a marine organism made up of many multicellular organisms, could we decide between claims like "there are many individuals here" (counting all the component organisms) or "there is one individual"?
3 To sum up, all these explanations embed some form of Hamilton's rule, which says that a behaviour taking place between a focal actor and another entity

whose relatedness to the first is r (often measured by the numbers of genes in common, in addition to the average number of genes in common in the population), and with a cost c for the actors and benefit b for the others, evolves iff c<br (West and Gardner 2013).
4 See Haber (2013) about the notion of a "paradigm approach" to biological individuality.
5 Where the placenta meets the fetus, a layer emerges, called syncytiotrophoblast; cells that are able to fuse together do it because they produce a protein called syncytin, only made in those cells. This envelope plays a crucial role in the gestation of the fetus. Genes that encode such proteins in humans have been shown to come from viruses that infected the germ line in early hominids (Mi et al. 2000). But the story is much more complicated. All primates actually fixed varieties of the syncytin genes (Mallet et al. 2004), labelled syncytin-1 and -2, and then it later appeared that other varieties of syncitin genes, integrated from different viruses, also characterise other lineages among mammals (carnivorans, mice, etc.) (Dupressoir et al. 2012).
6 "Proven examples [of positive contributions of prions to biological functioning of metazoans] include self/nonself recognition, stress defense and scaffolding of other (functional) polymers. The role of prion-like phenomena in memory has been hypothesized. As an additional mechanism of heritable change, prion formation may in principle contribute to heritable variability at the population level" (Inge-Vechtomov et al. 2007, 228).
7 Janzen (1977) famously argued that notwithstanding the possibility of a plurality of distinct phenotypes, such as dandelions in a field, what makes for an individual is the existence of a given genotype that is shared; all bearers of this genotype are the same individual.
8 Notice that niche construction was first investigated by functional ecologists such as Clive Jones (e.g., Jones et al. 1994) under the name "ecosystem engineering" – the phrase "niche construction" being only later used in the evolutionary perspective developed by Odling-Smee et al. (2003). "Ecosystem engineering" denotes all transformations of the environment induced by a species' activity, such as spiders building webs or earthworms modifying the pH of soils through continued osmotic exchanges.
9 In community ecology, the "niche" of a species is a volume in the hyperdimensional space whose axes are the relevant environmental parameters (pressure, temperature, light, predators, etc.) within which a given species can thrive (Hutchinson 1957). When the fundamental niches of two species overlap, the overlapping region ends up being inhabited by the species which is the best competitor, the two resulting non-overlapping niches being called "realised niches": this is called the "competitive exclusion principle", and the extent of its explanatory power has been discussed by ecologists.

References

Alvarez-Ponce, D., Lopez, P., Bapteste, E. and McInerney, J. (2013) 'Gene Similarity Networks Provide Tools for Understanding Eukaryote Origins and Evolution', *Proceedings of the National Academy of Sciences of the USA* 110(17), pp. 1594–1603.

Bird, A. (2007) *Nature's Metaphysics*, Oxford: Oxford University Press.

Bonner, J. T. (2009) *The Social Amoebae: The Biology of Cellular Slime Molds*, Princeton, NJ: Princeton University Press.

Bouchard, F. (2010) 'Symbiosis, Lateral Function Transfer and the (Many) Saplings of Life', *Biology and Philosophy* 25, pp. 623–641.

Bouchard, F. and Huneman, P. (Eds.) (2013) *From Groups to Individuals*, Cambridge, MA: MIT Press.

Bruno, J., Stachowicz, J. and Bertness, M. (2003) 'Inclusion of Facilitation into Ecological Theory', *Trends in Ecology and Evolution* 18(3), pp. 119–125.

Chakravarty, A. (2013) 'On the Prospects of Naturalized Metaphysic', in: Ross, D., Ladyman, J. and Kincaid, H. (Eds.), *Scientific Metaphysics*, New York: Oxford University Press, pp. 27–50.

Clarke, E. (2014) 'The Multiple Realizability of Biological Individuals', *Journal of Philosophy* 110(8), pp. 413–435.

Clements, F. (1916) *Plant Succession: An Analysis of the Development of Vegetation*, Washington, DC: Carnegie Institution.

Costello, E., Stagaman K., Dethlefsen, L., Bohannan, B. J. and Relman, D. A. (2012) 'The Application of Ecological Theory Toward an Understanding of the Human Microbiome', *Science* 336, pp. 1255–1262.

Damuth, J. and Heisler, L. (1988) 'Alternative Formulations of Multi-level Selection', *Biology and Philosophy* 3, pp. 407–430.

Davidson, D. (1974) 'On the Very Idea of a Conceptual Scheme', *Proceedings and Addresses of the American Philosophical Association* 47, pp. 5–20.

Dedeine, F., Vavre, F., Fleury, F., Loppin, B., Hochberg, M. and Boulétreau, M. (2001) 'Removing Symbiotic *Wolbachia* Bacteria Specifically Inhibits Oogenesis in a Parasitic Wasp', *Proceedings of the National Academy of Sciences of the USA* 98(11), pp. 6247–6252.

Doolittle, W. F. and Bapteste, E. (2007) 'Pattern Pluralism and the Tree of Life Hypothesis', *Proceedings of the National Academy of Sciences of the USA* 104(7), pp. 2043–2049.

Dunne, J. E., Williams, R. J. and Martinez, N. D. (2002) 'Food Web Structure and Network Theory: The Role of Connectance and Size', *Proceedings of the National Academy of Science of the USA* 99, pp. 12917–12922.

Dupressoir, A., Lavialle, C. and Heidmann, T. (2012) 'From Ancestral Infectious Retroviruses to Bona Fide Cellular Genes: Role of the Captured Syncytins in Placentation', *Placenta* 33(9), pp. 663–671.

Dupré, J. and O'Malley, M. A. (2009) 'Varieties of Living Things: Life at the Intersection of Lineage and Metabolism', *Philosophy, Theory and Practice in Biology* 1(3), doi:10.3998/ptb.6959004.0001.003.

Eldredge, N. (1985) *Unfinished Synthesis: Biological Hierarchies and Modern Evolutionary Thought*, New York: Oxford University Press.

Featherston, J. and Durand, P. M. (2012) 'Cooperation and Conflict in Cancer: An Evolutionary Perspective', *South African Journal Science* 108(9/10), pp. 1–5.

Folse, H. J. III and Roughgarden, J. (2010) 'What is an Individual Organism? A Multilevel Selection Perspective', *Quarterly Review of Biology* 85, pp. 447–472.

Gardner, A., West, S. A. and Wild, G. (2011) 'The Genetical Theory of Kin Selection', *Journal of Evolutionary Biology* 24, pp. 1020–1043.

Gilbert, S. F., Sapp, J. and Tauber, A. I. (2012) 'A Symbiotic View of Life: We Have Never Been Individuals', *The Quarterly Review of Biology* 87(4), pp. 325–341.

Godfrey-Smith, P. (2009) *Darwinian Populations and Natural Selection*, New York: Oxford University Press.

Goodnight, C. J. (2013) 'Defining the Individual', in: Bouchard, F. and Huneman, P. (Eds.), *From Groups to Individuals*, Cambridge, MA: MIT Press, pp. 37–54.

Haber, M. (2013) 'Colonies Are Individuals: Revisiting the Superorganism Revival', in: Bouchard, F. and Huneman, P. (Eds.), *From Groups to Individuals*, Cambridge, MA: MIT Press, pp. 196–217.

Hamilton, W. D. (1964). 'The Genetical Evolution of Social Behaviour I and II', *Journal of Theoretical Biology* 7 (1), pp. 1–52.

Hennig, W. (1966) *Phylogenetic Systematics,* transl. by D. Davis and R. Zangerl, Urbana, IL: University of Illinois Press.

Hull, D. L. (1980) 'Individuality and Selection', *Annual Review of Ecology and Systematics* 11, pp. 311–332.

Huneman, P. (2013) 'Adaptation in Transitions', in: Bouchard, F. and Huneman, P. (Eds.), *From Groups to Individuals*, Cambridge, MA: MIT Press, pp. 141–172.

Huneman, P. (2014a) 'Individuality as a Theoretical Scheme 1. Formal and Material Concepts of Individuality', *Biological Theory* 9(4), pp. 361–337.

Huneman, P. (2014b) 'Individuality as a Theoretical Scheme 2. About the Weak Individuality of Organisms and Ecosystems', *Biological Theory* 9(4), pp. 374–381.

Hutchinson, G. E. (1957) 'Concluding Remarks. Cold Spring Harbor Symposium', *Quantitative Biology* 22, pp. 415–427.

Inge-Vechtomov, S. G., Zhouravleva G. A. and Chernoff, Y. O. (2007) 'Biological Roles of Prion Domains', *Prion* 1(4), pp. 228–235.

Janzen, D. H. (1977) 'What are Dandelions and Aphids?', *The American Naturalist* 111, pp. 586–589.

Jones, C., Lawton, J. and Shachak, M. (1994) 'Organisms as Ecosystem Engineers', *Oikos* 69, pp. 373–386.

Kerr, B. and Godfrey-Smith, P. (2002) 'Individualist and Multi-level Perspectives on Selection in Structured Populations', *Biology and Philosophy* 17, pp. 477–517.

Ladyman, J. and Ross, D. (2009) *Everything Must Go: Metaphysics Naturalized*, Oxford: Oxford University Press.

Lewis, D. (1986) *On the Plurality of Worlds*, Oxford: Blackwell.

Lowe, E. J. (2001) *The Possibility of Metaphysics. Substance, Identity, and Time*, Oxford: Clarendon Press.

Lowe, E. J. (2016) 'Non-individuals', in: Guay A. and Pradeu T. (Eds.), *Individuals Across the Sciences*, New York: Oxford University Press, pp. 49–60.

Mallet, F., Bouton, O., Proudhomme, S., Chaynet, V., Oriol, G., Bonnaud, B., Lucotte, G., Duretand, L. and Mandrand, B. (2004) 'The Endogenonous Retroviral Locus ERVWE1 Is a Bona Fide Gene Involved in Hominoid Placental Physiology', *Proceedings of the National Academy of Sciences of the USA* 101, pp. 1731–1736.

Margulis, L. (1970) *Origin of Eukaryotic Cells*, Boston, MA: Yale University Press.

Maynard Smith, J. and Szathmáry, E. (1995) *The Major Transitions in Evolution*, New York: Oxford University Press.

Meheust, R., Lopez, P. and Bapteste, E. (2015) 'Metabolic Bacterial Genes Built all High-level Composite Lineages', *Trends in Ecology and Evolution* 30(3), pp. 127–129.

Mi, S., Lee, X., Li, X., Veldman, G. M., Finnerty, H., Racie, L., LaVallie, E., Tang, X. Y., Edouard, P., Howes, S., Keith, J. C. J. and McCoy, J. M. (2000) 'Syncytin Is a Captive Retroviral Envelope Protein Involved in Human Placental Morphogenesis', *Nature* 403, pp. 785–788.

Mishler, B. and Theriot, E. (2000) 'A Defense of the Phylogenetic Species Concept (sensu Mishler and Theriot): Monophyly, Apomorphy, and Phylogenetic Species

Concepts', in: Meier J. and Wheeler, Q. (Eds.), *Species Concept and Phylogenetic Theory. A Debate*, New York: Columbia University Press, pp. 179–184.

Montoya, J. M. and Solé, R. V. (2002) 'Small World Patterns in Food Webs', *Journal of Theoretical Biology* 214, pp. 405–412.

Nathan, M. (2014) 'Molecular Ecosystems', *Biology and Philosophy* 29(1), pp. 101–122.

Noe, R. and Hammerstein, P. (1994) 'Biological Markets: Supply and Demand Determine the Effect of Partner Choice in Cooperation, Mutualism and Mating', *Behavioral Ecology and Sociobiology* 35(1), pp. 1–11.

Nowak, M. A., Tarnita, C. E. and Wilson, E. O. (2010) 'Evolution of Eusociality', *Nature* 466, pp. 1057–1062.

Odling-Smee J., Laland K. and Feldman M. (2003) *Niche Construction. The Neglected Process in Evolution*, Princeton, NJ: Princeton University Press

Okasha, S. (2006) *Evolution and the Levels of Selection*, New York: Oxford University Press.

Pepper, J. W. and Herron, M. D. (2008) 'Does Biology Need an Organism Concept?', *Biological Review of the Cambridge Philosophical Society* 83, pp. 621–627.

Pimm, S. L. (2002) *Food Webs*, Chicago, IL: University of Chicago Press.

Pradeu, T. (2010) 'What Is an Organism', *History and Philosophy of Life Sciences* 32(2/3), pp. 247–267.

Queller, D. C. and Strassmann, J. E. (2009) 'Beyond Society: The Evolution of Organismality', *Philosophical Transactions of the Royal Society London B Biological Sciences* 364, pp. 3143–3155.

Quine, W. V. O. (1948) 'On What There Is', *Review of Metaphysics* 2(5), pp. 21–36.

Scadden, D. (2006) 'The Stem-cell Niche as an Entity of Action', *Nature* 441(7097), pp. 1075–1079.

Shorter, J. and Lindquist, S. (2005) 'Prions as Adaptive Conduits of Memory and Inheritance', *Nature Reviews Genetics* 6, pp. 435–450.

Simon, H. (1980) *The Sciences of the Artificial*, Cambridge, MA: MIT Press.

Slater, M. and Yudell, Z. (2017) *Metaphysics and the Philosophy of Science*, New York: Oxford University Press.

Sober, E. and Wilson, D. S. (1998) *Unto Others: The Evolution and Psychology of Unselfish Behavior*, Cambridge, MA: Harvard University Press.

Strassmann, J. E. and Queller, D. C. (2010) 'The Social Organism: Congresses, Parties, and Committees', *Evolution* 64, pp. 605–616.

Strawson, P. (1959) *Individuals: An Essay in Descriptive Metaphysics*, London: Methuen.

Turner, J. S. (2000) *The Extended Organism: The Physiology of Animal-Built Structures*, Cambridge, MA: Harvard University Press.

Umesono, Y. and Agata, K. (2009) 'Evolution and Regeneration of the Planarian Central Nervous System', *Development, Growth and Differentiation* 51, pp. 185–195.

van Baalen, M. (2014) 'Adaptation, Conflicting Information, and Stress', *Biological Theory* 9(4), pp. 431–439.

Walsh, D. M. (2010) 'Not a Sure Thing: Fitness Probability and Causation', *Philosophy of Science* 77, pp. 141–171.

West, S. A. and Gardner, A. (2013) 'Adaptation and Inclusive Fitness', *Current Biology* 23, pp. R557–R584.

West, S. A., Griffin, A. S. and Gardner, A. (2008) 'Social Semantics: How Useful has Group Selection Been', *Journal of Evolutionary Biology* 21, pp. 374–385.

Wilson, J. (1999) *Biological Individuality: The Identity and Persistence of Living Entities*, Cambridge: Cambridge University Press.

Woodward, J. (2003) *Making Things Happen: A Theory of Causal Explanation*, Oxford: Oxford University Press.

4 What is the problem of biological individuality?

Eric T. Olson

4.1 Biological individuals

One of the fundamental questions of biology is how to characterise life. We know, for example, that life involves complex structures that are intrinsically unstable and in need of constant renewal. And this renewal comes from within. Life builds and maintains itself: we don't need to take it to the shop for repairs. We know that this building and maintenance is accomplished through metabolism: life takes in materials, imposes a characteristic form on them, then expels them in a less ordered state. We know that life grows and reproduces, making more life of the same sort. This too is not merely a result of external circumstances: fires proliferate as temperature, fuel, and oxygen allow, but life expands according to an internal plan. These and other features distinguish the living from the non-living world.

But knowing what life is raises a second large question: how does it divide into units? Life does not simply occur here and there, like wind or gravity. It comes in ecosystems, species, herds, lineages, generations, organisms, and cells.[1] For there to be life is for there to be living things, just as for there to be movement is for there to be things that move. The jargon calls these units *biological individuals*.

The concept of a biological individual plays a vital role in the life sciences. A mature Portuguese man-of-war looks just like a single large animal, but the way it develops leads most zoologists to consider it a colony of many small ones. This is a claim about how life divides into organisms. The theory of evolution, to take another example, is about the spread of traits or genes through populations – their change in frequency from one generation to the next. And the very idea of a generation involves a distinction between parents and their offspring, and thus a division of life into individuals (Clarke 2010, p. 313, Godfrey-Smith 2015). For that matter, whether a gene becomes more or less frequent depends on how many individuals it occurs in. The number of occurrences within a single organism doesn't count towards its frequency: you don't spread your genes in the evolutionary sense by putting on more cells with those genes, but only by producing more organisms.[2] This implies that biological individuals must be countable (Clarke 2010, p. 313, Godfrey-Smith 2014, p. 67).

Knowing what life is and where it occurs does not tell us what individuals it consists of. We could know which spacetime regions contain life and why, yet have no idea how to distinguish colonies from single organisms, parents from offspring, or reproduction from growth.

The question of how life divides into units is called the *problem of biological individuality*. My interest is not in solving the problem, but in stating it. What exactly is the question that theories of biological individuality are supposed to answer? What would count as a solution to the problem? Actual statements of the problem have been vague and incomplete. What's more, proposed theories of biological individuality are not detailed enough to solve the problem even if they are correct. In many cases they have entirely the wrong form. The root of these troubles, I believe, is that philosophers of biology have not recognised the metaphysical claims presupposed in the discussion. These claims are implicit not only in proposed solutions to the problem, but in the statement of the problem itself. Making these claims explicit will enable us to see better what the problem is and what form a solution to it would need to have.

4.2 Boundary-drawing and its limitations

I will focus on organisms and leave aside other biological individuals. Organisms are the units that develop from a single cell in ontogeny, that fall ill and fight infections, that engage in reproduction, and that occupy the roles in ecological systems (Clarke 2010, p. 313). I take them to be the paradigmatic individuals in biology. But nearly everything I say will apply equally to others.

Thinking of the problem of biological individuality as how life divides into organisms suggests that it has to do with boundaries within the regions where life is present: boundaries around organisms or between one organism and another. These boundaries need not be precise – some atoms might be neither definitely parts of me nor definitely not, for example – but there could not be an organism with no boundary at all.

We can ask about organisms' spatial boundaries – their size and location at a given time – as when we ask whether we have a single complex organism or a colony. Or we can ask about their temporal boundaries: when they begin and end. What counts as a generation has to do with the temporal boundary between parent and offspring. Sometimes both questions operate at once: to ask whether a trait has increased in frequency is to ask both about the temporal boundaries between organisms of different generations and about the spatial boundaries between organisms within each generation.

But not all questions of biological individuality are about the location of boundaries. Many animals have symbiotic bacteria in their gut that are needed for healthy digestion and tolerated by their immune system. We can ask whether this makes them parts of the animal (giving it smaller organisms as parts), or whether the gut is simply a habitat for them, so that they

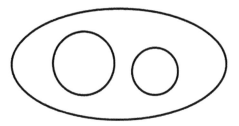

Figure 4.1

and the animal form a colony. This is a question about how life divides into organisms, but it's not about the location of boundaries. Knowing where the boundaries lie – the boundary of each bacterium and the outer boundary of the animal – would not tell us whether the bacteria are parts of the animal. Learning whether they are parts of it would not enable us to draw any further boundary. It would only tell us whether the boundary of each bacterium is also part of the boundary of the animal: whether the bacteria, like grains of sand, exclude the animal from their location.

We can also ask whether my gut bacteria themselves make up or compose an organism – where by definition some things, the xs, compose something y just if each of the xs is a part of y and every part of y shares a part with one or more of the xs.[3] This is analogous to asking whether the members of an ant colony compose a "superorganism" (Wilson and Sober 1989). These too are questions about individuality, but not about where boundaries lie. They are about whether the boundary of each individual insect or bacterium (when they're not touching) is also part of the boundary of a larger organism.

So we could know the location of all the boundaries around organisms without knowing what organisms there are – that is, which things they are boundaries *of*. Suppose Figure 4.1 shows all the boundaries around organisms. (That it shows just two dimensions is due to my department's limited art budget. Imagine it showing three dimensions of space and one of time.) Suppose life is present throughout the oval region and nowhere else. What would this tell us about what organisms there are?

It would tell us that there are at least two: one whose outer boundary is oval and one located entirely within it. But it would not imply that there are only two. In fact it would not tell us the boundaries or the size of any organism. Given our assumptions, the information in the diagram is consistent with these descriptions:

- There is an organism occupying the whole of the oval region apart from the two round regions, an organism occupying the larger round region,

and an organism occupying the smaller round region: three in all. (By "occupy" I mean "exactly occupy". A thing occupies a region when its boundaries are the same as the region's.)
- There is an organism occupying the whole of the oval region including the round regions, and two round organisms. In this case, in contrast with the first, the round organisms are parts of the oval organism.
- There is an organism occupying the whole of the oval region apart from the larger round region, and two round organisms. The smaller round organism but not the larger one is a part of the oval organism.
- As before, only the larger round organism but not the smaller is a part of the oval organism.
- There is an organism occupying the whole of the oval region apart from the round regions and an organism occupying the sum of the round regions that is not a part of the oval organism: two in all. There are no round organisms.
- There is an organism occupying the whole of the oval region and an organism that is a part of it occupying the sum of the round regions.

There will be more possibilities if an organism occupying the sum of the round regions could have round organisms as parts, giving three "levels" of organisms: round, composed of several round ones, and oval. In that case there could be an oval organism, an organism occupying the sum of the round regions, and two round organisms. Any of the smaller organisms may or may not be parts of the oval one.

This shows that knowing the boundaries around organisms barely begins to tell us what organisms there are. The problem of biological individuality is not just about where these boundaries lie, or what determines them, but also about what determines which regions are occupied by an organism. Given that things are located where their parts are, this is more or less equivalent to asking what makes things parts of a single organism.[4]

The situation will be yet more complicated if (as many philosophers believe) two things can occupy the same spacetime region and be made up of the same matter throughout their careers. Jack Wilson takes organisms to come in different kinds: there are, among others, *genetic individuals*, defined in terms of genetic homogeneity and common ancestry, and *functional individuals*, defined in terms of causal integration (1999, pp. 86–99). A genetic and a functional individual can coincide exactly in both space and time (1999, p. 47). But nothing can belong to both kinds, as they entail different modal properties, such as the conditions in which they would continue existing. In that case, even knowing which regions are occupied by organisms would not tell us what organisms there are. There could be any number of them in the diagram.

To simplify matters I will set this possibility aside. That enables us to state the problem of biological individuality as asking what determines which spacetime regions are occupied by organisms.[5] To say that something *determines* what regions are occupied by organisms is to say that what regions these are follows from this something together with the "underlying" facts about non-organisms that the discussion takes for granted, such as the distribution of matter and the laws of nature. A theory of biological individuality will be a principle or set of principles answering this question. Given our simplifying assumption, this will also tell us what determines how many organisms there are and what distinguishes one from another.

4.3 The problem of psychological individuality

The problem of biological individuality is presumably one problem of individuality among others. There is individuality in other phenomena besides life. Psychology, for example – *mind* in the mass-noun sense – does not simply occur here and there, like wind or gravity. It comes in units. The most important units are individual mental subjects: thinking or conscious beings, *minds* in the count-noun sense. For there to be thought or consciousness is for there to be thinking or conscious beings. So we should expect there to be a question of how mind divides into mental individuals analogous to the question of how life divides into organisms.

And in fact there are puzzles about psychological individuality very like those concerning biological individuality: about what mental individuals there are and what distinguishes one from another. We can know where mind is present without knowing what mental individuals there are. In cases of conjoined twinning, multiple personality, or commissurotomy, for example, we know that there is consciousness, but there is dispute over the number of conscious beings (Puccetti 1973, Olson 2003, 2014, Campbell and McMahan 2010). There is also dispute over the temporal boundaries of mental individuals, manifest in the literature on personal identity over time. There are even views analogous to Wilson's, saying that distinct mental individuals can occupy the same spacetime region (Baker 2000, p. 103, Hawthorne and McGonigal 2008). The parallels between the two problems will be instructive.

4.4 "Defining the biological individual"

I have argued that the problem of biological individuality is the question of what determines the spatiotemporal locations of organisms. An answer will tell us what determines how many there are and what distinguishes one from another. The problem of psychological individuality is analogous.

But philosophers of biology have stated the problem in a very different way. They take it to be what it is to be an organism, as opposed to a non-organism: to "define the biological individual".[6] By their lights, a theory of biological individuality would be a completion of the formula:

x is an organism *iff*...x....

(It may need some sort of necessity operator to ensure that it's not true only by accident: even if every organism contains potassium, that's no part of what it is to be an organism and should not appear in the definition.)

This is puzzling. How could an account of what it is to be an organism tell us what determines an organism's spacetime location, or how many there are, or what distinguishes one from another? An account of what it is to be an F does not generally tell us what determines the locations of Fs. Suppose we define "artefact" as "object fashioned by an intelligent being for a purpose". It's hard to see how this, by itself, could tell us anything at all about what determines an artefact's location or how many there are. Yet that doesn't make the definition wrong.

Think of the problem of psychological individuality again. No one expects a definition of "mental individual" to tell us what determines the spatiotemporal locations of mental individuals or how many there are. You and I might agree perfectly about what it is to be a mental individual and about all the underlying facts concerning the distribution of matter and the like, yet disagree wildly about the boundaries of mental individuals and even whether there are any at all.[7] Most disputes about psychological individuality have nothing to do with the definition of anything.

Though philosophers of mind have said rather little about psychological individuality in general, they've written volumes about the special case of people (or "persons"). Most say that to be a person, as opposed to a non-person, is to have certain special mental properties, such as intelligence and self-consciousness. Locke, for instance, famously defined "person" as "a thinking intelligent being that has reason and reflection, and can consider itself as itself, the same thinking thing in different times and places" (1975, p. 335).

It's easy to see that such definitions cannot tell us what determines a person's spatiotemporal location.[8] Suppose Plato's being a person amounts to his being intelligent and self-conscious. That leaves entirely open when he began and ended. His having those properties in his prime is consistent with his beginning at fertilisation, at birth, or when he was first intelligent and self-conscious. Nor does the definition imply that a person comes to an end when he ceases to be intelligent and self-conscious: for all it says, Plato might have continued to exist after his death as a corpse. All we can infer is that he was not then a person. (Unless he was a person essentially; but that does not follow from Lockean definitions.) Or at least this is so if we take "x is a person *iff* x is intelligent and self-conscious" to mean "x is a person at time t *iff* x is intelligent and self-conscious at t" – as all those who offer such definitions do. Lockean definitions do not even rule out a person's arbitrarily ceasing to exist without any interruption in her intelligence or self-consciousness.[9] They allow that each of us might come to an end right now and be imperceptibly replaced by an exactly similar but numerically different person.

Nor do they tell us anything about people's spatial boundaries. Most of us take Plato to have extended just as far as the surface of his skin. But even if that's right, it doesn't follow from his intelligence and self-consciousness, or from any other proposed definition of "person" or "mental individual". Those who think Plato was smaller – brain-sized, perhaps (Puccetti 1973, Hudson 2007) – or larger – as the "extended self" thesis has it (Clark and Chalmers 1998, p. 18, Olson 2011) – can and usually do accept a Lockean definition of personhood. For that matter, so can those who think that Plato was entirely immaterial and had no spatial location at all.

Lockean definitions of "person" are no more help in thinking about what determines how many people there are at any one time: whether there are one or two in cases of conjoined twinning, for example. They don't tell us what distinguishes one person from another. In this respect they're just like my definition of "artefact". How, then, could it be any different with definitions of "organism"?

Here is another way of expressing my puzzlement. No proposition about the number of organisms follows logically from any proposition about what it is to be an organism together with propositions about non-organisms (a description of the "underlying facts" about the distribution of matter and laws of nature, say). The claim that to be an organism is to be F can tell us nothing about what organisms there are unless we know something about what Fs there are: what material things with biological properties are *candidates*, so to speak, for being organisms. A definition of "organism" cannot solve the problem of biological individuality without a principle about the existence of the candidates to which it is to be applied. And the definition itself cannot provide such a principle.

The point has nothing to do with the details of the definition. It makes no difference, for example, if the definition of "person" can be known a priori, whereas an adequate account of what it is to be an organism must incorporate empirical discoveries. That no proposition of the form "there are n Gs" follows from any proposition of the form "x is G *iff* x is F" (together with a description of the underlying facts) is a simple matter of logic.

Why have philosophers of biology not noticed this or been troubled by it? The answer, I think, is that they have simply assumed the existence of the candidates (the Fs). To work out whether a Portuguese man-of-war is a single organism, it may seem that we need only know whether it satisfies the definition of "organism". If it doesn't, it's a colony of smaller polyps. But this presupposes that there *is* something composed of the polyps, which satisfies either the definition of "organism" or the definition of "colony". This is assumed to be one of the "underlying facts" taken for granted in the discussion. But it is a substantive metaphysical claim. Philosophers of biology have assumed a number of claims about the ontology of material things, often without realising it. Given these claims, a definition of "organism" really could be a theory of biological individuality. Yet the claims are never stated. What's more, they are highly controversial among metaphysicians. And they have important implications about what form a definition of "organism" needs to have in order to count as a theory of biological individuality – implications that have not been recognised. I will show this using some proposed theories of biological individuality.

4.5 A test case: the genetic theory

Let's examine an account of what it is to be an organism and see what would have to be the case for it to be a theory of biological individuality. We need one with enough detail to enable us to see what follows from it. Its plausibility is less important. (It's hard to find an account of biological individuality that is both detailed and plausible. The more detail you add to any theory, the more objectionable consequences you get.) Consider, then, the view that an organism is an entity with a uniform genotype.[10] I take this to mean a thing composed of *cells* with the same genotype. It's not clear what it could mean to speak of the genotype of fluids that are not parts of cells, such as blood plasma. A thing composed of certain entities is a *mereological sum* of those things. So the proposal is that an organism is a sum of cells with the same DNA.

The metaphysician's term "sum", and its synonyms "fusion" and "aggregate" are sometimes understood in different ways, but on the standard definition, "sum of the xs" means simply "thing composed of the xs" (where, again, the xs compose y just if each of the xs is a part of y and every part of y shares a part with one or more of the xs). This does not imply that a sum of things is *essentially* composed of those things, or rule out a sum's being composed of different things at different times (van Inwagen 2006). Nor does it imply that the mere existence of certain things suffices for there to be a sum of them: that's a substantive metaphysical claim (more on this later). The definition also says nothing about how a sum's parts must relate to one another. For all it says, things might compose a sum only if their parts interact in a special way – the way characteristic of biological life, for instance (van Inwagen 1990a, Section 9). Mereological sums, on the standard definition, are not things of a special kind. Everything is a sum.

I use the technical term "sum" because the ordinary word "collection" is notoriously slippery. "The collection of the xs" can mean either the xs in the plural or a single thing that the xs compose, obscuring this vital distinction. A sum of things is emphatically something distinct from any of those things (except in the degenerate case in which there is only one of them: because everything is a part of itself, everything is a sum of itself).

Do not confuse sums with *sets* in the mathematical sense. There are deep formal differences between sets and sums. The set whose only member is x is distinct from x itself, but again the sum of x is just x. And although the sets $\{x, y\}$, $\{x, \{x, y\}\}$, $\{\{x, y\}, y\}$, and so on are all different, there is no analogous distinction among sums. Nor do sums have anything analogous to the null set that is a subset of all sets. No one thinks that organisms are sets.

The claim, then, is that an organism is a sum of cells (that is, a thing composed of cells) with the same genotype. Presumably the cells must also have a common ancestry with that genotype: if, by sheer coincidence, some cells in another galaxy were genetically identical to mine, no one would say that all these cells together composed an organism, scattered across millions of light-years. I will ignore worries about how the ancestry requirement should be specified, or what it is for cells to have the same genotype (Dupré 2014, p. 10). Call this the *genetic theory* of biological individuality.

Almost no one thinks it's a *good* theory. It has the absurd implication that red blood cells, having no DNA, can never be parts of an organism. Nor can atoms that are not parts of cells. And although multicellular organisms commonly contain cells with mutant genes, the genetic theory implies that such cells are never parts of them, but are themselves organisms. There are many further objections (see e.g., Santelices 1999, p. 152). But even the critics of the genetic theory usually take it to be a genuine theory of biological individuality. It has the right form. It answers the questions that the problem of biological individuality consists in, even if it answers them wrongly. (Compare: "Ten" is an answer to the question, "How many English home counties are there?", though a wrong one; "Bananas" is not an answer at all.)

To see what follows from the genetic theory, we need a more precise statement of it. It is usually put in a loose and informal way, and this has discouraged questions about exactly how it would solve the problem of biological individuality. For example, it's no good saying

> *x* is an organism *iff x* is composed of cells with the same genotype and the right ancestry.

Everyone takes the genetic theory to imply that what we call asexual reproduction is not really reproduction at all, but growth: when an amoeba splits, it does not produce two organisms, but merely comes to be composed of two detached parts (Wilson 1999, pp. 87f., Godfrey-Smith 2014, p. 69). Yet this does not follow from the statement just given. If anything, the statement implies that the two free-swimming fission products are themselves organisms, as each is a sum of cells with the same genotype and the right ancestry. (Remember that everything is a sum of itself.) So the splitting must increase the number of organisms. It also counts my left thumb as an organism, composed as it is of cells with the same genotype and the right ancestry – an implication that no one would attribute to the genetic theory.

Not just any sum of cells with the right genotype and ancestry should count as an organism, but only one that is *maximal* – that is, not a part of any larger such sum. But although this is an improvement, it too lacks the implications ascribed to the genetic theory. It may seem to imply that the cells resulting from amoebic fission are not organisms, as they're not maximal sums of homogeneous cells, but parts of a larger such sum composed of the two cells. But this is so only on the assumption that there *is* something composed of those cells. Otherwise each cell will itself be a maximal sum of homogeneous cells and thus an organism. Likewise, the account rules out my thumb's being an organism only given that there is something composed of its cells together with my other cells. And nothing in the genetic theory implies either of these claims about composition. Call this the *composition problem*.

Let me say a brief word on this point. It is no tautology that whenever there are two things – even two things of the right sort arranged in the right way – there is a third thing that they compose: a sum of them. No principle of logic can take us from the proposition that there are two cells, for example, however arranged, to the proposition that there is a thing that is not a cell, but has two cells as parts. That requires a claim about when smaller things compose something bigger. And such claims are a subject of metaphysical debate.

You may wonder how anyone could accept that there is an oxygen atom and two hydrogen atoms attached to it by covalent bonds, yet deny that there is a water molecule. Isn't a water molecule by definition an oxygen atom and two hydrogen atoms so attached? But a molecule is not some atoms attached in a certain way. It's something that is not an atom, but rather *made up of* atoms.[11] To say that there is a crowd of people may be only a loose way of saying that there are people crowded together, but to say that there is a molecule is not just a loose way of saying that there are atoms appropriately attached. And the same goes for organisms: a dog is not some cells related in a canine way, but something that is not a cell. A dog is (according to the genetic theory) a thing composed of cells, and cells are not composed of cells. At any rate this is presupposed, as we saw earlier, in the claim that questions of biological individuality are questions about which entities satisfy the definition of "organism". Such a definition says what it is for *a thing* to be an organism. It takes the form "*x* is an organism *iff...x...*", where "*x*" is a singular variable. Only a thing composed of many cells can satisfy it (unless it is a unicellular organism). The cells themselves cannot. Definitions alone cannot bridge the gap from atoms to molecules, or from cells to multicellular organisms.

I will return to the composition problem in the next section. But there is another reason why the genetic theory as stated lacks the right implications. Suppose the two products of an episode of amoebic fission do compose something, and the theory counts that thing as an organism. Still, nothing in the theory implies that that organism is the original parent. For all it says, fission may destroy the parent and create a new organism composed of the two resulting cells. Although this would not increase the number of organisms, it would create a new generation and thus count as reproduction and not growth. No one would take that to be compatible with the genetic theory.

More generally, the theory says nothing about what determines an organism's temporal extent. It allows an organism to cease to exist arbitrarily without any genetic change, and be instantly replaced with a new and numerically different organism.

Nor does it rule out an organism's surviving a genetic change. It tells us only what cells an organism must have at any given time. Or at least this is so if it's understood as saying what it is for something to be an organism at a time (just as Lockean definitions say what it is for something to be a person at a time):

> x is an organism at t iff x is composed, at t, of cells with the same genotype and the right ancestry, and x is not a part, at t, of any larger such entity.

This says at most that an organism must be composed of homogeneous cells at each time when it exists.

In fact it says even less: only that an organism must be composed of homogeneous cells at each time *when it is an organism*. It doesn't rule out a thing's being an organism at one time and a non-organism at another, just as Lockean definitions don't rule out a thing's being a person at one time and a non-person at another. It allows an organism to become composed of heterogeneous cells, if this happens when it's no longer an organism. (Unless an organism must be an organism essentially. But that doesn't follow from the genetic theory.) The genetic theory as stated does not say what determines the spatiotemporal region an organism occupies, or where one leaves off and another begins, for the same reason that Lockean definitions don't say this about people.

The genetic theory is clearly intended to avoid these shortcomings by specifying what cells an organism is composed of not merely at a given time, but without temporal qualification. It should define an organism as a maximal sum of homogeneous cells that exist at *any* time. And it should say not what it is to be an organism at a time, but what it is to be an organism *simpliciter*. It should look like this:

> x is an organism *iff* x is composed of cells with the same genotype and the right ancestry – whatever their spacetime location – and x is not a part of any larger such entity.

Qualifications such as "at a time" don't come into it. This rules out a thing's being an organism at one time and a non-organism at another (though not a thing's being an organism in some possible worlds and not in others).

This looks more like the right sort of thing. It seems to imply, as the genetic theory should, that no new organism comes into being when an amoeba divides; rather, the original amoeba continues to exist in divided form.

The parent cell and its daughters, having the same genotype and the right ancestry, must all be parts of a single organism that persists through the division. Most of the things we ordinarily call amoebas – individual free-swimming cells – are not organisms, but parts of a larger organism composed of many cells scattered across space and time. Or at least this is so on the assumption that all amoebic cells with the same genotype compose something – that is, assuming an appropriate solution to the composition problem.

4.6 Temporal parts and material plenitude

So construed, the genetic theory implies that organisms have their parts without temporal qualification. But that's rather puzzling. Do I not have different parts at different times? Few of the cells that made me up in 1970 are parts of me now, and the atoms that composed me then are now scattered to the four corners of the earth. If I have my parts without temporal qualification, these cells and atoms must somehow be both parts of me and not parts of me. How could that be? How, if parthood is timeless, can things *change* their parts?

Metaphysicians usually answer by appealing to *temporal* parts. A temporal part of something is a part of it that takes up "all of that thing" whenever the part exists. Socrates's nose is a part of him, but not a temporal part, as it doesn't take up all of him while it exists. It's too small, spatially speaking. His adolescent portion, if we may so speak, would be a temporal part. His temporal parts are exactly like him when they exist. They eat and dring and ask awkward questions. They differ from him only in their shorter temporal extent.

Friends of temporal parts say that talk of temporary or temporally qualified parthood is a loose manner of speaking (Sider 2001, p. 57). For an atom to be a part of me now is for its current temporal part to be a part, without temporal qualification, of my current temporal part. More generally:

> x is a part of y at t *iff* the temporal part of x located at t (exactly located then) is a part of the temporal part of y located at t.

Strictly speaking, then, I'm not composed of atoms – no atom is a part of me without temporal qualification – but of temporal parts of atoms (roughly those located entirely within my spatiotemporal boundaries). And on the genetic theory I am also composed of temporal parts of cells. If we call a temporal part that exists for only an instant (or if there are no instants, a very brief period) a "stage", the genetic theory should say that an organism is a maximal sum of cell-stages with the same genotype and the right ancestry.

This implies that organisms, atoms, cells, and presumably all persisting things are composed of temporal parts. Or better, of *arbitrary* temporal parts. For every period when I exist, long or short, there is a temporal part of me located then and only then. This follows from the principle about temporary parthood just given: because atoms can be parts of me for any period, both they and I must have temporal parts of any length.

That all persisting things are composed of temporal parts (and that things have their parts without temporal qualification) is highly contentious.[12] The genetic theory presupposes a second contentious metaphysical claim as well, namely *unrestricted composition*. This is the view that for any things whatever, there is something composed of them: a sum.[13] If some entities did not compose anything, there would be no guarantee that all the cells (or cell-stages) of a given genotype composed anything, or even any apparent reason to suppose it. Think of all the amoebic cells (or cell-stages), scattered across space and time, with the same genotype as the one on this microscope slide, and the right ancestry. What reason could there be to suppose that there is a vast, disconnected material thing composed of those cells, other than the thought that *any* things, whatever their nature or arrangement, must compose something?

But unless those cells do compose something, they cannot compose an organism, even according to the genetic theory. If there is no sum of them at all, there is none that can satisfy the definition of "organism". The theory would *allow* that any amoebic cells (or cell-stages) with the right genotype and ancestry that are not parts of any larger sum of such cells compose an organism. But it would also allow that some such cells compose organisms and others don't. It would even allow that none of them do (except in the trivial sense that each composes itself), and that each individual cell is itself an organism. The theory says, in effect, that anything such cells compose is an organism. But it doesn't say whether they do compose anything.

No one discussing the genetic theory considers these possibilities. Both its advocates and its critics assume that the cells (or cell-stages) in question compose something. Their disagreement is about whether that thing is an organism or a non-organism. Debates over biological individuality are about classification: about assigning individuals to sorts. That follows from their statement of the problem as asking for a definition of "organism". Questions about what individuals there are – about which homogeneous cells or cell-stages compose anything, for example – do not arise. This can only be because the debate presupposes a "generous" account of composition: one implying that any entities that anyone might take to compose an organism or other biological individual compose something or other. And the only such account that has ever been proposed is unrestricted composition.

Assuming unrestricted composition thus avoids the composition problem. It rules out the possibility that the two cells resulting from an amoeba's division are each maximal sums of homogeneous cells and thus organisms themselves. And it rules out my left thumb's being an organism. It does so by implying that each of these entities is a part of a larger sum of homogeneous cells.

The temporal-parts ontology and the doctrine of unrestricted composition are nearly always held together.[14] Their conjunction implies that every matter-filled spacetime region is occupied by a material thing.[15] As Quine once put it, a physical object "comprises simply the content, however heterogeneous, of some portion of space-time, however disconnected and gerrymandered".[16] Call this the *principle of material plenitude*.

4.7 The functional-integration theory

I have argued that both friends and enemies of the genetic theory presuppose important metaphysical claims: that biological individuals are composed of temporal parts and that any cell-stages whatever compose something. The debate is about which sums of cell-stages are organisms and which are not, never about whether the sums exist in the first place. The only systematic principle that would secure these claims, or at least the only one that anyone has ever actually proposed, is the principle of plenitude. All parties to the debate take it for granted.

In a way this is not surprising. As we saw earlier, no claim about what it is to be an organism (together with propositions about atoms, cells, and the like) can entail a claim about what organisms there are or how many. To reach a conclusion about what organisms there are from the premise that all and only organisms are F, we need to know what Fs there are. We need an account of the "candidates" to which the definition can be applied. The principle of plenitude supplies this. But any such account will be independent of a definition of "organism". It follows that no definition can be a theory of biological individuality by itself, but at best in conjunction with a metaphysical principle about what material things there are. Or, to put it differently, a definition can be such a theory only if some such metaphysical principle is presupposed.

The point has nothing to do with the genetic theory in particular, but is a simple matter of logic. Consider the more appealing view that an organism is a sort of "functionally integrated whole" (e.g., Sober 1993, pp. 150–153, Wilson 1999, p. 89, Pradeu 2010, p. 252). What makes things parts of the same organism is not any sort of similarity, but something to do with their causal interrelations: metabolic and immune activities, for instance.

Suppose we could specify these activities at the level of atoms. (Specifying them at the level of cells would be easier, but a definition of "organism" based on such a specification would imply that organisms are composed of cells and cannot include extracellular fluids.) Then we could say that an organism is a sum of atoms, each of which interacts in this way with every other – or better, any two of which either interact in this way or stand to each other in a chain of such interactions. Call such atoms *I-related*. An organism would then be a sum of I-related atoms. Or rather a maximal such sum: one that is not a part of any larger one. Otherwise my left thumb would count as an organism.

(That is presumably why organisms are called integrated *wholes*.) Call this the *functional-integration theory* of biological individuality.

Does it tell us what determines which spatiotemporal regions are occupied by organisms? Take the spatial case first. Suppose my symbiotic gut bacteria are I-related to my animal cells because my immune system interacts with both in the same way.[17] More precisely, the atoms composing these bacteria and the rest of my atoms are I-related. Does the proposal imply that those atoms compose an organism? Well, only if they compose something at all. Otherwise they will not compose a maximal sum of I-related atoms, even though they are all I-related to each other and not to any other atoms – maximally I-related, we might say. If instead my "animal" atoms that are not parts of my gut bacteria compose something, *it* will be a maximal sum of I-related atoms, even though those atoms are not maximally I-related. In that case the functional-integration theory will imply that this smaller entity is an organism and there is none composed of it together with my gut bacteria. There will be no such organism because there is no such entity at all.

No one would take the functional-integration theory to have this consequence. Everyone assumes that maximally I-related atoms compose a maximal sum of I-related atoms, which the theory classifies as an organism. But for this to be the case, such atoms must compose something. What could ensure this? The most obvious answer is, again, that any atoms whatever compose something, no matter what their nature or arrangement. Those discussing biological individuality simply don't worry about when atoms or other small things compose something. This insouciance is appropriate only on the assumption of unrestricted composition (or some other "generous" ontology of material things).

Does the theory tell us what determines organisms' temporal extent? Not if it says only what it is for something to be an organism at a given time – that something is an organism at a time just if it is composed, at that time, of atoms that are then I-related and it is not then a part of any larger such sum. That would allow that an organism might arbitrarily cease to exist without any interruption of its metabolic or immune activity and be replaced with a new organism. Or it might continue after all such activity stops and cannot be restarted: the theory would imply only that it was not an organism then. It might carry on after its death as a former organism, just as graduates become ex-students.

Like the genetic theory, the functional-integration theory is clearly meant to tell us not which parts an organism is composed of at a given time, but which it is composed of *simpliciter*: that an organism is a maximal sum of I-related entities that exist at any time. As we saw in discussing the genetic theory, this implies that organisms are not composed of atoms, but of temporal parts of atoms. So the functional-integration theory too presupposes that atoms and other persisting objects have temporal parts – arbitrary temporal parts in fact, as an atom can be a part of an organism for any period. Combining this with unrestricted composition yields the principle of material plenitude.

Similar remarks will apply to any theory of biological individuality that takes the form of a definition. It cannot do its job without a generous ontology of material things.

4.8 What a definition needs to say

Philosophers of biology never argue for the principle of plenitude or any other account of what material things there are. They rarely even mention the point.[18] For the most part they appear unaware that their discussions of individuality presuppose such a principle. My purpose is not to criticise them for this or to argue against the principle, but only to show that this is an unavoidable feature of any theory of biological individuality taking the form of a definition.

This matters because the principle of plenitude (or any other generous ontology) implies that not just any definition of "organism" can be a theory of biological individuality, right or wrong, but only one having a special form. The definition must say what determines an organism's spatiotemporal boundaries. This is because a theory of biological individuality needs to say what distinguishes a colony from a single complex organism, copies of a gene in one organism from copies in many organisms, and members of different generations. It needs to say what determines how many organisms there are and what distinguishes one from another. And the principle of plenitude makes this task more difficult by providing an awkward surplus of things with biological properties – of candidates, so to speak, for being organisms.[19] It implies that my office contains a vast number of entities made up entirely of living matter: me; my left thumb; my northern half; sums of a human being and certain dust mites, and of arbitrary portions of man and mite; my current stage; the thing composed of my stages located up to now and yours located thereafter; and so on. No one would take more than a trivial proportion of these things to be biological individuals of any sort. A theory of biological individuality will have to tell us what determines which ones they are – what distinguishes the organisms from the arbitrary pieces of ontological rubbish. To do that, it needs to say what determines the spatiotemporal boundaries of organisms.[20]

I have never seen a definition that does this. Accounts in terms of reproductive capacities, for instance (Clarke 2010, p. 317), or in terms of autonomy and self-sufficiency (Santelices 1999, pp. 152f., Boden 2008), do not say what determines organisms' boundaries, and thus, given plenitude, which of the many things with biological properties count as organisms. Accounts in terms of the nature of the life cycle, such as the "bottleneck" view characterising an organism as what develops from a single cell (Wilson 1999, pp. 99–101, Clarke 2010, pp. 317f., Dawkins 2016, pp. 334–341), may tell us when an organism begins, but say little about its other boundaries. These proposals may contribute towards a theory of biological individuality, but they cannot be theories of biological individuality by themselves. They simply have the wrong form.

Even the genetic theory falls short. My best formulation in Section 4.5 implies that any homogeneous cells (or cell-stages) with the right ancestry are parts of a single organism. It follows that if a bramble scratches off some of my skin, I don't lose any cells. The detached cells remain parts of me. I merely become disconnected. If a violent accident destroys all my cells but one, I still exist (as I am located where my parts are). Almost no one takes the theory to have these consequences. (Wilson is again an exception: 1999, pp. 87f.) As normally understood, it requires an organism's cells to stand in a special causal relation in addition to their having the right genotype and ancestry – one that does not hold between the cells currently within my skin and those that have been detached. Genetic theories have not specified this relation.

Functional-integration theories seem explicitly designed to solve this problem: they're all about the causal relations among an organism's parts. But existing proposals of this kind fall short. Thomas Pradeu's version, for example, says only that what determines an organism's temporal extent is "the spatiotemporal continuity of the interactions between components of the being involved" (2012, p. 249). That sounds right, but what follows from it? Spatiotemporal continuity comes in infinite varieties. When an amoeba divides, there is plenty of spatiotemporal continuity between the interactions of the parent's atoms and those of the atoms composing each daughter cell. But is it the right sort? Does the theory imply that the original amoeba survives the split? If so, are both daughters identical to it (contrary to the transitivity and symmetry of identity, which are theorems of standard logic)? Or does the original become composed of the two daughter cells? If it doesn't survive, where is the spatiotemporal *dis*continuity? Without further detail, there's no saying. Wilson's account says more (1999, pp. 89–99), but still leaves many questions unanswered.

4.9 Individuality without definitions

I have argued that no definition of "organism" can be a theory of biological individuality on its own, but only in conjunction with a substantive claim about the ontology of material things providing the candidates to which the definition is applied. This claim is rarely stated: it is a hidden assumption that those engaged in debates over biological individuality are often unaware of. And it requires a definition far more detailed than any of those actually proposed.

Let me make one more point. Philosophers of biology give the impression that a theory of biological individuality must take the form of a definition, or at least include one. It needs to complete the formula

$(x)x$ is an organism *iff*...x....

This is not so. It's possible to say what determines the spatiotemporal locations of organisms without defining "organism". And it can be done without presupposing any metaphysical claim about the existence of the candidates – the values of "x" in the formula. Rather than asking what it is for something to be an organism as opposed to a non-organism, we can ask instead under what circumstances *there is* an organism as opposed to there not being one. Or better, we can ask when things compose an organism. (All philosophers of biology that I know of agree that organisms have parts: they're not mereological simples.) We can try to complete the formula

$(ys)(\exists x) x$ is an organism and the ys compose x *iff*...the ys....

How are the two formulas different? Well, because the variable "x" occurs on the right-hand side of the first formula, its blanks need to be filled with a condition on organism-candidates – complex material things with biological properties. If we complete it with "x is a maximal sum of I-related atoms", for example, we presuppose that there *are* such sums – that maximally I-related atoms always compose something. Otherwise the theory will tell us nothing about what determines how many organisms there are. It will not imply that there are any organisms at all, even given the "underlying facts" about the distribution of matter and so on mentioned in Section 4.2 – including those about which atoms are maximally I-related. It will not do what a theory of biological individuality is supposed to do. That's why a definition of "organism" needs to be combined with a claim about the ontology of material things, such as the principle of material plenitude (the one most discussions of biological individuality appear to presuppose).

The second formula, by contrast, does not presuppose any metaphysical claim about the existence of the candidates. The variable occurring on the right-hand side, "the ys", ranges not over candidates for being organisms, but over candidates for being their parts: atoms, cells, or the like. Filling in the blanks will specify what nature and arrangement such things need to have in order to compose an organism – as opposed to composing either a non-organism or nothing at all. (By "their nature" I mean their intrinsic properties and by "their arrangement" I mean the spatiotemporal and causal relations they bear to one another and perhaps to their surroundings.) Given that things are located where their parts are (and that composition is defined in terms of parthood), it will tell us what determines which spatiotemporal regions are occupied by organisms.[21]

We might call this an "existential" statement of the problem of biological individuality, and a solution to it an existential theory – as opposed to the "definitional" statement or theory usual in the philosophy of biology. The definitional statement presupposes the existence of organism-candidates and says that an organism is one having the right features. The existential statement presupposes only the existence of smaller things such as atoms, and says that an organism exists just when those smaller things have the right features.

It's a trivial exercise to derive a definition of "organism" from an existential theory: an organism will be anything composed of entities satisfying the condition got by filling in the blanks in the second formula – atoms related in the right way, as it may be. But such a definition is no part of the theory, and is not needed in order to say what determines the parts of an organism and thus what region one occupies. An existential theory of biological individuality requires no definition of "organism". (Of course, we'll need to know what the word "organism" means in order to understand the theory, just as we need to understand the other words occurring in it. But understanding a word is not the same as knowing its definition.) So we can think of the problem of biological individuality not as what makes something an organism as opposed to a non-organism, but rather as what has to be the case for there *to be* an organism – or more specifically, what nature and arrangement atoms or other things need to have in order to compose one.

If composition is timeless and something is or is not an organism without temporal qualification, such an account will also tell us what determines organisms' temporal boundaries – though as we saw in Section 4.6, that requires persisting things to be composed of temporal parts. Otherwise it will need a separate clause covering persistence – a completion of the formula:

If x is an organism at t and y exists at t^*, then $x=y$ iff $...x...t...y...t^*....$

The existential statement of the problem is not quite equivalent to the one I gave in Section 4.2: what determines which spacetime regions are occupied by organisms. Knowing what determines which things compose an organism and what it takes for one to persist will tell us what determines its spacetime location. But not vice versa: knowing an organism's location will not tell us what composes it. It won't tell us whether the neutrinos now passing through me are among my parts, for instance. That said, the difference is unimportant. We are unlikely to know the answer to the original question, about organisms' locations, without knowing the answer to the new one, about their parts. We could find out what determines which region an organism occupies by discovering what it takes for things to compose one. It's hard to see how we could discover it in any other way.

In any event, the existential statement of the problem has an important advantage over the definitional statement: it does not presuppose the principle of plenitude or any other controversial metaphysical claim. It requires no independent list of candidates to which the definition can be applied, and makes no tacit assumptions about when atoms or other entities make up larger things. It has no hidden metaphysical commitments. Of course, it will be uninteresting unless there are atoms or other entities whose nature and arrangement are responsible for the existence of organisms. But this commitment is evident on the surface. Nor is it much disputed.

The existential statement also has the advantage of not presupposing, as the definitional statement does (Section 4.5), that an organism must be an organism throughout its existence. This does not imply the opposite – that something could be an organism at one time and a non-organism at another. That's left open, as it should be.

The definitional statement, by contrast, has no evident advantage over the existential statement. I don't know how much practical difference it would make to discussions of biological individuality if the problem were formulated existentially rather than definitionally. Maybe all parties are happy with the metaphysical presuppositions of the definitional statement – though I doubt it. And adopting the existential statement would remind us that theories of biological individuality need to specify what determines the spatiotemporal boundaries of organisms. By obscuring this fact, the definitional statement makes the problem appear easier than it is. And it's surely better for contentious metaphysical claims to be recognised and open for discussion than tacitly presupposed by the terms of the debate.

Acknowledgements

I am grateful to Arthur Carlyle, Ellen Clarke, John Dupré, Charles Jansen, Anne Sophie Meincke, Will Morgan, and Karsten Witt for generous comments on earlier versions.

Notes

1 See Dupré and O'Malley (2012) for more examples.
2 Bouchard (2008) argues that there are exceptions.
3 Most definitions also require the xs not to overlap, but we can ignore this. The word "compose" is sometimes used in other, more tendentious ways. I will use it exclusively in the way defined here.
4 Godfrey-Smith says the problem is "what collects the parts…of a system into a living individual" (2015, p. 85; see also Sober 1993, pp. 149–153). I will return to the relation between a thing's parts and its location in Section 4.9.
5 If an organism can have imprecise boundaries, they may be vague regions, which certain areas are neither definitely parts nor definitely not parts of.
6 Clarke (2013, p. 414). See also Santelices (1999), Wilson (1999, p. 1), Pradeu (2010, p. 248), Pradeu (2012, pp. 227, 244), Godfrey-Smith (2014, pp. 66–80), Wilson and Barker (2017, section 2). Clarke says explicitly that the problem is not about organisms' spatiotemporal boundaries (2013, pp. 414f. fn.).
7 On the view that there are no mental individuals, see van Inwagen (1990, pp. 72f.), Merricks (2001, chapter 5), Olson (2007, chapter 8).
8 Do not confuse Lockean definitions of "person" with Lockean theories of personal identity over time. These theories do say what determines a person's temporal boundaries: it has to do with psychological continuity. But they don't follow from Lockean definitions of "person".
9 Shoemaker (1999) disagrees, but his view is unique in this respect.

10 This seems to be the thought behind the view that there are "clonal organisms" composed of parts that look like organisms themselves: individual aphids or aspen trees, for example (Janzen 1977, Bouchard 2008, p. 563). But advocates of this view rarely commit themselves to an explicit principle of individuality.
11 Van Inwagen (1990a, §§2, 9, 10, 1994); see also the references in note 13.
12 See e.g., Thomson (1983), Lewis (1986, pp. 202–204), van Inwagen (1990b), Sider (2001). Some use the language of temporal parts in a less contentious way (e.g., Shoemaker 1984, pp. 74f.). By a "stage" of an organism they mean only a temporal part of its history, which they take to be distinct from the organism. My statement of the genetic theory requires organisms themselves to have temporal parts.
13 Advocates include Lewis (1986, pp. 212f.), Sider (2001, pp. 120–139), and Hudson (2007, pp. 223–228); critics include van Inwagen (1990a – the classic discussion of theories of composition) and Merricks (2001).
14 Without temporal parts, unrestricted composition rules out a thing's having different parts at different times (van Inwagen 1990a, p. 78, Olson 2007, p. 230). And much of the theoretical work the temporal-parts ontology is designed for requires unrestricted composition (Sider 2001, pp. 120–139).
15 Or every "occupiable" region (van Inwagen 1981, pp. 135f., fn. 3). It may be metaphysically impossible for a material thing to occupy a region of fewer than three spatial dimensions, for example.
16 Quine (1960, p. 171). Like Quine, I don't mean anything special by "object" or "thing". I use them as completely general count nouns. "x is a thing" is equivalent to "($\exists y)y=x$". Everything is, by definition, a thing.
17 Pradeu (2010, pp. 259f.), Dupré and O'Malley (2012, p. 224); see Pradeu (2012, p. 248) for further references.
18 Wilson is a commendable exception: "all earthly life or any span of it", he says, "can rightly be considered a particular" (1999, p. 61). This seems to mean that any part of the spacetime region where life is present is occupied by a material thing. But even he does not think the claim needs any defence.
19 This may be what Hull means when he speaks of "the welter of individuals that clutter our conceptual landscapes" (1992, p. 183).
20 It needn't be completely precise. Material plenitude implies that many beings differ from me only trivially, by a few atoms or milliseconds. That's not the problem here. We can count these things as if they were one (Lewis 1993). It's such things as my thumb and my first half that the theory needs to classify as non-organisms.
21 This is van Inwagen's approach (1990a, §§2, 9).

References

Baker, L. R. (2000) *Persons and Bodies*, Cambridge: Cambridge University Press.
Boden, M. (2008) 'Autonomy: What Is It?', *BioSystems* 91, pp. 305–308.
Bouchard, F. (2008) 'Causal Processes, Fitness, and the Differential Persistence of Lineages', *Philosophy of Science* 75, pp. 560–570.
Campbell, T. and McMahan, J. (2010) 'Animalism and the Varieties of Conjoined Twinning', *Theoretical Medicine and Bioethics* 31, pp. 285–301.
Clarke, A. and Chalmers, D. (1998) 'The Extended Mind', *Analysis* 58, pp. 7–19.
Clarke, E. (2010) 'The Problem of Biological Individuality', *Biological Theory* 5, pp. 312–325.
Clarke, E. (2013) 'The Multiple Realizability of Biological Individuals', *Journal of Philosophy* 110, pp. 413–435.

Dawkins, R. (2016) *The Selfish Gene, 40th Anniversary Edition*, Oxford: Oxford University Press.

Dupré, J. and O'Malley, M. (2012) 'Varieties of Living Things. Life at the Intersection of Lineage and Metabolism', in: Dupré, J. (Ed.), *Processes of Life: Essays in the Philosophy of Biology*, Oxford: Oxford University Press, pp. 206–229.

Dupré, J. (2014) 'Animalism and the Persistence of Human Organisms', *Southern Journal of Philosophy* 52, Spindel Supplement, pp. 6–23.

Godfrey-Smith, P. (2014) *Philosophy of Biology*, Princeton, NJ: Princeton University Press.

Godfrey-Smith, P. (2015) 'Individuality and Life Cycles', in: Guay, A. and Pradeu, T. (Eds.), *Individuals Across the Sciences*, Oxford: Oxford University Press, pp. 85–102.

Hawthorne, J. and McGonigal, A. (2008) 'The Many Minds Account of Vagueness', *Philosophical Studies* 138, pp. 435–440.

Hudson, H. (2007) 'I am Not an Animal!', in: van Inwagen, P. and Zimmerman, D. (Eds.), *Persons: Human and Divine*, Oxford: Oxford University Press, pp. 216–236.

Hull, D. (1992) 'Individual', in: Keller, E. F. and Lloyd, E. A. (Eds.), *Keywords in Evolutionary Biology*, Cambridge, MA: Harvard University Press, pp. 181–187.

Janzen, D. (1977) 'What Are Dandelions and Aphids?', *The American Naturalist* 111, pp. 586–589.

Lewis, D. (1986) *On the Plurality of Worlds*, Oxford: Blackwell.

Lewis, D. (1993) 'Many, but Almost One', in: Campbell, K., Bacon, J. and Reinhardt, L. (Eds.), *Ontology, Causality, and Mind*, Cambridge: Cambridge University Press, pp. 23–38.

Locke, J. (1975) *An Essay Concerning Human Understanding*, 2nd ed., Oxford: Oxford University Press. (Original work 1694).

Merricks, T. (2001) *Objects and Persons*, Oxford: Oxford University Press.

Olson, E. T. (2003) 'Was Jekyll Hyde?', *Philosophy and Phenomenological Research* 66, pp. 328–348.

Olson, E. T. (2007) *What Are We? A Study in Personal Ontology*, New York: Oxford University Press.

Olson, E. T. (2011) 'The Extended Self', *Minds and Machines* 21, pp. 481–495.

Olson, E. T. (2014) 'The Metaphysical Implications of Conjoined Twinning', *Southern Journal of Philosophy* 52, Spindel Supplement, pp. 24–40.

Pradeu, T. (2010) 'What is an Organism? An Immunological Answer', *History and Philosophy of the Life Sciences* 32, pp. 247–268.

Pradeu, T. (2012) *The Limits of the Self: Immunology and Biological Identity*, Oxford: Oxford University Press.

Puccetti, R. (1973) 'Brain Bisection and Personal Identity', *British Journal for the Philosophy of Science* 24, pp. 339–355.

Quine, W. V. O. (1960) *Word and Object*, Cambridge, MA: MIT Press.

Santelices, B. (1999) 'How Many Kinds of Individual Are There?', *Trends in Ecology and Evolution* 14, pp. 152–155.

Shoemaker, S. (1984) 'Personal Identity: A Materialist's Account', in: Shoemaker, S. and Swinburne, R. (Eds.), *Personal Identity*, Oxford: Blackwell.

Shoemaker, S. (1999) 'Self, Body, and Coincidence', *Aristotelian Society Supplementary Volume* 73, pp. 287–306.

Sider, T. (2001) *Four-Dimensionalism: An Ontology of Persistence and Time*, Oxford: Oxford University Press.

Sober, E. (1993) *Philosophy of Biology*, Boulder, CO: Westview.
Thomson, J. J. (1983) 'Parthood and Identity Across Time', *Journal of Philosophy* 80, pp. 201–220.
Van Inwagen, P. (1981) 'The Doctrine of Arbitrary Undetached Parts', *Pacific Philosophical Quarterly* 62, pp. 123–137.
Van Inwagen, P. (1990a) *Material Beings*, Ithaca, NY: Cornell University Press.
Van Inwagen, P. (1990b) 'Four-Dimensional Objects', *Noûs* 24, pp. 245–255.
Van Inwagen, P. (1994) 'Composition as Identity', in Tomberlin, J. (Ed.), *Philosophical Perspectives* 8, *Logic and Language*, Atascadero, CA: Ridgeview, pp. 207–220.
Van Inwagen, P. (2006) 'Can Mereological Sums Change Their Parts?', *Journal of Philosophy* 103, pp. 614–630.
Wilson, D. S. and Sober, E. (1989) 'Reviving the Superorganism', *Journal of Theoretical Biology* 136, pp. 337–356.
Wilson, J. (1999) *Biological Individuality: The Identity and Persistence of Living Entities*, Cambridge: Cambridge University Press.
Wilson, R. A. and Barker, M. (2017) 'The Biological Notion of Individual', in: Zalta, E. (Ed.), *The Stanford Encyclopedia of Philosophy* (Spring 2017 Edition), https://plato.stanford.edu/ archives/spr2017/entries/biology-individual/.

5 The role of individuality in the origin of life

Alvaro Moreno

5.1 Introduction

In all sciences, the search for the most fundamental classificatory concepts is of paramount importance. In biology, the concept of individuality is probably one of those that has triggered most debates of this kind. Generically, the term "individuality" refers to a system or entity that is distinct from other neighbouring systems or entities. In biology, however, the concept of individuality has different uses and is understood at many levels: in addition to linking it to the concept of organism, many authors speak of species or populations as individuals, and the term is also widely accepted as a means of referring to ecological, developmental or immune forms of individuality. What is it, then, that constitutes an individuated entity in biology? And what is the functional role of the formation of individuated entities in the evolution of life?

When searching for the roots of life's impressive capacity to proliferate, to create an enormous variety of forms and, particularly, to act and modify different environments, we tend to focus on those cohesive and spatially well-localised entities that we call organisms. However, no agreement exists regarding what actually constitutes a full-fledged organism. Moreover, as mentioned above, the problem with the concept of individuality is not only the difficulty inherent in clarifying what is meant by an individuated organism, but also the fact that individuality may be manifested in many different dimensions which do not necessarily coincide: physiological, developmental, immune, ecological, anatomical or evolutionary (Clarke 2010, 2013, Godfrey-Smith 2013, Pradeu 2016a, 2016b, Love and Brigandt 2017, DiFrisco 2019).

In light of the above, the difficulties involved in moving towards the formulation of a satisfactory answer to the question of biological individuality seem overwhelming. And yet, despite all these undeniable obstacles, the question requires a clear answer. For how else can we hope not only to classify in a reasonable way but also – and more importantly – to understand biological phenomenology? In other words, how can we hope to determine which type of entities populate the biological domain and what are the real categories behind the huge amount of data provided by the life sciences? Which forms of individuality are more fundamental? How did they emerge? Why do they relate to each other in the way they do?

Nowadays, it is widely accepted that, right from the very start, life has developed a multitude of associations, with different degrees of cohesion. For example, multicellular individuals on Earth are closely associated with a multitude of microorganisms, which are of fundamental importance for their survival (McFall-Ngai et al. 2013, McFall-Ngai 2015). Some associations between a eukaryotic multicellular host and all the elements of its symbiotic microbiome are so deep and cohesive that, for many, they even constitute a new form of biological individuality (Gilbert, Sapp and Tauber 2012, Gilbert and Tauber 2016, Skillings 2016, Gilbert, Rosenberg and Zilber-Rosenberg 2017, Roughgarden et al. 2017, Stencel and Proszewska 2018). And if we look back in time, we see that multicellular organisms are the result of certain processes of association between unicellular eukaryotes (which in other cases have formed other types of associations). For their part, these unicellular eukaryotes live in an ocean of prokaryotes (with the latter often living literally inside the formers) which are of fundamental importance for the survival of eukaryotic life. Unicellular eukaryotes themselves are the result of a special history of symbiosis between several unicellular prokaryotes. Ultimately, unicellular prokaryotic entities form many associations, some of which are so cohesive that they have become a new composite form of individuality.

Thus, even at its most simple expression, life appears in a variety of forms and dimensions: diachronically, as lineages of self-reproducing organisms, surrounded by and merging with a host of genetically autonomous entities (i.e., viroids, virus, plasmids), in continuous evolution; synchronically, as a large diversity of groups harnessing the flow of energy and matter through themselves and their abiotic environment, so as to create ecological networks. Moreover, most of the time, prokaryotic cells organise themselves in cohesive colonies, often constituted by a variety of different species, losing autonomy in exchange for increasing their chances of survival under adverse conditions. And, as we shall see, there are good reasons to go even further back in time and argue that the progressive constitution of a multidimensional system, organised at different temporal and spatial scales, began much earlier, during prebiotic evolution. If we accept this, how and when did all this complementary process of individuation and association start? What was the minimal form of individuality and what was the functional role of the formation of individuated entities in prebiotic evolution? How did the different dimensions of individuality emerge?

In this chapter, I will address these questions on the basis of the hypothesis that the study of how the earlier forms of individuality originated will help us gain a better understanding of the nature of biological individuality. Life is certainly a highly complex and multifarious phenomenon, but its historical unfolding in many different dimensions and forms of organisation cannot be explained without elucidating the fundamental role that the creation of forms of individuality has played in this process. My main claim, therefore, is that the different forms and levels of biological organisation ultimately date back to – and are rooted in – the appearance of a minimal form of individuality, understood as a (selectively permeable) encapsulated self-maintaining organisation, which constitutes the core of what we now mean by organismality.

5.2 Origins

The process of biogenesis – also known as prebiotic evolution – began when, under highly favourable boundary conditions, certain sets of molecules started a self-sustained process of chemical complexification. This initial complexification was supported by two types of processes: the formation of far-from-equilibrium autocatalytic self-maintaining reaction networks, and the formation of near-to-equilibrium self-assembling structures, which, soon, began to merge or "cooperate" in different ways.[1]

How could such a process of cooperation have begun? Since the spontaneous formation of self-assembled vesicles in an aquatic environment favours the formation and maintenance of self-maintaining reaction networks (Fallah-Araghi et al. 2014, Walde et al. 2014), it is likely that, at a given moment, both processes became functionally associated. This association, however, was by no means a trivial achievement and was probably the result of a long process of repeated attempts and mutual transformations. As Shirt-Ediss (2016) points out:

> Numerous times in prebiotic chemistry, there may have occurred the chance insertion of complex reaction systems, developed in different chemical contexts, into various self-assembled compartments (e.g., into the internal aqueous phase and/or membrane of lipid vesicles). However, such structures would have been unlikely to result in stable functional protocell systems since the membrane would probably not possess the correct composition, nor contain the correct molecular machinery (selective channels, carriers, ion pumps, mechanisms to transduce external energy into a form usable by chemical reactions, etc.) in order to meet the permeability, catalytic and energy requirements to keep the metabolism running, and at the same time avoid e.g., osmotic burst. Equally, the metabolic processes would probably not be able to synthesize all of the key system components enabling division into equally functional offspring. The first systems with the ability to robustly maintain themselves far-from-equilibrium (and with the ability to divide in a controlled way into equally functional progeny) could only have existed as such if they were encapsulated and their internal (proto-metabolic) organization and compartment were tightly integrated.
>
> (Shirt-Ediss 2016, p. 85)

In other words, encapsulation emerges as being closely linked to the appearance of a proto-metabolic organisation. Compartments create a phase separation between the inside and outside of the system, thus generating concentration gradients (Mitchell 1961, Harold 1986), and pH and oxidation-reduction differences (Morowitz 1981, 1992, Chen and Szostak 2004). Interestingly, these differences could be used as energy storage mechanisms, supporting endothermic transformations and active solute transport. Hence, encapsulation emerges as a necessary condition for the evolution of a primitive metabolic organisation, capable of both managing the energy flows required for the maintenance of the system and developing internal organisational differentiation (Ruiz Mirazo and Moreno 2004). In turn, the encapsulated proto-metabolic organisation would be a tool for the evolution of a compartment, providing catalysts and other compounds that enable the emergence of selective permeability.

A cyclic, self-maintaining chemical system that produces its own physical border, which in turn contributes to the maintenance of the enclosed chemical network, is called an autopoietic system (Maturana and Varela 1980). The first point I would like to emphasise here is that an autopoietic entity is not only an epistemic category, a set of related features that *we* build to differentiate something from other domains of the world. An autopoietic system is also *an ontologically self-identifying entity*: it is a system that itself dynamically produces and maintains a differentiated identity. And this, I believe, is the first aspect that characterises the appearance of a form of biological individuality.

Moreover, if an autopoietic system is organised in such a way that the deployment of its continuous process of self-production generates growth, the compartment and all that contained inside may divide into two new units. In fact, when certain conditions occur, the growth and division of a compartmentalised self-maintaining system becomes an inevitable consequence of its own self-maintaining dynamics: autocatalysis causes the production and accumulation of different types of components within the system, and the cellular nature of this causes division to occur spontaneously once a critical size has been reached and the system has become unstable. Thus, ultimately, reproduction is a consequence of a certain time relation between the production and decay of the constitutive components in a self-maintaining organisation. If the rate of replacement of the constitutive components is faster than their decay, the system's self-maintaining cycles will prompt it to establish reproductive cycles: the system will either grow and reproduce; or disintegrate (Zepik, Blöchliger and Luisi 2001, Mavelli and Ruiz-Mirazo 2013).

Two important and related points should be added here. First, the process of reproduction includes the conservation/maintenance of a specific identity (see Section 5.2); and second, that the very organisation of the system is the source of the reproductive process. The first claim is supported by the fact that reproductive enclosed proto-metabolic systems, devoid of sequence-based biopolymers, could through reproduction transmit certain features of their specific identity (see Section 5.3). The second claim is based on the fact that, as explained earlier, reproduction is a form of autopoiesis. Together, these two facts lead us to qualify the process as "self-reproduction".

If we compare this form of organisation with the preceding scenario, several interesting differences emerge. Let us remind the reader. We have said that biogenesis started when, in certain places and under very specific boundary conditions, different sets of reactions gave rise to cyclic processes, namely to *self-maintaining* networks, and that these networks further evolved when they became associated with self-assembling vesicles. Admittedly, the use of the term "self" in this context implies a certain form of active distinction. For example, a self-maintaining chemical network is a cyclic series of component production (and destruction) processes, thus generating a certain form of organisational identity; and a self-assembling structure could also become a specific template, making copies of itself. However, none of these "selves" induces an active and stable distinction between themselves and other neighbouring systems or entities.

By contrast, the formation of populations of self-re-producing autopoietic entities radically changes the situation, giving rise to a scenario in which two (spatially and temporally) different types of processes begin to appear. First, a not well-bounded macroscopic set of chemicals, driven by some external sources of energy and favoured by a set of highly special physical conditions, slowly and heterogeneously yields progressively complex compounds.[2] And second, within this messy domain, some microscopic spatially localised areas emerge where far-from-equilibrium chemical processes become entangled with self-assembling processes, thus creating autopoietic entities capable of actively harnessing the flows of matter and energy necessary for their own maintenance. Some of these entities become capable not only of growing and dividing but also (at least to a certain degree) of controlling these processes so as to self-re-produce. Needless to say, the first process provides the necessary environment for the development of the second.

As a global result, these autopoietic entities emerged as microscopic local "centres" of complexity within the – by comparison – large macroscopic ocean of connected matter undergoing prebiotic evolution. This process of encapsulation led to the creation of what I like to describe as the second fundamental aspect of the origin of individuation, namely the appearance of *organisationally integrated* units. I use the term "organisational integration" as opposed to (mere) "organisational coordination" in the following sense: whereas a set of different systems are held to be organisationally *coordinated* when they enter, or establish, a functional network in which they maintain their identity and specific organisation, they become organisationally *integrated* when they merge together, bringing forth a new, unique, cohesive and spatially localised entity, whose identity is the result of a re-organisation of the former constitutive entities. Thus, organisational integration implies not only that the functional parts of the former constitutive entities either disappear or change, so as to become functional parts of the new entity (thereby, ensuring global viability), nor simply that the former functional organisations become merely coordinated. Rather, it implies that new functions be created and that the older functional organisations be transformed so as to enter into a new more encompassing and hierarchically structured cohesive organisation. Thus, if we compare this scenario with the preceding one, in which self-assembling vesicles and self-maintaining networks began to cooperate, we find that these new (autopoietic) entities created a minimal form of integrated organisation.

Let's see why. Since the vesicle acts globally (affecting for example the concentration of components in the network), it could be said that it exerts a global control on all the remaining constraints of the system (such as catalysts). It is, in fact, the compartment which controls the selective forms of permeability, the energy flows required for the maintenance of the system and the mechanisms of its growth and reproduction. The compartment is what ensures the global cohesion of the system, because it affects the ensemble of different local constraints within it. The different roles played by the compartment, such as the selective forms of permeability and the mechanisms of its shrinkage or growth (as well as reproduction when it occurs), exemplify this global control over the system, and this very fact establishes a sharp distinction between the whole set of encapsulated processes and the rest of the world.

For these reasons, one could argue that the beginning of a minimal form of individuality in terms of cohesion, physical boundedness and organisational asymmetry with its environment was the encapsulation of a proto-metabolic system, which progressively took over the production and maintenance of its own compartment (Moreno 2016). Let us now analyse the implications of this.

5.3 Earlier implications

We will now assess the consequences of the appearance of this basic form of individuality. As we have seen, some of these earlier individuated entities were capable of self-reproduction. However, the establishment of a long-term process enabling a cumulative retention of innovative events also requires some form of inheritance in the reproductive steps. How could a mechanism of inheritance have emerged before the appearance of more sophisticated mechanisms that include genetic components? One likely example of primitive inheritance may be, as Segré and co-workers have suggested (Segré and Lancet, 2000, Segré et al. 2001), the so-called "compositional genomes", namely compositionally biased catalytic networks, devoid of sequence-based biopolymers, capable of transferring their compositional specificity through reproduction. Segré and Lancet's hypothesis is based on the idea of a "lipid world", which has been criticised for its limitations. But more recently, Vasas et al. (2012) and Hordijk and Steel (2014) have shown that a world of polypeptides, as opposed to the lipid world, could also ensure a form of statistical inheritance (i.e., a non-reliable way of transmitting the specific identity of a given system through reproduction).[3] Hence, self-enclosed peptide-based proto-metabolic networks in lipid compartments may transmit specific features across generations.

Once encapsulated autopoietic organisations capable of reproducing with a certain form of inheritance had developed, the appearance of populations of autopoietic entities may have led to a primitive form of evolution based on some kind of competitive dynamics. In this sense, Evelyn Fox Keller (2009, 2010) has argued that a process of intergenerational accumulation of complexity may have occurred before any form of evolution by natural selection. As she points out, all that is needed for a primitive form of evolution are systems with properties that contribute to their persistence, since that would trigger a selective process favouring the more stable systems. Hence, the generation of different degrees of stability is enough to trigger simpler evolutionary-competitive dynamics that result in different degrees of maintenance of those same stabilities.

This process in turn gives rise to a new scenario. Intergenerational entailments between types of autopoietic individuals (which we have qualified as a primitive form of "evolution"[4]) create a trans-individual diachronic process, operating at a much larger spatial and temporal scale than that of the individual entities. This evolution, allowing a cumulative process, creates the conditions for the appearance of increasingly complex forms of individual organisations.

5.4 Towards a second step of individuation in prebiotic evolution

The scenario described in the previous section enabled a process of complexification, in which the individualised systems gradually generated and incorporated more complex compounds. This complexification would have resulted in the appearance of polymers of some kind, capable of displaying both template and sequentially dependent catalytic activity, similar to present-day RNA. Indeed, it is commonly accepted that one key stage in biogenesis was the emergence of sequentially specific polymers, acting both as templates that in some sense reliably "recorded" the specific metabolic organisation of each system during reproduction; and as highly specific catalysts, controlling their constitutive ongoing organisation (i.e., their metabolism). This stage is usually known as the "RNA world".[5] Before long, these components would have been recruited by the proto-metabolic organisation of primitive autopoietic systems, leading to the origin of genetically controlled metabolic organisations, or, in other words, to autopoietic systems whose self-maintaining organisation was driven by RNA. Let us call them "genetic autopoietic systems".

The appearance of genetically instructed autopoietic systems would have had many important consequences, which we shall explore in more detail in the sections below. They can, however, be summarised as follows. First, this new form of organisation enabled a new type of evolution, much closer to Darwinian evolution, which in turn reinforced individual complexity and diversity. Second, as a consequence of this increasing diversity, a new form of collective – this time synchronic – organisation was developed, which led to the creation of the material conditions required for long-term sustainability of the increasingly diverse populations of prebiotic systems. Third, at the same time, this genetically based evolutionary stage triggered the emergence of a large and diverse population of new *quasi-autonomous individuals*, modifying both the evolutionary and the organismic framework. This set of changes would have triggered a new process of individuation, which we shall analyse in Section 5.5.

5.4.1 *The appearance of a new form of evolution*

Given that the catalytic function of RNA depends entirely on the specificity of its sequence, the preservation of this specific order is of vital importance. The fact that these components are also templates is crucial because, as a result of this property, they ensure the intergenerational transmission of this specific order. In other words, they are hereditary constraints, which can reliably preserve organisational changes from one system to another. Yet, at the same time, over longer periods (i.e., many generations), the sequential order of these hereditary constraints may statistically undergo changes.

All these changes led to an exploration of the sequential space linked to a correlative selective retention of the organisational forms. However, instructed metabolisms enabled the appearance of a much richer phenotype variety, which in turn involved differential fitness. This gave rise to a scenario in which natural selection could operate, a development which, along with reliable inheritance, accelerated the evolutionary process.

Thus, genetically driven autopoietic systems could coherently and consistently link the individual dimension of their activity (related to their constitutive organisation) to a progressively larger temporal and spatial dimension (related to their long-term maintenance and evolution as a whole population). This articulation was enabled by the inherited sequential structure of the special functional components (i.e., RNAs) of each individual entity. And at the individual system level, this permitted the complexification of both the metabolic organisation and its behaviour, as well as (most likely) a deeper articulation between the two.

5.4.2 *The appearance of proto-ecosystems*

Another consequence of this stage was the development of a progressive differentiation among the set of evolving genetic autopoietic entities, generating different hereditary *types* of individuals. The most salient form of differentiation likely took place in the type of metabolic organisation. Consequently, the emergence of metabolically differentiated populations (i.e., proto-species[6]) constituted the basis for the establishment of metabolic complementarities among them. The establishment of these metabolic complementarities was of paramount importance for the sustainability of the very process of biogenesis, because, as Guerrero pointed out (1995, Guerrero, Piqueras and Berlanga 2002), this complementarity between different types of prebiotic metabolising entities would resolve the need to "clean" the environment of an increasing amount of non-digestible organic waste. It is therefore a sensible hypothesis that, over time, some individuated systems, if forming dense and diverse enough proto-species, would have established *synchronic* functional interactions among different groups of proto-species, thus constituting the early ecological networks.[7]

The importance of these *synchronic* networks of interactions lies in the fact that they enabled the long-term sustainability of a diversified community of proto-species in both energy and material terms. The metabolic activity of each type of proto-species would have constrained the flux of energy and matter necessary for the maintenance of the other types of proto-species, which in turn would have done the same for others, and so on, until a cyclic process was established. This cyclic process could be indefinitely maintained provided certain geological and astronomical conditions were met (for example, the network was ultimately driven by a stable external energy source, such as the sun). This type of cyclic process could be qualified as "proto-ecological", since ecological interactions (such as the synthesis and recycling of organic compounds, pedogenesis and niche construction) were the inter-community relations that regulated the flow of energy and matter through both themselves and the abiotic environment, so as to ensure their long-term maintenance as a diversified community (Nunes Neto, Moreno and El-Hani 2014). Thus, in a certain sense, an ecological system would be a kind of "biologically constructed environment", as Dagg (2003) has indeed pointed out.

5.4.3 *The appearance of new quasi autonomous individuals*

Another important consequence of a stage dominated by populations of genetically driven autopoietic entities was the appearance of a new type of "autonomous" entity. Once autopoietic entities had developed genetic capacities, certain genetic structures (most of which probably originated within them) were able to develop autonomous activities. Now, on the one hand, we have individuated autopoietic systems whose organisation is driven (instructed) by a certain type of self-replicating polymer; and on the other, we have many of these latter entities spreading outside the former ones. This situation would have led to opportunities for all kinds of encounters, some of which would have given rise to non-viable associations, while others would have resulted in new, viable ones. But before considering these associations, we should first analyse how these former entities could have become, *in this context*, "autonomous agents" of some kind, with the capacity to functionally sequester the organisation of the autopoietic entities they successfully penetrated.

It is important to clarify some features of these "autonomous" genetic entities.[8] First, they were genuinely (proto)biological, because they were necessarily generated within the autopoietic protocells. Second, they could become active, because they could penetrate autopoietic entities and manage their host's metabolic organisation to work for their own survival. And third, despite this capacity to act autonomously, these entities were entirely dependent on more complex entities, namely the evolved (i.e., genetically instructed) autopoietic individualities, and cannot, therefore, be considered either minimal forms of biological individuals or genealogically older entities. However, in an environment of genetic autopoietic individualities, their capacity to display a form of autonomous activity enables them, in a certain sense, to qualify as individualities also. I will therefore use the term "quasi-individuals" to refer to these entities.

This fact is paradoxical since, as I have already argued, only a self-encapsulated proto-metabolic entity can be considered a (minimal) biological individual. Yet, these entities could perform functional activities insofar as they became embedded in the autonomous organisation of the host. These activities were functional, not for the host but rather for the infectious "agent". How could the invading entity deploy its own functional domain? Paradoxically, by becoming in some way a part (usually a pathological one) of this organisation, modifying it in its own interests, and in so doing, deploying new functions (such as capacities to compensate for the obstacles deployed by the host).

It is hard to hypothesise how complex early autonomous replicative structures must have been to be capable of infecting primitive genetic autopoietic protocells. Present-day viruses are molecular robots, made up not only of nude replicators but also including a protein coat, which surrounds and protects the genetic material (and in some cases an envelope of lipids that surrounds the protein coat). Viroids, however, which are composed solely of a short strand of circular, single-stranded RNA, may constitute "living relics" of the widely assumed ancient RNA world (Diener 1989, Flores et al. 2014). Thus, viroids may represent the most plausible form of these hypothetical autonomous genetic entities, which parasitised earlier genetically driven protocells. Whatever the case, these primitive parasitic replicators must have developed ways to penetrate their "prey's" membrane. Once within the cell, infectious replicators were able to deploy self-assembling and self-replicating capacities, which means that they had the ability to modify the organisational environment of genetically driven autopoietic systems in order to generate copies of *themselves*. Thus, they were able to evolve autonomously, not in the sense that they could evolve *by* themselves (they needed the organisation of their host), but rather in the sense of evolving *for* themselves: they were selected so as to maximise their own fitness, not that of their host. In this sense, they can be considered, albeit in a minimal sense, evolutionary individuals.

But however, on some occasions the penetration of an autonomous genetic agent could become also functional for the host organisation (or at least, not harmful). Interestingly, this could generate a viable metabolic reorganisation of the host, leading to a new functional autopoietic entity. This association could therefore lead to the formation of new species through changing the gene expression of the host. Along these lines, Cazzolla Gatti (2016) has also recently argued that pieces of genetic material could form symbiotic associations when they become included in the host's nuclear genetic structure. Yet, strictly speaking, it is hard to qualify this relation as a form of symbiosis, since insofar as the relation becomes beneficial for both entities, the former autonomous genetic agent loses its autonomous identity. Only in the case of a parasitic relation could we speak of "unlike individualities" living together.

5.5 Consequence: the origin of a new form of individuality

As explained earlier, this massive exchange of genetic material (viral genetic material, plasmids, viroids and other potentially parasitic entities) may have posed a serious threat to genetically instructed proto-organisms, since in the majority of cases, the acquisition of foreign genetic material implied the possible destruction of the invaded system. Thus, in a context of widespread "promiscuity" between autonomous genetic entities and genetically driven autopoietic entities, it would have been practically impossible for the latter to have thrived and evolved without the creation of some mechanism to protect them against harmful infections. This leads us to think that a form of protection must have been developed before the appearance of prokaryotic life. What could have been the earlier mechanisms for protecting genetically based autopoietic systems against harmful infections? On most occasions the membranes of these autopoietic entities would have offered a certain degree of protection against the invasion of harmful genetic agents. They may also have developed collective strategies, such as collective rapid mutation. However, infectious replicators would likely have overcome these strategies, since they mutate faster.[9] Hence, unless an individualised mechanism could be developed, capable of operating within an ontogenetic timeframe and monitoring the continuation of the "right" metabolic functions to protect them against alien genetic takeover, the high risk of harmful infection would have made any evolution of genetic autopoietic individuals towards more complex forms of organisation very difficult. Such a mechanism was therefore the origin of the earliest immune systems.

Present-day immune defences of bacteria and archaea are quite sophisticated. In addition to innate defensive mechanisms, such as superinfection exclusion (Sie) systems, which block phage DNA injection into the cell, restriction-modification (R-M) systems, which can destroy invading DNA, or abortive infection (Abi) systems, which cause the death of the infected cell to protect the surrounding clonal population, many prokaryotic cells also possess an adaptive immune system called CRISPR-Cas. This complexity and diversity of mechanisms points to a long period of evolution, suggesting that much more simple forms of immune defence must have emerged during prebiotic evolution. Obviously, an RNA world could not have achieved such a high degree of sophistication and complexity as did the mechanisms of genetically controlled recognition and regulation which exist in a world of DNA and proteins. But RNA has been found to have important recognition and sequential manipulation (i.e., processing) capacities, which are key tasks in immune defence, and this fact supports the likelihood of the appearance in RNA-based autopoietic entities of an early form of immune mechanism. Let us clarify what exactly we mean by this.

Nowadays, the immune system works as a general supervisor of the constitutive organisation of the organism, tying together its identity. As Chiu and Eberl (2016) have recently pointed out, "in addition to the management of energy flows, immunity is emerging as one important way living systems resist degeneration in the face of continuous external invasion and perpetual injury, cellular death, and tumors" (p. 822). Likely, immune functions emerged as a meta-control over global metabolic functions (including adaptive regulatory tasks). Whereas metabolic functions are primarily related to the *physical-chemical processes* (i.e., the material and energy interactions) required for the production and maintenance of the constitutive identity of a biological (or even proto-biological) system, immune functions deal instead with *already biological* interactions which affect (or challenge) forms of constitutive identity. In other words, the immune mechanism controls these interactions so as to maintain the metabolic identity of each autopoietic individuality. In a prebiotic scenario, an immune mechanism should deal with the alien biologically – or rather, proto-biologically – based activity, preventing it from threatening the constitutive identity of genetic autopoietic systems.[10] More specifically, the early immune task would have been to counteract the action of alien biological (or, rather, biologically generated) *agents*, by first detecting the precise source of their functional activity, and then trying to block this activity. For example, Pradeu (2012, p. 31) defines immunity in prokaryotic systems as "a group of molecular mechanisms that allow specific interactions with pathogenic patterns (especially viral patterns) and the elimination and inactivation of these pathogens". Also, much of the work carried out by the immune system consists of correctly identifying alien individualities as they affect the ongoing dynamic constitutive identity, a task which is far from easy.[11]

Any system endowed with immune capacities (even in their most basic form) will have a higher degree of complex hierarchical organisation and integration than those we have analysed so far. These individuated systems have the infrastructure required for monitoring their constitutive organisation. In other words, they have the capacity to exert a global regulatory control over themselves. This capacity is supported by off-line mechanisms operating through selective processing of the genetic components of the system (Bechtel 2007, Griesemer and Szathmáry 2009).[12] In other words, this requires a subsystem that operates *dynamically decoupled* from the process network that it regulates. As Bechtel has pointed out,

> What is required for an information-based control system that goes beyond direct negative feedback loops, then, is a property that is sufficiently independent of the processes of material and energy flow such that it can be varied without disrupting these basic processes (which may themselves be maintained by negative feedback), but still be able to be linked to parts of the mechanism to enable the modulation of their operations.
> (Bechtel 2007, p. 290)

> [...] independent control can only be achieved by a property not directly linked to the critical stoichiometry of the system.
> (Bechtel 2007, p. 289)

In sum, this is a much more integrated form of individuality than that of basic, non-genetic autopoietic systems.

Let us now explore the other implications of the emergence of immune-based individuated entities.

5.6 Implications: the prebiotic origin of colonies and symbiosis

The establishment of stable, intimate relations among similar individuals, as we understand them, requires said individuals to be relatively complex and capable of regulating their constitutive organisation according to the interactive actions of their neighbours. However, intimate associations with emergent properties which may have contributed to the survival of their components probably appeared fairly early on. For example, some experiments (Carrara, Stano and Luisi 2012, Stano et al. 2014) have shown the formation of protocell "associations". Interestingly, these aggregates increase the survival of their integral entities (i.e., the associated stage fosters solute capture and vesicle fusion). Of course, the entities featured in these experiments are extremely simple, and it would not, therefore, be justified to view the association formed as an *intimate* association between *individuals*. But once primitive protocells became more developed autopoietic systems, capable of detecting neighbouring signals and triggering flexible functional actions, then these actions could have been coordinated and a collective, mutually beneficial interaction may have been generated.[13]

The hypothesis of a prebiotic origin of a kind of colonial organisation is supported by the fact that the most ancient known forms of life on Earth are colonies growing on the surface of (or inside) a solid medium.[14] This is only logical, since these associations increase individual survival capacities.

For example, in such colonies, which generally consist of layers of individual organisms held together by polymers secreted by them, the collective action creates locally more friendly environments (neighbouring organisms live in a wide range of internal chemical environments, so they can feed on or at least tolerate the dominant chemicals at their level). Obviously, this type of close association requires not only a considerable and complex interactive capacity among members, but also a highly mouldable metabolic organisation.

Today, primitive forms of colonial association (bacterial mats and biofilms) require significant capacities, such as motility, sensing capabilities, regulatory modulation of secretions and the ability to trigger changes in the physiological organisation of their members, supported by a host of regulatory genes, for example quorum sensing (i.e., a change in the gene expression of each individual according to the detected changes of the density of the community). It is likely that these capacities would have emerged gradually, and many of them probably became available progressively as genetically instructed autopoietic entities evolved. Genetically instructed autopoietic individuals may have modulated the expression of certain genes so as to generate phenotypes enabling closer forms of collaborative interactions, such as sharing the chemicals they secreted, etc.

In some cases, probably favoured by circumstances where close ecological complementarities occurred, different proto-species may have evolved towards increasingly close forms of association. This idea is supported by the fact that in the case of most ancient *biomats* different species occupy different levels. Little by little, the ecological interactions would have led to a more intimate interaction between the different proto-species. And when they were capable of developing rich interactive capacities, it would have been possible for a group of different proto-species to have entered into a process of increasingly intimate association. Genetically instructed autopoietic individuals may have modulated the expression of certain genes so as to allow closer collaborative interactions, such as sharing the chemicals segregated by them, which in turn would have paved the way for earlier forms of co-metabolism or co-aggregation.[15]

Interestingly, due to the close proximity of different types of cells, these multispecies associations may have stabilised viable horizontal genetic exchanges. In fact, present-day multispecies biofilms create a more-than-adequate environment for horizontal gene transfer because they sustain high bacterial density and provide cell-cell contact (Sorenson et al. 2005). The role of multispecies biofilms as facilitators of novel genotypes is further illustrated by the fact that the availability of plasmid recipients (also known as plasmid permissiveness) is, among other things, affected by community composition (Burmølle et al. 2014). Obviously, to stabilise such an intimate cooperative process would have required immune individualities.

In sum, this type of association, although less intimate than that of a formerly autonomous replicator establishing a "beneficial" relationship with its host cell, arises from different forms of true individualities. Indeed, in primitive multispecies associations cooperation is unlikely to have occurred within one of the partners;[16] but the different types of individuals may have established relations of co-metabolism (Elias and Banin 2012) and even co-aggregation (op. cit., Rickard et al. 2003), as found in present-day biofilms. Hence, in this hypothetical form of prebiotic association, the partners would have been full-fledged autonomous individualities, and, to a large extent, would have maintained their identities throughout their intimate and durable association. Thus, they can be considered a true form of symbiosis. Not surprisingly, the result of this association would have been a mixture of cooperation and conflict. For example, as we see in present-day multispecies biofilms, cooperative interactions are essential for the overall biofilm fitness: cooperative interactions generate specific spatial organisation of different species, which results in an efficient diffusion of organic compounds within the whole system. However, on the other hand, a certain degree of conflict still persists, since different species compete for nutrients and try to inhibit the growth of other species (Yang et al. 2011).

5.7 Concluding remarks

By the end of prebiotic evolution, hierarchically and cohesively organised individuated systems would have emerged. These systems would have used and fought a host of alien genetic agents (ultimately generated by them) and, in close association with similar individuated systems, would have generated new cohesive associated entities. They would also have belonged to spatially and temporally larger networks, constituted by different types of individuals, establishing metabolic complementarities and ensuring that the flux of energy and matter necessary for their own maintenance was constrained so as to be indefinitely maintained. Before their appearance, an even more primitive form of self-reproducing individuality would have existed, yielding a long-term process of intergenerational lineages to which these individuals belonged, and from which they would have inherited crucial material components.

What, then, is the meaning of individuality in this somewhat messy scenario? One could view the entire process as a global unfolding of an increasingly complex system, organised at different temporal and spatial scales. Yet, at the same time, during this process, different forms of individuality emerge as places where peaks of organisational complexity become embodied and, only in this way, are able to develop new properties and interactive capacities. From a global perspective, the emergence of different instances of individuality can be seen as the materialisation of a distributed organisation in spatially bounded, cohesive and integrated units. In this sense, the focus is on the global, which carries the explanatory load. But the problem with this view is that it does not account for the appearance of the most radically new and emergent capacities and properties in the process of the origin of life.

If we shift perspective, however, and focus instead on these centres of cohesive and integrated organisations, then it becomes much easier to understand how they could have generated new relationships, new spatially and temporally larger systems, new inter-individual associations, etc., which in turn would have helped them maintain these individuals, increase their complexity and, ultimately, produce new ones. This would always have required a period of time in which some different types of entities managed to establish a stable association, while still maintaining their differentiated identities over a long period. For this reason, such an associated entity cannot yet be considered an individual. But sometimes, after many generations of intimate complementary dependency, the interactions between different associated individuals may have slowly transformed and given rise to a new functionally integrated individual.

Throughout the process of biogenesis, the creation of a basic form of individuality (and its further evolution) represented, in spatially and temporally small instances of actual organisation, the materialisation of the consequences of spatially and temporally much larger processes. Thanks to the spatially and temporally wider scaffold they created, these instances of actual organisation evolved and became capable of generating new forms of individuality. The earlier individual proto-organisms have progressively created a set of collective forms of organisation, some of them displaying their own cohesiveness and identity, and even, in certain cases, harbouring inside other individuals.

However, this raises a very fundamental question: should the individuals that belong to these more encompassing organisations still be considered as such? This problem shows that the evolution of individuality makes it to some extent context-dependent what counts as an individual. This happens, for example, when one individual becomes a part of another more integrated and cohesive individual.

Hence, since what defines the basic form of individuality is the capacity of a system to dynamically distinguish itself from its environment, and since this later could be constituted in turn by active agents, the organisational structure necessary to ensure individual agency in face of these alien agents would be more demanding. That is why, as we have seen, in the context of massive exchange of autonomous genetic agents, the only form of individuality that can survive and strive is the one which incorporates immune mechanisms. The lesson is, therefore, that to determine whether a given biological entity is or is not an individual will depend not only on its autopoietic organisation but also on whether that system is itself its own main source of normative control, or instead, whether the system is under the control of something external.

Acknowledgements

The author acknowledges the Research Projects of the Basque Government IT 590-13, and of MINECO FFI2014-52173-P, as well as a Salvador de Madariaga Fellowship PRX17/00379. Also, the author wants to acknowledge the IHPST (Paris) for hosting his research stay during the first semester of 2018, and the useful comments by Dr Mossio.

Notes

1 Of course, replication matters; but to move towards a deeper understanding of life, we have to think within a framework in which populations of molecules, instead of competing for faster replication, have diverse catalytic effects on each other, as a means of coordinating the particular locations, times and speeds at which their chemical transformations occur. This implies gathering together different reactions, i.e., embedding the processes of synthesising new structures (and degrading others) within a self-maintaining organisation. Indeed, many advocates of the primacy of replication and selection establish links with, or include replicators in, a self-maintaining organisation, thereby implicitly admitting that such an organisation is a necessary condition for making any progress in terms of complexity (Moreno and Ruiz-Mirazo 2009).
2 This scenario includes not only chemical processes but also other types of macroscopic and microscopic geological and physical processes (Nisbert and Sleep 2001).
3 The idea of a "statistical inheritance" is based on the model of "compositional genomes" proposed by Segré and co-workers (2000, 2001). According to these authors, certain non-genetic catalytic networks can grow and reproduce, transmitting, however, some of their specific compositional features from generation to generation (that is the idea behind what they call a "compositional genome"). In fact, it is the actual organisational dynamics of the system that cause the growth and subsequent division-multiplication, implying a certain probability that some of the components and features of the "mother system" get produced once and again, and thus transmitted statistically to the offspring.
4 Although based on the individual reproduction of each autopoietic entity, evolution is understood as the intergenerational change of populations, because the process necessarily involves large numbers of individuals and therefore what matters is the statistical average of the hereditary changes.
5 The RNA world hypothesis holds that the appearance of systems based on DNA as a means of support for genetic information, and of proteins as highly specific catalysts, was preceded by a stage of systems in which both functions were supported by RNA molecules. This hypothesis is supported by RNA's capacity to both store and transmit information (just like DNA), and to act as a sequentially dependent catalyst (just like proteins).
6 Since, in a prebiotic scenario, it is likely that neither hereditary lineages nor ecological specificities would have been as clearly delimited as in the present day, I would adopt here a pluralist concept of species (namely, as denoting an ecologically specific group of (proto)organisms, able to maintain this specificity through time and space); and, moreover, I would use it in a very rough sense. A more accurate term to describe this concept would therefore be "proto-species".
7 This new trans-individual dimension differs from the evolutionary one because it is an actual, synchronic organisation that, similarly to the individuals which comprise it, functions in far-from-equilibrium conditions. In fact, these interspecies networks are in organisational terms similar to the intermolecular self-maintaining catalytic networks (Cazzolla Gatti, Hordjik and Kauffman 2017). However, they are less cohesive and integrated than these.
8 Autonomous not in the full sense of being organisationally complete agents, since they could only deploy functional operations within the organisation of the true agents (the protocells), but rather in the sense that they could deploy functional operations for themselves. Hence, I use here the terms "autonomy" and "agency" in a loose sense. By qualifying as "autonomous agents" entities like prebiotic viroids, I just want to say that they are capable of using certain parts of their environment (precisely those parts which are highly organised,

i.e., the genetic protocells) for their own survival and propagation. Yet, this "activity" is neither autonomous nor even strictly speaking an active behaviour, since it depends on the organisation of the protocell, without which the viroid is purely inert (see more on this question in Moreno 2018).
9 For example, present-day prokaryotic cells prevent virus adsorption by modifying the receptor structure in their membrane through mutation and by concealing receptors with an additional physical barrier.
10 The constitutive identity of basic individuated systems lies essentially in their metabolic specificity, which is roughly transmitted in the reproductive steps. Yet, since the metabolisms of these new individuated systems are genetically specified (and reliably transmitted in reproduction), their constitutive identity lies essentially in their genetic specificity.
11 Since the immune system deals specifically with inter-individual relations that challenge the constitutive identity, it has evolved to admit intimate inter-individual symbiotic-like relations (namely, inter-individual intimate relations affecting constitutive identities in non-harmful ways), rather than just engaging in systematic and invariant rejection.
12 At the same time, and based also on the development of off-line mechanisms (for example, by modulating the expression or repression of certain functional parts of their genetic material), these systems could engage in regulatory control of other type of processes, such as interactive relations (i.e., adaptive agency, Bich and Moreno 2016).
13 Conceptually, such a scenario is not that far from present-day bacterial mats and biofilms, where cells of different species secrete polymeric substances (EPSs) that adhere to each other and/or to a surface. Moreover, the most ancient recorded form of life on Earth (i.e., stromatolites) is also a kind of colony.
14 The earlier forms of full-blown biological entities were probably small, single-species associations of chemotrophs forming biomats in hydrothermal vents. In addition, bacterial mats were the most important members and maintainers of the planet's ecological system.
15 The formation of present-day multispecies biofilms is based on surface pili, flagella and their mediated motilities, as well as on certain cell surface proteins (Yang et al. 2011). But cooperative behaviour between species requires also an exchange of a set of signalling molecules.
16 Whether proto-organisms could have successfully lived within other proto-organisms is difficult to tell. Endosymbiosis in prokaryote cells is a rare phenomenon, because cells with a wall cannot engulf other cells through phagocytosis.

References

Bechtel, W. (2007) 'Biological Mechanisms: Organized to Maintain Autonomy', in: Boogerd, F., Bruggerman, F., Hofmeyr, J. H. and Westerhoff, H. V. (Eds.), *Systems Biology: Philosophical Foundations*, Amsterdam: Elsevier, pp. 269–302.
Bich, L. and Moreno, A. (2016) 'The Role of Regulation in the Origin and Synthetic Modelling of Minimal Cognition', *BioSystem* 148, pp. 12–21.
Burmølle, M., Ren, D., Bjarnsholt, T. and Sørensen S. J. (2014) 'Interactions in Multispecies Biofilms: Do they Actually Matter?', *Trends in Microbiology* 22, pp. 84–91.
Carrara, P., Stano, P., and Luisi, L. (2012) 'Giant Vesicles "Colonies": A Model for Primitive Cell Communities', *ChemoBioChem* 13, pp. 1497–1502.
Cazzolla Gatti, R. (2016) 'A Conceptual Model of New Hypothesis on the Evolution of Biodiversity', *Biologia* 71(3), pp. 343–351.

Cazzolla Gatti, R., Hordijk, W. and Kauffman, S. (2017) 'Biodiversity is Autocatalytic', *Ecological Modelling* 346, pp. 70–76.

Chen, I. A. and Szostak, J. W. (2004) 'Membrane Growth can Generate a Transmembrane pH Gradient in Fatty Acid Vesicles', *Proceedings of the National Academy of Sciences of the USA* 101(21), pp. 7965–7970.

Chiu, L. and Eberl, G. (2016) 'Microorganisms as Scaffolds of Biological Individuality: An Eco-immunity Account of the Holobiont', *Biology and Philosophy* 31, pp. 819–837.

Clarke, E. (2010) 'The Problem of Biological Individuality', *Biological Theory* 5(4), pp. 312–325.

Clarke, E. (2013) 'The Multiple Realizability of Biological Individuals', *The Journal of Philosophy* 110(8), pp. 413–435.

Dagg, J. (2003) 'Ecosystem Organization as Side-Effects of Replicator and Interactor Activities', *Biology and Philosophy* 18, pp. 491–492.

DiFrisco, J. (2019) 'Kinds of Biological Individuals: Sortals, Projectability, and Selection', *The British Journal for the Philosophy of Science* 70(3), pp. 845–875.

Diener, T. O. (1989) 'Circular RNAs: Relics of Precellular Evolution?', *Proceedings of National Academy Science USA* 86(23), pp. 9370–9374.

Elias, S. and Banin, E. (2012) 'Multi-species Biofilms: Living with Friendly Neighbors', *FEMS Microbiology Review* 36, pp. 990–1004.

Fallah-Araghi, A., Meguellati, K., Baret, J. C., El Harrak, A., Mangeat, T., Karplus, M., Ladame, S., Marques, C. M. and Griths, A. D. (2014) 'Enhanced Chemical Synthesis at Soft Interfaces: A Universal Reaction-Adsorption Mechanism in Microcompartments', *Physical Review Letters* 112(2), pp. 28301–28305.

Flores, R., Gago-Zachert, S., Serra, P., Sanjuán, R. and Elena, S. F. (2014) 'Viroids Survivors from the RNA World?', *Annual Review of Microbiology* 68, pp. 395–414.

Gilbert, S. F., Sapp, J. and Tauber, A. I. (2012) 'A Symbiotic View of Life: We have Never been Individuals', *The Quarterly Review of Biology* 87(4), pp. 325–341.

Gilbert, S. F., Rosenberg, E. and Zilber-Rosenberg, I. (2017) 'The Holobiont with its Hologenome is a Level of Selection in Evolution', in: Gissis, S. B., Lamm, E. and Shavit, A. (Eds.), *Landscapes of Collectivity in the Life Sciences*, Cambridge, MA: MIT Press, pp. 305–324.

Gilbert, S. F. and Tauber, A. I. (2016) 'Rethinking Individuality. The Dialectics of the Holobiont', *Biology and Philosophy* 31(6), pp. 839–853.

Godfrey-Smith, P. (2013) 'Darwinian Individuals', in: Bouchard, F. and Huneman, P. (Eds.), *From Groups to Individuals: Evolution and Emerging Individuality*, Cambridge, MA: MIT Press, pp. 17–36.

Griesemer, J. and Szathmáry, E. (2009) 'Ganti's Chemoton Model and Life Criteria', in: Rasmussen, S., Bedau, M., Chen, L., Deamer, D., Krakauer, D. C., Packard, N. H. and Stadler, P. F. (Eds.), *Protocells: Bridging Nonliving and Living Matter*, Cambridge, MA: MIT Press, pp. 481–513.

Guerrero, R. (1995) 'Vida Arcaica y Ecopoyesis', in: Moran, F., Pereto, J. and Moreno, A. (Eds.), *Los Orıgenes de la Vida*, Madrid: Editorial Complutense, pp. 225–243.

Guerrero, R., Piqueras, M. and Berlanga, M. (2002) 'Microbial Mats and the Search for Minimal Ecosystems', *International Microbiology* 5(4), pp. 177–188.

Harold, F. M. (1986) *The Vital Force: A Study of Bioenergetics*, New York: Freeman.

Hordijk, W. and Steel, M. (2014) 'Conditions for Evolvability of Autocatalytic Sets: A Formal Example and Analysis', *Origins of Life and Evolution of Biospheres* 44(2), pp. 111–124.

Keller, E. F. (2009) 'Self-Organization, Self-Assembly, and the Inherent Activity of Matter', in: Otto, S. H. (Ed.), *The Hans Rausing Lecture 2009*, Uppsala University, Disciplinary Domain of Humanities and Social Sciences, Faculty of Arts, Department of History of Science and Ideas, pp. 1–29.

Keller, E. F. (2010) 'It is Possible to Reduce Biological Explanations to Explanations in Chemistry and/or Physics', in: Ayala, F. J. and Arp, R. (Eds.), *Contemporary Debates in Philosophy of Biology*, Oxford: Wiley, pp. 19–31.

Love, A. C. and Brigandt, I. (2017) 'Philosophical Dimensions of Individuality', in: Lidgard, S. and Nyhart, L. K. (Eds.), *Biological Individuality: Integrating Scientific, Philosophical and Historical Perspectives*, Chicago, IL: University of Chicago Press, pp. 318–348.

Maturana, H. and Varela, F. (1980) *Autopoiesis and Cognition. The Realization of the Living*, Dordrecht: Reidel Publishing.

Mavelli, F. and Ruiz-Mirazo, K. (2013) 'Theoretical Conditions for the Stationary Reproduction of Model Protocells', *Integrative Biology* 5(2), pp. 324–341.

McFall-Ngai, M., Hadfield, M. G., Bosch, T. C.G., Carey, H. V., Domazet-Loso, T., Douglas, A. E., Dubilier, N., Eberl, G., Fukami, T., Gilbert, S. F., Hentschel, U., King, N., Kjelleberg, S., Knoll, A. H., Kremer, N., Mazmanian, S. K., Metcalf, J. L., Nealson, K., Pierce, N. E., Rawls, J. F., Reid, A., Ruby, E. G., Rumpho, M., Sanders, J. G., Tautz, D. and Wernegreen, J. J. (2013) 'Animals in the Bacterial World, a New Imperative for the Life Sciences', *Proceedings of the National Academy of Sciences of the USA* 110(9), pp. 3229–3236.

McFall-Ngai, M. (2015) 'Giving Microbes Their Due – Animal Life in a Microbially Dominant World', *The Journal of Experimental Biology* 218, pp. 1968–1973.

Mitchell, P. (1961) 'Coupling of Phosphorylation to Electron and Hydrogen Transfer by a Chemi-osmotic Type of Mechanism', *Nature* 191, pp. 144–148.

Moreno, A. and Ruiz-Mirazo, K. (2009) 'The Problem of the Emergence of Functional Diversity in Prebiotic Evolution', *Biology and Philosophy* 24(5), pp. 585–605.

Moreno, A. (2016) 'Some Conceptual Issues in the Transition from Chemistry to Biology', *History and Philosophy of the Life Sciences* 38(4), pp. 1–19.

Moreno, A. (2018) 'On Minimal Autonomous Agency: Natural and Artificial', *Complex Systems* 27(3), pp. 289–313, doi:10.25088/ComplexSystems.27.3.289.

Morowitz, H. J. (1981) 'Phase Separation, Charge Separation and Biogenesis', *BioSystems* 14(1) pp. 41–47.

Morowitz, H. J. (1992) *Beginnings of Cellular Life: Metabolism Recapitulates Biogenesis*, New Haven, CT and London: Yale University Press.

Nisbert, E. and Sleep, N. (2001) 'The Habitat and Nature of Early Life', *Nature* 409, pp. 1083–1091.

Nunes Neto, N., Moreno, A. and El-Hani, C. (2014) 'Function in Ecology: An Organizational Approach', *Biology and Philosophy* 29(1) pp. 123–141.

Pradeu, T. (2012) *The Limits of the Self: Immunology and Biological Identity*, New York: Oxford University Press.

Pradeu, T. (2016a) 'The Many Faces of Biological Individuality', *Biology and Philosophy* 31, pp. 761–773.

Pradeu, T. (2016b) 'Organisms or Biological Individuals? Combining Physiological and Evolutionary Individuality', *Biology and Philosophy* 31, pp. 797–817.

Rickard A. H., Gilbert, P., High, N. J., Kolenbrander, P. E. and Handley, P. S. (2003) 'Bacterial Coaggregation: An Integral Process in the Development of Multispecies Biofilms', *Trends in Microbiology* 11(2), pp. 94–100.

Roughgarden, J., Gilbert, S. F., Rosenberg, E., Zilber-Rosenberg, I. and Lloyd, E. A. (2017) 'Holobionts as Units of Selection and a Model of their Population Dynamics and Evolution', *Biological Theory*, doi:10.1007/s13752-017-0287-1.

Ruiz Mirazo, K. and Moreno, A. (2004) 'Basic Autonomy as a Fundamental Step in the Synthesis of Life', *Artificial Life* 10(3), pp. 235–259.

Segré, D. and Lancet, D. (2000) 'Composing Life', *EMBO Reports* 1(3), pp. 217–222.

Segré, D., Ben-Eli, D., Deamer, D. and Lancet, D. (2001) 'The Lipid World', *Origin of Life and Evolution in the Biosphere* 31(1–2), pp. 119–145.

Shirt-Ediss, B. (2016) *Modelling Early Transitions Toward Autonomous Protocells*, PhD Dissertation, University of the Basque Country, San Sebastian.

Skillings, D. (2016) 'Holobionts and the Ecology of Organisms: Multi-species Communities or Integrated Individuals?', *Biology and Philosophy* 31, pp. 875–892.

Sorensen, S. J., Bailey, M., Hansen, L. H., Kroer, N. and Wuertz, S. (2005) 'Studying Plasmid Horizontal Transfer *in situ*: A Critical Review', *Nature Reviews Microbiology* 3, pp. 700–710.

Stano, P., de Souza, T. P., Carrara, P., Altamura, E., D'Aguanno, E., Caputo, M., Luisi, P. L. and Mavelli, F. (2014) 'Recent Biophysical Issues about the Preparation of Solute-Filled Lipid Vesicles', *Mechanics of Advanced Materials and Structures* 22(9), pp. 748–759.

Stencel, A. and Proszewska, A. M. (2018) 'How Research on Microbiomes is Changing Biology: A Discussion on the Concept of the Organism', *Foundations of Science* 23, pp. 603–620, doi:10.1007/s10699-017-9543-x.

Vasas, V., Fernando, C., Santos, M. and Kauffman, S. (2012) 'Evolution before Genes', *Biology Direct* 7, pp. 1–14.

Walde, P., Umakoshi, H., Stano, P. and Mavelli, F. (2014) 'Emergent Properties Arising from the Assembly of Amphiphiles. Articial Vesicle Membranes as Reaction Promoters and Regulators', *Chemical Communications* 50, pp. 10177–10197.

Yang, L., Wu, H., Hóiby, N., Molin, S. and Song, Z. (2011) 'Current Understanding of Multi-species Biofilms', *International Journal of Oral Science* 3(2), pp. 74–81.

Zepik, H., Blöchliger, E. and Luisi, P. L. (2001) 'A Chemical model of Homeostasis', *Angewandte Chemie International Edition* 40(1), pp. 199–202.

6 The being of living beings
Foundationalist materialism versus hylomorphism

Denis M. Walsh and Kayla Wiebe

> ... in the case of living things, their being is to live.
> —Aristotle *de Anima* B 4, 415b13

6.1 Introduction

Domains of scientific discourse have ontologies. An ontology is a theory of being for the domain of discourse, a kind of inventory for a science. It specifies the phenomena that a science ranges over, the roles they play, and what their existence consists in. Ontologies are often implicit; they go unarticulated or unscrutinised. However, they exert an influence on our thinking. One of the principal desiderata of a theory of being is that it should "stay out of the way". Staying out of the way implies, at minimum, that the theory of being should not constrain a scientific theory from saying about the things in its domain what it ought to be able to say.

The entities in the domain of biology are principally organisms. At the very least, the existence of organisms provides biology with its subject matter. Furthermore, the phenomena that biology sets itself to explain—fit, form, function—are primarily phenomena of organisms.[1] Any theory of being for organisms faces special challenges. On the one hand, organisms are material entities like anything else. On the other, organisms are unlike anything else in that they are self-building, self-maintaining, processual, purposive, and emergent. Given the principle that our ontologies "stay out of the way", our theory of being for organisms should encompass these unique features, or at the very least not prevent us from according them their full significance. Throughout much of 20th- and early 21st-century biology, the emphasis in any given theory of being for organisms has fallen primarily on stressing the commonality between organisms and other material objects. The tacit supposition seems to be that any theory of being that suffices for non-living matter should be adequate for living things as well. But things were not always so. Vitalisms of the 17th to the 20th century, as well as organicism of the early 20th century, are united by the conviction that living beings are fundamentally different from non-living things in ways that demand a different theory of being.

The question of what it is to be a living thing has resurfaced of late. This stems perhaps from rising discontent with the concept of the organism as it appears in the orthodox Evolutionary Synthesis theory of evolution (Laland et al 2015, Huneman and Walsh 2017); perhaps it is a result of advances in the study of self-organising systems (Moreno and Mossio 2015); perhaps it is due to a rekindling of interest in the metaphysics of processes (Nicholson and Dupré 2018, Meincke 2019). However it arose, the mode of being of living beings—and, subsequently, the question of what the corresponding theory of living beings ought to be—is once again a live issue.[2]

In this chapter, we survey two ontologies and interrogate their aptness for the scientific study of organisms. The first we consider has no commonly accepted name, but, nevertheless, seems to be the default theory of being for many of the natural sciences, and much of our common-sense parlance about material objects, both living and non-living. For want of a better name, we call it "Foundationalist Materialism".[3] The second we consider is "hylomorphism". While it is routinely considered as one of the more deserving victims of the "Aristotelian purge" of the natural sciences that took hold during the scientific revolution, in this chapter we seek to give it a second hearing.[4] The principal difference between these two ontologies is that where Foundationalist Materialism represents objects as exclusively constituted of matter, hylomorphism represents its objects as unities of matter and form. We argue that Foundationalist Materialism constrains what a scientific study of organisms might fruitfully say about the objects in its domain; it gets in the way. In particular, it prevents us from according the self-building, self-maintaining, processual, and emergent capacities of organisms their full due.[5] Faced with the failure of Foundationalist Materialism, the options seem to be to pursue a biology that diminishes the distinctiveness of organisms, as much of 20th- and 21st-century evolutionary biology has sought to do, or to canvass the prospects of an alternative theory of being better suited to the job. We take the latter option and call for a re-evaluation of the merits of hylomorphism for biology.[6]

We proceed in the following way. We begin with a brief discussion of the mode of being of organisms as "processual-emergents", arguing that any theory of being for organisms ought to account for this distinctive, foundational feature (Section 6.2). We then lay out what we take to be the commitments of Foundationalist Materialism, and the difficulty it has in acknowledging organisms as processual-emergents. By way of illustrating this inadequacy, we examine an influential argument from Jaegwon Kim, which if successful entails that characterising organisms as emergent is incoherent. The objective is not to endorse Kim's argument or to refute it, but to point out some of its tacit metaphysical assumptions, and how these limit the scope of his argument. Specifically, we show how Kim's argument assumes from the start the commitments of Foundationalist Materialism. Importantly, this demonstrates only that Foundational Materialism is inconsistent with emergence, not that the kind of emergence that characterises organisms (processual-emergents) is incoherent on its own terms. Foundationalist Materialism is incapable of countenancing the emergent (and *a fortiori*, the processual-emergent) nature of organisms, and it therefore "gets in the way" (Section 6.3). Finally, we outline an interpretation of Aristotle's hylomorphism and demonstrate that it is perfectly suited, indeed tailor-made, for the task of describing the distinctive way of being of organisms. Hylomorphism does not "get in the way", because it accommodates the processual-emergent nature of organisms (Section 6.4).

6.2 Organisms as processual-emergents

Organisms, like all primary substances, are material things.[7] They derive their characteristic properties from the material components of which they are constituted. Furthermore, like everything else in nature, an organism is *exclusively* material. There is no other non-material, vital component that goes into the make-up of an organism. It is natural to suppose, then, that the relationship between an organism and the matter that composes it is just like that between any material object and its matter. For run-of-the-mill objects, their identity and persistence conditions are intimately tied to their material composition, so much so that if they undergo no change in the matter that makes them up, then they persist. Organisms are manifestly not like that. Quite the contrary, organisms persist only through the constant exchange and orchestration of their material constituents. In this sense, and unlike common-sense material entities, organisms are constitutively "processual" things. They do not merely persist through change in their material constitution; they *subsist in change*.

The processual nature of organisms is well documented.[8] As C. H. Waddington suggests: an organism "is more nearly comparable to a river than to a mass of solid rock" (1957, p. 2). Ludwig von Bertalanffy describes living things as not merely existing, but as happening: "living forms are not *in being*; they are *happening*; they are the expression of a perpetual stream of matter and energy which passes the organism and at the same time constitutes it" (1952, p. 124; emphases in original).[9] As such, organisms share a mode of constitution common to a range of processual things like flames, cyclones, waves, or convective cells. Like cyclones, flames, and convective cells, organisms exist far from thermodynamic equilibrium. Their stability derives from their capacity to dissipate energy efficiently. But organisms are not mere dissipative processes. Whereas non-living dissipative systems run themselves down, organisms uniquely build and maintain structures that restore the very conditions that their existence would otherwise destroy. Their continued existence as living things depends upon their ability to do so.

Like any system, an organism is governed by the Second Law of Thermodynamics; it is subject to the tendency to increase disorder. Yet the construction and maintenance of spontaneous order appears to be a contravention. An organism "pays its debt to the Second Law" (to borrow Schrödinger's

(1944) apt phrase) by increasing the entropy around it. Both the internal order and the external disorder are achieved through metabolism. Metabolism is an organism's capacity to build stable structures that increase the efficient exchange of energy with its environment.

> In biology, [metabolism] [...] accounts for many characteristics of living systems that have appeared to be in contradiction to the laws of physics, and have been considered hitherto as vitalistic features.
> (von Bertalanffy 1950, p. 23)

Organisms thus face a thermodynamic predicament.[10] "Its 'can' is a 'must', since its execution is identical with its being. It can, but it cannot cease to do what it can without ceasing to be" (Jonas 1966, p. 83).

Negotiating its thermodynamic predicament imposes certain architectural demands on an organism (Moreno and Mossio 2015). Perhaps the most important of these is "the closure of constraints".

> [...] biological systems are, as von Bertalanffy [...] had already emphasised, thermodynamically open (dissipative) systems, traversed by a continuous flow of matter and energy; yet [...] they realise *closure*, which refers to mutual dependence between a set of constituents which *could not exist in isolation, and which maintain each other through their interactions*. [...] In a word, biological systems self-determine because they are organised, and they are organised because they realise closure.
> (Montévil and Mossio 2015, p. 180; emphases added)

Closure of constraints is a property of a system as a whole. It consists in the capacity of the system to construct parts—and to orchestrate their activities—whose existence and function simultaneously both *depend upon* and *sustain* the existence of the system. The elements of a causally closed system function to secure the unity of the system, by engaging in activities that each by itself could not perform.

Further, causally closed systems are autonomous:

> Unlike a merely physical dissipative structure (e.g., a hurricane), which maintains its identity as long as certain specific boundary conditions are met, *a system is autonomous if it actively maintains its identity*: for example, by modulating its internal, constitutive organization, in accordance with environmental changes.
> (Moreno 2019, p. 291)[11]

Autonomy signifies the freedom of the organism from the vicissitudes of its external conditions. More importantly, autonomy consists in the fact that organisms exhibit a significant measure of control over their own material constitution and function (Moreno 2019). Like any material entity, the structures, capacities, and activities of an organism as a whole are causally realised by the properties, relations, and interactions of its constituents. But, *unlike* run-of-the-mill objects, the converse relation also holds. The structures, capacities, and activities—indeed the existence and maintenance—of the constituents are the direct result of the activities of the organism as a whole.

This closure and autonomy are reminiscent of Kant's depiction of the organism as "a thing that is cause and effect of itself" (1793 [2000], p. 371). Closure and autonomy secure the reciprocity between part and whole that, according to Kant, marks out organisms as a special category of being.

> The definition of an organic body is that it is a body, every part of which is there for the sake of the other (reciprocally as an end, and at the same time, means). [...] An organic (articulated) body is one in which each part, with its moving force, necessarily relates to the whole (to each part in its composition).
>
> (*Opus postumum*, quoted in Guyer (2005, p. 104))

It is, moreover, this reciprocity between system and component, which allows for causal autonomy, that marks organisms out as *emergent* entities.

The characteristic processual nature of organisms is non-dissociable from their emergent nature. The processes and structures by which an organism subsists both constitute and are constituted by the life of the organism as a whole. Organisms are, to coin a phrase, *"processual-emergents"*. This marks them out as entirely distinctive sorts of material entities for which a distinctive theory of being is required.

> [I]n living things, nature springs an ontological surprise in which the world-accident of terrestrial conditions brings to light an entirely new possibility of being: systems of matter that are unities of a manifold [...] in virtue of themselves, for the sake of themselves and continually sustained by themselves.
>
> (Jonas 1966, p. 79)

An ontology that is adequate for biology must be able to accommodate this uniquely organismal mode of being, and any that fails to do so therefore "gets in the way". Having set out what is distinctive of organisms, we can now turn to examine both Foundationalist Materialism (Section 6.3) and hylomorphism (Section 6.4), to see which ontology better accommodates the processual-emergent nature of organisms.

6.3 Foundationalist materialism

On the general theory of being that emerged during the 16th and 17th centuries, the constitution of the world comprises a single, basic, fundamental kind of stuff: matter. Everything in the world is an aggregate of matter. Much has changed since the inception of this way of thinking, of course, but certain basic features of this theory of being survive into our own current folk and scientific worldview. In particular, the idea that "matter" is a primary substance with its own principles of change and stasis is a central stay. Even the so-called "Standard Model" in particle physics supports this metaphysical picture: "All matter around us is made of elementary particles, the building blocks of matter"; "[...] everything in the universe is found to be made from a few basic building blocks called fundamental particles, governed by four fundamental forces."[12] Further up the scale of being, there are, of course, more complex entities of different kinds, all entirely built out of matter. At last count, there are 118 elements made from the fundamental building blocks (94 of which occur naturally on Earth).[13] These, in turn, are capable of entering into further, more complex combinations as proteins. Proteins combine to make cells, and cells make up organisms. All the way up, the capacities of non-fundamental entities are determined by the properties of their constituents.[14] This is the basis of the theory of being we are calling "Foundationalist Materialism".

Foundationalist Materialism is materialist in the sense that it takes matter to be a substance, and further, it is basically the only substance. It is foundationalist in that it holds that the properties and activities of all complex non-basic entities are ultimately determined by the properties and activities of the basic components out of which they are made. The obverse of being foundationalist is being compositional. Matter combines with other matter to make new material objects, and when it does, the components confer on these combinations their properties. In this way, the nature of a complex entity is vouchsafed by the nature of the basic material that constitutes it. For composition to work in this way the properties of matter must be context-insensitive; the nature of matter is not transformed by combining with other matter.

Furthermore, according to Foundationalist Materialism, the nature of a material entity is tied to its causal powers (Ellis 2000, Bird 2007). Causal powers, in turn, are fixed by an entity's material constitution. Because causal powers of material entities are tied exclusively to their constitution, and the material constitution of an entity is intrinsic to it, it follows that the causal powers by which an entity is individuated are intrinsic properties. For this reason, the causal powers of material entities do not change with changing contexts.

6.3.1 Analytic Mechanism

Foundationalist Materialism has its own scientific method, "Analytic Mechanism", that reflects its foundationalist and compositional commitments. Its essence is nicely captured by Nancy Cartwright:

> [...] to understand what happens in the world, we take things apart into their fundamental pieces; to control a situation we reassemble the

pieces, we reorder them so they will work together to make things happen as we will. You carry the pieces from place to place, assembling them together in new ways and new contexts. But you always assume that they will try to behave in new arrangements as they have tried to behave in others. They will, in each case, act in accordance with their nature.

(Cartwright 1999, p. 83)[15]

The passage reveals much about the ways in which the metaphysical presuppositions underlying Foundationalist Materialism issue in its proprietary methodology. Taking "things apart into their fundamental pieces" is only an effective way of understanding the features of complex entities if nothing too much about the parts is changed by dissociating them. That, in turn, is only a means to understanding complex material things if nothing too much is altered by putting the parts back together.

The commitment to compositionality inherent in Foundational Materialism requires that the relation of composition is fairly transparent. John Stuart Mill's principle of the composition of causes nicely encapsulates this feature: "I shall give the name of the Composition of Causes to the principle which is exemplified in all cases in which the joint effect of several causes *is identical with the sum of their separate effects*" (Mill 1843, p. 243, emphasis added). Mill avers that the composition of causes is sufficiently widespread to make Foundationalist Materialism worth pursuing: "There are no objects which do not, as to some of their phenomena, obey the principle of the Composition of Causes; none that have not some laws which are rigidly fulfilled in every combination into which the objects enter" (Mill 1843, p. 244).[16]

The assumption that (as Cartwright calls them) "pieces" "[...] behave in new arrangements as they have tried to behave in others" further reflects the supposition of context insensitivity. The insensitivity of "pieces" derives from their natures. Each piece behaves "according to its nature", and the nature of each piece, being intrinsic, is not altered by being moved from context to context, or by being combined with other pieces into a complex entity. Compositionality, intrinsicality, and context-insensitivity together imply that the capacities of complex entities are inherited *exclusively* from the causal powers of their parts.

Analytic Mechanism is an example of what Michael Strevens (2017) calls a "compositional theory".[17] Such a theory

> [...] divides the system into parts and assigns dynamical properties to these components that, when aggregated, predict the behavior of the system as a whole. If you have a compositional theory of a certain kind of system, then, you do not need to theorize anew for each instance of that system.
>
> (Strevens 2017, p. 43)

Strevens asserts that "[c]ompositional theories are, quite obviously, greatly desirable; indeed, it is hard for science to make much headway against the world's complexity without them" (p. 43). In the case of Analytic Mechanism, the "components" are spatiotemporal parts, and the relation of aggregation represents the mechanical interaction of spatiotemporal parts.[18]

Analytic Mechanism trades extensively in a special kind of explanation: mechanistic explanation. Mechanistic explanations advert to entities and their "characteristic behaviours" to account for regular occurrences (Machamer et al. 2000, Bechtel 2006). In particular, we can explain the properties and activities of complex entities by citing the mechanical interactions among their parts: "...one needs to add that mechanistic explanations are constitutive. They explain the behavior of the mechanism as a whole in terms of the activities of its component parts" (Craver 2007, p. 161).

> [...] the behavior of a system [is explained] in terms of the functions performed by the parts and the interactions between these parts. [...] A mechanistic explanation identifies these parts and their organization, showing how the behavior of the machine is a consequence of the parts and their organization.
>
> (Bechtel and Richardson 2010, p. 17)

The constitutive nature of mechanistic explanation implies that it is reductive in a special sense (Nicholson 2018). While mechanism allows one to explain the properties of a complex entity by adverting to the properties of its parts, one cannot explain the properties of the parts by adverting to the nature of the whole.

To recap, then, the features of Foundationalist Materialism are:

Materialism: matter is basic and fundamental, and all complex entities are composed exclusively of matter.

Foundationalism: there are basic units of matter, and the nature of every compound entity is grounded exclusively in the nature of its constitutive matter.

Compositionality: matter composes. The natures of complex entities can thus be understood by analysing the natures of their constituent parts.

Intrinsicality: the natures and powers of material entities are intrinsic to them. Context- insensitivity, in fact, follows from intrinsicality.

Context-insensitivity: the natures of material entities are context-insensitive. A material entity behaves according to its nature, and its nature is constant across contexts.

Mechanism: the natures of material entities are their causal powers. The causal powers of a complex entity are inherited exclusively from the powers of its constituent parts.

Reduction: the properties of complex entities are explained by the properties of their parts, but not *vice versa*.

Certainly, Foundationalist Materialism faces certain challenges as a generalised ontology for non-living things; it is subject to a battery of objections.[19] The question we want to address in this chapter, however, is whether Foundational Materialism provides a satisfactory theory of being for living things (organisms), given their distinctive processual-emergent properties.

6.3.2 Emergence

As discussed above, organisms are processual-emergents. That is to say, the processes and properties of an organism are the consequence of its components, *and* the properties of the component parts and processes are the consequence of the organism as a whole. The reciprocity that holds between an organism and its component parts and processes is the most significant challenge that a theory of being for living things faces. Reciprocity entails emergence, the idea that the natures or causal powers of the parts are derived from the system as a whole. The metaphysical status of this kind of emergence—generative emergence—is an open question, to say the least.[20]

There is an influential argument from Jaegwon Kim (1999, 2006) to the effect that this sort of emergence is incoherent. In its outline the argument goes as follows. Kim believes that for a complex entity C to have an emergent property it must have the capacity to bring about an effect E that the collection of its parts $<c_1, c_2, c_3, \ldots c_n>$ —its "microphysical base"—does not already have.[21] But C could not have the causal power to bring about E unless its microphysical base $<c_1, c_2, c_3, \ldots c_n>$ also has the power to bring about E. So, according to the emergentist position, C must have the capacity to confer on its microphysical base, $<c_1, c_2, c_3, \ldots c_n>$, some causal properties that it wouldn't otherwise have. This is the phenomenon that Kim calls "reflexive downward causation" (RDC) (Kim 1999). According to Kim, RDC is incoherent.

> Emergentism cannot live without downward causation but it cannot live with it either. Downward causation is the raison d'être of emergence, but it may well turn out to be what in the end undermines it.
> (Kim 2006, p. 548)

To see why, we should ask where the causal powers of C come from. Here Kim appeals to a common and plausible-sounding intuition, the Causal Inheritance Principle (CIP), which says that complex entities inherit their causal powers exclusively from their microphysical bases. So C could not have its capacity to bring about E unless $<c_1, c_2, c_3, \ldots c_n>$ also does. Genuine emergence, then, requires *both* RDC and CIP. But taken together, these principles entail a contradiction. By CIP, $<c_1, c_2, c_3, \ldots c_n>$ must *have* the putative emergent causal property (otherwise C couldn't have it), and by RDC $<c_1, c_2, c_3, \ldots c_n>$ must *lack* it (otherwise, it wouldn't be emergent).

There have been multiple attempts to either resist or circumvent Kim's argument (Humphreys 1997, O'Connor 1994, Walsh 2013), or to settle for a weaker, workable conception of emergence (Bedau 1997, Huneman and Humphreys 2008). Our interest here lies less in the correctness or otherwise of Kim's opposition to emergence, than in the metaphysical presuppositions that motivate it. Reconstructing Kim's argument in a way that maximises its chances of success seems to require attributing to it the metaphysical precepts of Foundational Materialism: materialism, foundationalism, compositionality, intrinsicality, context-insensitivity, mechanism, and reduction. If so, then to the extent that Kim's argument is successful, it is less an indictment of the coherence of generative emergence *per se*, than a demonstration that generative emergence is incompatible with Foundationalist Materialism.

The first step in showing how generative emergence *per se* is not as incoherent as Kim suggests is to point out that CIP and RDC taken together do not by themselves entail a contradiction. CIP says that the properties of wholes are determined by the properties of the parts. RDC says that the properties of the parts are fixed by the properties of the whole. Taken together they imply a form of reciprocal constitution of parts and wholes in which the properties of the whole are wholly determined by the properties of the parts (CIP) *and* the constitutive properties of parts (by which they determine the properties of the whole) are determined by the whole (RDC). That, at least, seems to be a reasonable conception of what generative emergence is committed to, and it isn't evidently nonsense. The reciprocity that holds between an organism and its constituent parts and processes looks to be precisely of this sort.

It is easy to demonstrate the (in principle) consistency of this reciprocity.[22] We can imagine a substance in which this relation of reciprocal constitution holds, that is to say, where the properties of the parts constitute the whole *and* those properties are conferred on the parts by the whole. Take the fanciful, *ersatz* substance the connubium. A connubium is a married couple; it is a complex entity comprising two married persons. Of course, not any dyad of married persons is a connubium. Pythias the Elder and Harriet Taylor are both married, but do not constitute a connubium. Like any complex substance, a connubium's parts must have certain relations to one another that give the substance its unity; in this case the married persons have to be married to each other.[23] Take the connubium CharlesEmma, comprising Charles and Emma as its constituent parts. It is the property that the Charles part has of being married to Emma, along with the property that the Emma part has of being married to Charles, that makes it the case that CharlesEmma is a connubium. The relevant properties, of course, are (*inter alia*) causal properties. Emma has certain rights and powers, and Charles has certain rights and powers, that neither would have, except in the context of the connubium. Emma has the right to inherit Charles' wealth and Charles has the right to inherit Emma's. And the connubium itself has certain causal properties too. Couples have rights—e.g., to pool their taxable income, to share their parental leave—that mere pairs of persons don't have.

So, where do these properties of Charles, Emma, and CharlesEmma come from? There was a marriage ceremony that conferred on Emma the property of being married to Charles, and on Charles the property of being married to Emma. At the same time it created the connubium. In this instance, the parts could not instantiate the properties by which they constitute the composite entity unless and until the composite entity existed. This is the analogue—or an instance—of RDC. Conversely, CharlesEmma does not exist unless, and until such time as, Charles has the property of being married to Emma and Emma has the property of being married to Charles. This is the analogue—or an instance—of CIP. The conjunction of CIP and RDC does not entail a contradiction. This sort of reciprocal determination is not incoherent.

Therefore, Kim would need to add some further metaphysical principle about the make-up of complex substances that, when conjoined with CIP and RDC, entails a contradiction. The best motivated would be the principles of intrinsicality, and (consequently) context-insensitivity. With these principles in place, Kim could argue that the relevant properties by which parts constitute their wholes are intrinsic to the parts, and therefore context-insensitive. This means that the relevant properties by which the parts constitute the whole cannot be conferred on the parts by the whole; similarly, this means that the parts have their properties already, and not in virtue (or in the context) of being parts of the whole. Not only would the insistence on the intrinsicality of constituting properties have the salutary effect of ridding the world of connubia as substances, it would also secure Kim's claim of the incoherence of emergence. Where causal properties are intrinsic and context-insensitive, then the only possible source of an entity's causal properties is its internal constitution. The matter of which a complex entity is composed is its internal constitution, so a complex entity could inherit its causal properties from its component parts *and nowhere else*. Under the additional assumption of intrinsicality, CIP seems perfectly reasonable, indeed unimpeachable. And RDC seems to be false. For any part, the context in which it is a proper part of another entity could not confer on it its causal properties. So, RDC, CIP, and intrinsicality of causal powers make an inconsistent triad. Faced with this contradiction, RDC would appear to be the most dispensable of the three principles.

Where the constitutive properties of the parts that constitute the complex substances are causal properties, the addition of intrinsicality and context-insensitivity seems to be reasonable. It makes a certain amount of practical sense to individuate complex entities by those capacities conferred on them by the stable capacities of their parts. But note that the intrinsicality of the causal properties of matter (and their concomitant context-insensitivity), together with CIP, are cornerstones of Foundationalist Materialism. If Kim's argument is successful, then it successfully demonstrates *not* that generative emergence is incoherent, but that it is incompatible with Foundationalist Materialism.

One the one hand, then, the distinctive features of organisms seem to entail emergence. On the other, emergence is incompatible with our most commonly held theory of being for material objects. We are faced with a choice: either we deny that organisms are processual-emergents, or we deny that Foundationalist Materialism is an appropriate theory of being for organisms. The former strategy renders incoherent the precepts of autonomous systems theory, by which we understand the possibility of living things. This looks like a serious case of an ontology getting in the way of the workings of science.

This is not the only problem that Foundationalist Materialism has with organisms. Quite apart from their emergent properties, the dynamic nature of organisms appears to put them beyond the ken of Foundationalist Materialism too (Johnston 2006, Jaworski 2016, Meincke 2019).

> [M]ere structure in the sense of configuration of parts is far too *static* a concept to tell you all there is about the form of an animal: [...] *dynamic* notions have to be added to the relatively static structural notions to get us to something like an account of the form of a living thing.
> (Oderberg 2014, p. 177; emphases in original)

> A form or principle of unity may be [...] *dynamic*, in that [...] the parts it holds of vary over time [...]. A paradigm case is a living thing whose organic matter is unified into an organism [...]. The operation of that disposition *requires* the matter to be exchanged over time.
> (Johnston 2006, p. 663f; emphases in original)[24]

The discontent with Foundationalist Materialism motivates the search for an alternative theory of being capable of accommodating processual-emergent beings of the sort we take organisms to be.

6.4 Hylomorphism

Hylomorphism is famously Aristotle's theory of the being of primary substances. According to Aristotle, a primary substance is an interaction of form and matter. Quite how to represent this relation and its relata is a question of considerable discussion and dispute. Following others, we suggest that the relation between matter and form is a modal, synchronic, dynamic, reciprocal, and contextual one, but *not* a mereological one.[25] In particular, the matter and the form of a substance are not related as parts are to a whole, nor is it the relation that the matter of a complex substance bears to the whole according to Foundationalist Materialism.

6.4.1 *Matter and form*

Aristotle often represents the distinction between matter and form in modal terms, by reference to possibility and actuality: "the proximate matter and the form are one and the same, potentially, and the other actually" (Metaphysics 1045b; cited in Marmodoro 2013, p. 19). The distinction between a mere *potentiality* and an actual *capacity* is crucial here. The matter of a substance (taken in abstraction) consists in a range of potentialities. In taking on the form of a particular substance, the potentialities are actualised as capacities to act, or to be acted on in various ways: "[...] form *actualizes the potencies [potentialities] of matter* in the sense of being the principle that unites with matter to produce a finite individual with limited powers and an existence circumscribed by space and time" (Oderberg 2007, p. 66; emphases in original). So, for example, the wood of a table (*qua* wood, as it were) only has the properties required of a table potentially. It is only when the wood takes on the form of a particular table that the wood's latent *potentialities* to be a table are transformed into actual table *capacities*.[26]

Marmodoro (2018) stresses that understanding the modal nature of the relation between matter and form is the key to understanding much of Aristotle's metaphysics. It figures in his theory of universals and particulars, and also his theory of change. For our purposes, it can help prevent us falling into the trap of reifying matter and form as quasi-substances.

> The ultimate constituents of hylomorphic compounds are not matter and form held together by relations between them. The ultimate constituents are powers that become activated. More graphically, the bedrock of Aristotle's reality is not a two-tier hylomorphic "clasp" between matter and form; but a single tier of powers that are either in potentiality or activated. If we are to think of a "cosmic generator" of creation for Aristotle's world, it is not the coming together of matter and form into hylomorphic compounds; but rather, it is the activation of powers from potentiality to actuality.
>
> (Marmodoro 2018, p. 18)

The relation of matter and form is synchronic in the sense that neither the form nor the matter of a substance is temporally prior to the other ("matter and the form are one and the same", Aristotle, *Metaphysics* 1045b). The form of a substance is that which is realised in its matter, and the matter is that which realises the form. In the case of organisms, in particular, it is also a dynamic relation. Aristotle says: "Substance is, on the one hand, matter, on the other hand, form, *that is activity* [*energeia*]" (Aristotle, *Metaphysics* H 2, 1043a27-28, quoted in Kosman 2013). By activity [*energeia*] he means "[...] a process or action [...] whose end is nothing other than itself, a process that contains its own end and completion [...]" (Kosman 2013, p. 44). The form of an organism, considered dynamically, consists in a set of changes to the capacities of matter over time, whose end is the enactment or realisation of those vital processes that mark the organism out as the kind of organism it is. We may think of the form of an organism as a set of organising principles, or a set of goal-directed dispositions, to organise its matter in such a way that the organism is capable of performing the particular soul functions distinctive of its kind.

In keeping with the idea that matter and form are not quasi-substances, but consist in potentiality and actualisation, matter and form must also be relative concepts. Matter is matter *relative* to some form or other; form is the form *of* some assemblage of matter. One and the same entity may be both form and matter, relative to different hierarchical levels of arrangement. For example, flesh and bone may be forms that some aggregates of matter might take. But flesh and bone are not form *per se*. They are also the matter *of* a living organism.

> Roughly speaking, form is a restriction of matter [...]. And this restriction-relation can apply at a variety of different levels [...]. At each level we can intelligibly ask: what are the structural features (form) in virtue of which the stuff or thing is what it is (at the level in question); and in what and out of what material substrate has that form been generated?
>
> (Hankinson 2009, p. 216)

Form and matter are thus constitutively reciprocal. An individual organism has its actual capacities to fulfil its characteristic way of life (its form), in virtue of the matter in which it is realised. In turn, the matter of an organism has the actual powers it has because the organism's form has been realised in it. It is only in virtue of being enmattered in the way it is that the organism has the capacity to pursue the life activities that constitute its form. It is, moreover, only in virtue of being enformed in the way it is that an organism's matter has its actual powers.

On this schema, capacities are context-dependent and non-intrinsic. They are potentialities actuated by the context provided by the substance (Marmodoro 2018). Potentialities may well be context-insensitive and intrinsic, but capacities are neither. Yet it is the capacities of a substance's matter, rather than its mere potentialities, that account for the substance's characteristic activities, and its causal interactions with other substances. It is, moreover, these actual capacities, rather than mere potentialities, that we appeal to in explaining how substances are capable of doing what they do. These are Aristotle's "material" explanations. We appeal to the organism's form to explain why the matter actually has these capacities (rather than others it might have had). These are Aristotle's "formal" explanations.

However it is to be done, we must not think of form and matter as components and a substance as their mereological sum. To do so is to fall into the trap of reconceptualising hylomorphism as a kind of Foundationalist Materialism. Doing so would forfeit any advantage that the former might have over the latter for representing the nature of organisms. Marmodoro (2013) suggests that matter is "re-identified" through its participation in a substance. "A substance is *not* its parts, [...] and it is *not* its parts plus a form,

[...]. *A substance is all its parts, re-identified*" (2013, p. 18, emphases in original). The "re-identification" consists in the parts taking on the properties which they would not otherwise have had, all of which jointly constitute the whole.[27]

6.4.2 Hylomorphism and processual-emergence

Hylomorphism offers a satisfactory theory of being for organisms only if it can accommodate organisms as both processual and emergent. It is immediately obvious that hylomorphism nicely accommodates the *processual* nature of organisms: their dynamical self-maintenance, and the pursuit of those goals that constitute their natures. "According to Aristotle, that status is conferred on them in virtue of their possessing 'principles of activity' which allow them to persist as unified beings [...]" (Austin and Marmodoro 2018, p. 178). The very being of an organism consists in it partaking of the processes of life: "[...] in the case of living things, their being is to live" (Aristotle, *de Anima* B 4 415b13). Their being consists in an active principle of change in which the organism's form consists in the power to orchestrate the production of those structures and activities that realise the organism's distinctive way of life (Lennox 2010).

Not only does Aristotle's hylomorphism represent living things as irreducibly dynamical (processual), it also makes clear their status as *emergent* entities. The reciprocity of matter and form reflects the reciprocal relation between the capacities of an organism's component parts and processes and the integrated functioning of the organism as a whole. As we saw (in Section 6.2), the closure of constraints involves parts and processes whose existence and function simultaneously both *depend upon* and *sustain* the existence of the system. The closure of constraints mirrors the reciprocal relation between matter and form. An organism's "principles of activity" crucially involve the synthesising of materials—the matter—of which it is made. The "principles of activity" then confers on these material parts the capacities necessary for them jointly to realise the organism's form. Thus, the hylomorphic relation between matter and form meets the requirement of closure of constraints.

Closure of constraints, in turn, entails a form of generative emergence. Generative emergence requires that (CIP) and (RDC) hold simultaneously in a system. This, as we saw, is incompatible with Foundationalist Materialism. Hylomorphism, in contrast, provides the ground for the possibility of emergence. The relation of matter to form, by which the capacities of matter confer on the organism its ability to pursue its particular way of life, is for all intents and purposes, a case of CIP. Conversely, the relation of form to matter, by which the organism's pursuit of its life activities (its form) confers on its component parts and processes (its matter) their particular capacities, is an instance of RDC.

Aristotle's hylomorphism, then, has the conceptual resources to make sense of organisms as processual-emergents. Hylomorphism represents living things as thus fundamentally dynamic, processual entities. For Aristotle the being of living beings consists in the process by which the form of the organism synthesises, alters, and orchestrates its constituent matter in ways that equip the organism for the pursuit of its distinctive way of life. Reciprocally, the matter of an organism confers on it the capacity to pursue those vital processes appropriate to its form. Organisms are also emergent in the following way: the capacities of the organism as a whole to pursue its ways of life (form) are underwritten by the capacities and arrangements of the organism's material parts and constitution (matter). Yet, it is also true that the organism's pursuit of its way of life (form) confers on the organism's material constituents (matter) the properties they need to contribute to the organism as a whole. There is a dynamic reciprocity, a mutual determination, between the organism and its material constitution that is exactly mirrored in the reciprocity of matter and form encoded in Aristotle's hylomorphism.

6.5 Conclusion

To be an organism is to be a system that subsists through the constant exchange of matter and energy with its environment. This constant exchange requires that organisms build and maintain those structures necessary to resist thermodynamic decay. The ability to resist thermodynamic decay, in turn, requires that an organism's architecture manifests a closure of causal constraints in which there is a "mutual dependence between a set of constituents which could not exist in isolation, and which maintain each other through their interactions" (Montévil and Mossio 2015, p. 180). In this sense, the processes that constitute an organism are emergent. If these uniquely defining features of organisms are to be accorded their rightful place in our scientific investigations of life, then our theory of being for organisms must be capable of accommodating them.

We have argued that Foundationalist Materialism, the theory of being most often implicitly deployed in our scientific and common-sense thinking about material things, is incompatible with the processual-emergent nature of organisms. This is for the simple reason that Foundationalist Materialism is inconsistent with the reciprocity that holds between the capacities of an organism's constituents and the activities of the organism as a whole. It gives us one-way determination only: material constitution determines the properties of complex substances, and not the other way around. As such, Foundationalist Materialism "gets in the way"; it prevents us from saying what a science of organisms ought to be able to say. For its part (so to speak), Aristotle's hylomorphism nicely reflects the reciprocity between an organism and its constituent matter. Hylomorphism, then, is a more appropriate theory of being by which to capture the distinctive features of organisms *qua* organisms. It is hardly surprising that Aristotle's theory of being adequately represents the nature of organisms. It is often said that it is tailor made for Aristotle's biology (Leroi 2014, Kosman 2013). What is less often said is that hylomorphism is the theory of being that is most consonant with our best contemporary scientific account of the nature of living beings.

Acknowledgements

Early versions of this chapter were given as talks in Montreal, Salt Lake City, and London. We thank Anne Peterson and Anne Sophie Meincke for the invitations, and audiences at these conferences for helpful feedback. Jim Lennox, as ever, has been a source of astute insights into Aristotle's biology. Anne Sophie Meincke and John Dupré provided very helpful editorial comments. This chapter was completed during a resident fellowship (by DMW) at Institut d'Études Avancées and FMSH, Paris. We wish to thank these institutions for their generous support.

Notes

1 Much of 20th-century evolutionary biology pursued a programme in which the canonical entities in evolution are genes, rather than organisms. Walsh (2015, 2018) argues that organisms play little part in evolutionary theory largely because its ontology got in the way.
2 See, for example, Turner (2013, 2017) who explicitly endorses a form of "process" vitalism, and Walsh (2018) for an argument for "methodological" vitalism. See Nicholson and Dupré (2018) for various process ontological approaches to understanding the nature of organisms.
3 Our Foundationalist Materialism bears some resemblances to, and some differences from, John Dupré's (1993) "compositional materialism". See his discussion of compositional materialism and reductive materialism. We thank John for drawing this issue to our attention.
4 The term "Aristotelian purge" is borrowed from Taylor (1989). There are, to be sure, various attempts to rehabilitate various versions of hylomorphism (Fine 1999, Johnston 2006, Oderberg 2007, Koslicki 2008, Marmodoro 2013, Jaworski 2014, 2016, Austin 2017, Austin and Marmodoro 2018). Some of these (especially, Marmodoro 2013, Austin 2017, Austin and Marmodoro 2018) are wholly consonant with our own, others probably less so. For a different, but complementary argument for the utility of hylomorphism for understanding organisms, see Oderberg (2018).
5 There may be myriad other reasons for abandoning Foundationalist Materialism (see perhaps Wilson 2006, Jaworski 2016). Our principal aim is to argue for its inability to represent organisms as the kinds of things they are.
6 While we endorse hylomorphism for living things, we prescind from the question of the unity of theories of being. That is to say, we do not address the question whether hylomorphism is the appropriate theory of being for non-living things as well.
7 We use the terms "substance" and "primary substance" here in their most generic sense as that which is not predicable of anything else (as per Aristotle's *Categories*) (Robinson 2018).
8 See Dupré and Nicholson (2018) for an extended account of the processual nature of organisms. See also various papers in Nicholson and Dupré (2018).

9 Cited in Nicholson (2018). We thank Dan for drawing our attention to this passage.
10 We are grateful to Alex Djedovic for discussions on this topic.
11 The significance of autonomy as a condition (and a consequence) of life was first recognised by Claude Bernard: "The constancy of the internal environment is the condition for a free and independent life" (Turner 2017, p. 32, translation by Turner).
12 https://home.cern/about/physics/standard-model. The Standard Model accounts for only three of the four fundamental forces. But even forces consist in the exchange of particles (Bosons).
13 In the words of Tom Lehrer, "These are the only ones of which the news has come to Harvard. And there probably are others but they haven't been discarvard."
14 Oppenheim and Putnam's (1958) Unity of Science hypothesis is a striking example of this way of thinking.
15 We thank Rasmus Winther for drawing our attention to this.
16 Famously Mill makes special exception for living things. In living things, he argues, the Composition of Causes breaks down comprehensively.
17 See also Winther (2011).
18 Strevens is not, it should be noted, endorsing the universal applicability of this kind of spatiotemporal decomposition. There are other ways of decomposing a complex system.
19 Unger (1979) and van Inwagen (1990), for example, deny that there are any non-living non-basic entities. These theories are nicely surveyed in Koslicki (2018).
20 We take the term "generative emergence" from Seibt (2018).
21 In itself, this is a decidedly odd supposition. Cast in its own terms, generative emergence probably entails no such thing. But, as will become evident, Kim seems to be constrained to think of emergence in this way because of certain unarticulated assumptions about the nature of substances.
22 A version of the following argument occurs in Walsh (2015).
23 A connubium is not just a heap (see Koslicki 2008). It has internal structure.
24 Jaworski (2016) provides ample reason for supposing that the nature of organisms, and much else besides, cannot be captured by their material constitution alone.
25 In particular, we are following Shields (2007), Kosman (2013), and Marmodoro (2013, 2018).
26 The modal nature of the form/matter relation is a common theme for contemporary hylomorphists (e.g., Fine 1999, Oderberg 2007).
27 The idea that parts are re-identified when they constitute a substance may help elucidate the claim in Levins and Lewontin (1983) that, according to dialectical materialism, parts have no prior existence to the complex entities that they constitute.

References

Austin, C. (2017) 'A Biologically Informed Hylomorphism', in: Simpson, W. M. R., Koons, R. C. and Teh, N. J. (Eds.), *Neo-Aristotelian Perspectives on Contemporary Science*, London: Routledge, pp. 185–210.

Austin, C. and Marmodoro, A. (2018) 'Structural Powers and the Homeodynamic Unity of Organisms,' in: Simpson, W. M. R., Koons, R. C. and Teh, N. J. (Eds.), *Neo-Aristotelian Perspectives on Contemporary Science*, London: Routledge, pp. 169–184.

Bechtel, W. and Richardson, R. (2010) *Discovering Complexity. Decomposition and Localization as Strategies in Scientific Research,* Cambridge, MA: The MIT Press.

Bechtel, W. (2006) *Discovering Cell Mechanisms: The Creation of Modern Cell Biology*, Cambridge: Cambridge University Press.
Bedau, M. (1997) 'Weak Emergence', in: Tomberlin, J. (Ed.), *Philosophical Perspectives 11: Mind, Causation, and World*, London: Blackwell, pp. 375–399.
Bird, A. (2007) *Nature's Metaphysics: Laws and Properties*, Oxford: Oxford University Press.
Cartwright, N. (1999) *The Dappled World: A Study in the Boundaries of Science*, Cambridge: Cambridge University Press.
Craver, C. (2007) *Explaining the Brain: Mechanisms and the Mosaic Unity of Neuroscience*, Oxford: Oxford University Press.
Dupré, J. (1993) *The Disorder of Things: Metaphysical Foundations of the Disunity of Science*, Cambridge, MA: Harvard University Press.
Dupré, J. and Nicholson, D. J. (2018) 'A Manifesto for a Processual Philosophy of Biology', in: Nicholson, D. J. and Dupré, J. (Eds.), *Everything Flows: Towards a Processual Philosophy of Biology*, Oxford: Oxford University Press, pp. 3–47.
Ellis, B. (2000) *Scientific Essentialism*, Cambridge: Cambridge University Press.
Fine, K. (1999) 'Things and Their Parts', *Midwest Studies in Philosophy* 23, pp. 61–74.
Guyer, P. (2005) 'Organisms and the Unity of Science', in: Guyer, P. (Ed.), *Kant's System of Nature and Freedom: Selected Essays*, Oxford: Oxford University Press, pp. 86–111.
Hankinson, R. J. (2009) 'Causes', in: Agnastopoulos, G. (Ed.), *A Companion to Aristotle*, London: Blackwell, pp. 213–229.
Humphreys, P. (1997) 'How Properties Emerge', *Philosophy of Science* 64, S337–S345.
Huneman, P. and Humphreys, P. (2008) 'Dynamical Emergence and Computation: An Introduction', *Minds and Machines* 18, pp. 425–430.
Huneman, P. and Walsh, D. M. (2017) (Eds.) *Challenging the Modern Synthesis: Adaptation, Inheritance, Development*, Oxford: Oxford University Press.
Jaworski, W. (2014) 'Hylomorphism and the Metaphysics of Structure', *Res Philosophica* 91, pp. 179–201.
Jaworski, W. (2016) *Structure and the Metaphysics of Mind: How Hylomorphism Solves the Mind-Body Problem*, Oxford: Oxford University Press.
Johnston, M. (2006) 'Hylomorphism', *Journal of Philosophy* 103, pp. 652–698.
Jonas, H. (1966) *The Phenomenon of Life: Toward a Philosophical Biology*, Evanston, IL: Northwestern University Press.
Kim, J. (1999) 'Making Sense of Emergence', *Philosophical Studies* 95, pp. 3–36.
Kim, J. (2006) 'Emergence: Core Ideas and Issues', *Synthese* 151, pp. 547–559.
Koslicki, K. (2008) *The Structure of Objects*, Oxford: Oxford University Press.
Koslicki, K. (2018) *Form, Matter, Substance*, Oxford: Oxford University Press.
Kosman, A. (2013) *The Activity of Being*, Cambridge, MA: Harvard University Press.
Laland, K. N., Uller, T., Feldman, M. W., Sterelny, K., Müller, G. B., Moczek, A., Jablonka, E. and Odling-Smee, J. (2015) 'The Extended Evolutionary Synthesis: Its Structure, Assumptions And Predictions', *Proceedings of the Royal Society B: Biological Sciences* 282: 20151019, doi:10.1098/rspb.2015.1019.
Lennox, J. (2010) 'Βιος, Πραχεις, and the Unity of Life', in: Follinger, S. (Ed.), *Was ist Leben?*, Stuttgart: Steiner Verlag, pp. 239–257.
Leroi, A. (2014) *The Lagoon: How Aristotle Invented Science*, London: Bloomsbury.
Levins, R. and Lewontin, R. (1983) *The Dialectical Biologist*, Cambridge, MA: Harvard University Press.

Machamer, P., Darden, L. and Craver, C. (2000) 'Thinking about Mechanisms', *Philosophy of Science* 57, pp. 1–25.
Marmodoro, A. (2013) 'Aristotle's Hylomorphism Without Reconditioning', *Philosophical Inquiry* 37, pp. 5–22.
Marmodoro, A. (2018) 'Potentiality in Aristotle's Metaphysics', in: Engelhard, K. and Quante, M. (Eds.), *Handbook of Potentiality*, Dordrecht: Springer, pp. 15–43.
Meincke, A. S. (2019) 'Autopoiesis, Biological Autonomy and the Process View of Life', *European Journal for Philosophy of Science* 9: 5, doi:10.1007/s13194-018-0228-2.
Mill, J. S. (1843) *A System of Logic Ratiocinative and Inductive*, London: Harper and Brothers. www.Gutenberg.Org/Files/27942/27942-Pdf.Pdf.
Montévil, M. and Mossio, M. (2015) 'Biological Organisation as Closure of Constraints', *Journal of Theoretical Biology* 372, pp. 179–191.
Moreno, A. (2019) 'On Minimal Autonomous Agency, Natural and Artificial', *Complex Systems* 27: 3, doi:10.25088/ComplexSystems.27.3.289.
Moreno, A. and M. Mossio (2015) *Biological Autonomy: A Philosophical and Theoretical Enquiry, History, Philosophy, and Theory of the Life Science*, Vol. 12, New York: Springer.
Nicholson, D. J. (2018) 'Reconceptualizing the Organism from Complex Machine to Flowing Stream', in: Nicholson, D. J. and Dupré, J. (Eds.), *Everything Flows: Towards a Processual Philosophy of Biology*, Oxford: Oxford University Press, pp. 139–166.
Nicholson, D. J. and Dupré, J. (Eds.) (2018) *Everything Flows: Toward a Processual Philosophy of Biology*, Oxford: Oxford University Press.
O'Connor, T. (1994) 'Emergent Properties', *American Philosophical Quarterly* 31, pp. 91–104.
Oderberg, D. (2007) *Real Essentialism*, London: Routledge.
Oderberg, D. (2014) 'Is Form Structure?', in: Novotný, D. and Novak, L. (Eds.), *Neo-Aristotelian Perspectives in Metaphysics*, New York: Routledge, pp. 164–180.
Oderberg, D. (2018) 'The Great Unifier: Form and the Unity of the Organisms', in: Simpson, W. M. R., Koons, R. C. and Teh, N. J. (Eds.), *Neo-Aristotelian Perspectives on Contemporary Science*, London: Routledge, pp. 211–234.
Oppenheim, P. and Putnam, H. (1958) 'Unity of Science as a Working Hypothesis', in: Feigl, H., Scriven, M. and Maxwell, G. (Eds.), *Minnesota Studies in the Philosophy of Science* 2, Minneapolis, MN: University of Minnesota Press, pp. 3–36.
Robinson, H. (2018) 'Substance', in: Zalta, E. (Ed.), *The Stanford Encyclopedia of Philosophy,* https://plato.stanford.edu/archives/win2018/entries/substance/.
Schrödinger, E. (1944) *What Is Life?* Cambridge: Cambridge University Press.
Seibt, J. (2018) 'Ontological Tools for the Process Turn in Biology', in: Nicholson, D. J. and Dupré, J. (Eds.), *Everything Flows: Towards a Processual Philosophy of Biology*, Oxford: Oxford University Press, pp. 113–137.
Shields, C. (2007) *Aristotle: An Introduction*, London: Routledge.
Strevens, M. (2017) 'Ontology, Complexity, and Compositionality', in: Slater, M. and Udell, Z. (Eds.), *Metaphysics and the Philosophy of Science: New Essays*, Oxford: Oxford University Press, pp. 41–54, doi:10.1093/acprof:oso/9780199363209.003.0003.
Taylor, C. (1989) *Sources of the Self*, Cambridge, MA: Harvard University Press.
Turner, J. S. (2013) 'Homeostasis and the Forgotten Vitalist Roots of Adaptation', in S. Normandin and C. Wolfe (Eds.), *Vitalism and the Scientific Image in Post-Enlightenment Life Science*, 1800–2010, Dordrecht: Springer, pp. 271–292.

Turner, J. S. (2017) *Purpose and Desire: What Makes Something "Alive" and Why Modern Darwinism Has Failed to Explain It*, Toronto: Harper Collins, Canada.
Unger, P. (1979) 'There Are No Ordinary Things', *Synthese* 41, pp. 117–154.
Van Inwagen, P. (1990) *Material Beings*, Ithaca, NY: Cornell University Press.
Von Bertalanffy, L. (1950) 'The Theory of Open Systems in Physics and Biology', *Science* 111, pp. 23–29.
Von Bertalanffy, L. (1952) *Problems of Life. An Evaluation of Modern Biological Thought*, London: Wiley.
Waddington, C. H. (1957) *The Strategy of the Genes*, London: Allen and Unwin.
Walsh, D. M. (2006). 'Evolutionary Essentialism', *British Journal for the Philosophy of Science* 57, pp. 425–448.
Walsh, D. M. (2013) 'Mechanism, Emergence, and Miscibility: The Autonomy of Evo-Devo', in Huneman, P. (Ed.), *Functions: Selection and Mechanisms*, Dordrecht: Springer (*Synthese Library* 363), pp. 43–65.
Walsh, D. M. (2015) *Organisms, Agency, and Evolution*, Cambridge, Cambridge University Press.
Walsh, D. M. (2018) 'Objectcy and Agency: Toward a Methodological Vitalism', in: Nicholson, D. J. and Dupré, J. (Eds.), *Everything Flows: Towards a Processual Philosophy of Biology*, Oxford: Oxford University Press, pp. 167–185.
Wilson, J. (2006) 'On Characterizing the Physical', *Philosophical Studies* 131, pp. 61–99.
Winther, R. (2011) 'Part-Whole Science', *Synthese* 178, pp. 397–427.

7 The origins and evolution of animal identity

Stuart A. Newman

7.1 Introduction

Any notion of biological identity will necessarily be a conditional one, dependent on which biological level one is considering (e.g., viruses, cells, multicellular organisms), what category within a level (e.g., phages vs. animal viruses, nucleated vs. nonnucleated cells, animals vs. plants), and how evolved from their points of origin are the entities under consideration. Assigning identity will also depend on the theory of the living state one brings to bear on the question. As an extreme (and naïve) example, if an organism's identity is taken to be defined by the information contained in its genome, then even one mutation will change this identity, since it would be, by this definition, a quasi-continuous property. If, in contrast, the type specificity of an organism is (as generally assumed) relatively autonomous of the details of its genome, then some genetic variation will be consistent with conserved biological identity.

To analyse identity within a category of organism and the range of variation consistent with membership in that category, it is necessary to distinguish that category from others. This is at base an evolutionary question – since however extensive the differences may be among present-day types, each type is constituted, if only incipiently, when it first diverges from other ones. Like the question of identity itself, the origination of types is a theory-laden issue. The standard model of evolution (the Darwinian modern synthesis), for example, considers new phylogenetic lineages to arise from the same microevolutionary processes that fine-tune established ones. Other models (e.g., some versions of evolutionary developmental biology), in contrast, assert that saltational processes play a role in origination events. Although gradualist and saltationist models both recognise the existence of genealogical nodes at which clades (consisting of an ancestral form and all its lineal descendants) are established, the ontological status of such bifurcations will differ depending on the mechanism by which they occur.

For many important examples of emergent biological entities – e.g., living vs. nonliving chemical systems, archaea vs. bacteria, prokaryotes vs. eukaryotes – origination scenarios, i.e., accounts of the mode and degree of initial separation from the ensemble of progenitors, are obscure. For others, such as the rise of the metazoans or animals from unicellular ancestors, key steps can be reconstructed. This chapter will address this case.

The idea that the identity of an organism can derive from its being an exemplar of a "natural kind" verges towards essentialism, a notion broadly rejected by modern philosophers of biology (Okasha 2002, Lewens 2012, Dupré 2014), although there are some exceptions (Wilson et al. 2007, Devitt 2010). Essentialism is widely seen to conflict with the reality of evolution, in which species identity is conditional and mutable. Even more alien to much of present-day thinking is the Kantian concept of "natural purpose," which implies a recursive self-definition that seems to defy natural explanation (Moss and Newman 2015). In this chapter, I argue that the concepts of natural kind, essence, and natural purpose are in fact applicable to living systems, but only if certain tenets of the modern synthesis are relinquished in favour of a "physico-genetic" perspective that emphasises the material properties of cell assemblages and their inherent properties (Newman 2012, 2018, 2019a). Rather than conflicting with naturalistic biological science, I propose that these concepts are necessary to it and enable an understanding of the relation between development and evolution, and organismal identity and individuality.

It may be impossible to formulate general criteria of organismal identity and individuality which would encompass cases as varied as bacterial biofilms, social amoebae, plants, and animals, or their prokaryotic and eukaryotic unicellular antecedents. However, it is reasonable to expect that intensive analysis of one example could inform consideration of others. Here I focus on the animals and the unicellular progenitors from which they arose, and the various phyla into which they have diversified. I argue that both genes and the physics of materials are causal factors in the constitution, diversification, and individuation of animal bodies. Further, I provide evidence that the inception of new categories of animals was usually associated with the appearance of novel genes which enabled, and in some cases made inevitable, the generation of unprecedented morphological motifs. The association of the genes in question with unique morphogenetic capabilities suggests that animal type-identities so defined are intrinsically stable, a conclusion at odds with gradualist and adaptationist evolutionary models. An important corollary is that individuality need not be in tension with or disruptive of categorical identity but will often serve to intensify it.

7.2 The constitution of animal identity

The first animals to emerge on Earth arose within a population of single-celled organisms that shared ancestry with present-day choanoflagellates (reviewed in Newman 2016b). The animals (metazoans), the choanoflagellates, and another unicellular group with extant members, the filastereans, are collectively termed "Holozoa." Unicellular holozoan species sometimes exhibit colonial forms, but the cell masses that constitute embryos and organ primordia of animals are very different, biologically and physically, from the cell clusters or aggregates formed by unicellular holozoans (reviewed in Newman 2019b). Unlike the latter's transient associations, the metazoan cells grip one another by cell attachment molecules (CAMs) that extend through their membranes and engage their internal cytoskeletons. Specifically, there was an evolutionary addition to a class of holozoan CAMs known as cadherins of a new cytoplasmic domain not found in any known protein of other unicellular or multicellular organisms, turning them into what are termed "classical cadherins." Additionally, the catenins, whose products link the cytoplasmic tails of classical cadherins to the cell's actin cytoskeleton, are unique to the animals among the holozoans (reviewed in Newman 2019a).

The resulting capacity of cells to extend random protrusions while remaining reversibly bound to their neighbours enables them to move between and past one another without disrupting the cohesion of the cellular mass. This endows developing animal tissues with *liquid-like* properties (Newman 2016a).[1] Although some other forms of multicellular life (e.g., social amoebae) have liquid-like properties (Nicol et al. 1999), none achieve it in precisely the same way as animal tissues.

In nonbiological liquids, the subunits (atoms or molecules) cohere by electrical forces while moving freely with respect to one another by random Brownian motion. Although, as indicated, cohesivity and random locomotion have entirely different bases in animal tissues, the same liquid-state physics applies. In the first place, the free exchange of cells between the periphery and interior causes these cell clusters to exhibit surface tension, and thus to assume a spherical morphology when unstressed, like nonbiological liquids. Second, any interior spaces will automatically fill in with cells, since simple liquids do not sustain lacunae. Third, if two parcels of the same liquid tissue are fused, eventually the cells will cross the interface and mix with each other, just like the molecules in drops of nonliving liquids (Foty et al. 1994).

In addition, if a liquid tissue, analogously to a binary liquid (e.g., a shaken mixture of oil and water) is composed of two different cell types, it can undergo phase separation, spontaneously sorting out (like the oil and water) into distinct layers. Such layering is the initial step in most modes of animal embryogenesis (Forgacs and Newman 2005). For cell mixtures to sort out and form layers the cells must differentiate from one another. Cells in a tissue mass can be induced to differentiate by *morphogens*, secreted diffusible proteins. There are more than a dozen of these, but the most fundamental is Wnt, a molecule produced by all animal embryos but in no other known organisms. Wnt modulates the expression of genes in the cells that produce it or those nearby, leading to heterogeneous cell mixtures (Loh et al. 2016).

In cases where their cells express a few additional components, liquid tissues can mirror the more elaborate organisational effects seen in nonliving liquids composed of, or containing, asymmetric subunits. The most important of these effects are also induced by Wnt, which in addition to its influence on gene expression can mobilise internal machinery inherited from unicellular ancestors to make cells nonuniform in surface properties ("apicobasal polarity") or anisotropic in shape ("planar polarity") (Karner et al. 2006a, 2006b). When apicobasally polarised cells rearrange, they can come to surround interior tissue spaces or lumens. Here they are behaving like molecules in nonbiological liquids with polar charge distributions, which spontaneously self-organise into closed micelles or vesicles. Alternatively, like the long polymers that comprise liquid crystals (droplets of which exhibit asymmetrical shapes), planar polarised cells can intercalate among one another and reshape tissues, causing them to narrow in one direction and elongate orthogonally to it.

Following the origination of animal life as Wnt-expressing liquid and liquid crystalline tissues, several structural motifs that characterise the metazoans emerged almost automatically due to their being inherent to this novel form of matter (Newman 2018, 2019a). The morphologically simplest animals, marine sponges and placozoans, which are the only metazoans that (with some exceptions) lack the planar polarisation mechanism induced by Wnt, are characterised by poorly defined cell layers and labyrinths and lumens. All the other, more complex, animal groups are termed "eumetazoans." These have sharply defined layers, elongated bodies, appendages, and, in more sophisticated forms, internal organs.

Contrary to the expectations of gradualist adaptationism, all these body plans emerged in a few time-compressed episodes (beginning in the Precambrian) (Rokas et al. 2005, Conway Morris 2006, Shen et al. 2008). They also have defining characteristics which (like the constitution of the metazoan liquid-tissue state itself) are attributable to the harnessing of previously inapplicable physical effects by novel molecular functionalities (Newman and Bhat 2009, Newman 2012). The implication is that the superphylum- and phylum-level features of animal life (what I refer to here as "phylotypic identity") emerged early, in broad strokes, and not as late products of repeated cycles of microevolution of simple forms. This has major implications for the notion of identity in this group of organisms.

7.3 The diversification of phylotypic identity

Identity is partly a matter of group membership. In the view of the modern evolutionary synthesis, boundaries between groups are porous, and every organismal category is only apparently stable. All types are on their way to becoming something else. The picture laid out in the previous section, however, points to constitutive physical properties and dispositions that define the metazoans and make the forms they assume predictable and, in important ways, inevitable. As evolution progressed, these properties were recruited for the elaboration of new structural motifs during body and organ development. The liquid (including liquid-crystalline) tissue state and its inherent morphogenetic effects are necessary and enabling for the origination and development of animals, and are generatively entrenched with respect to the subsequent evolution of these organisms (Wimsatt 2015).

The early-diverging sponges and placozoans, the so-called basal metazoans, lack one or more components of the planar polarisation pathway, as well as the ability to produce the stiff planar extracellular matrix, the *basal lamina*, essential for sharply separated tissue layers and complex morphogenesis (see below). However, the relevant genes may have been lost completely or in part in the evolution of these forms (one group of sponges, the homoscleromorphs, contain both the planar polarisation pathway and a basal lamina, for example, and the single extant placozoan species contains a truncated gene for the enzyme – peroxidasin – required to assemble the basal lamina) (Fidler et al. 2014, Degnan et al. 2015). The simple body plans of these animals thus may reflect a secondary paring down of the metazoan "morphogenetic toolkit" (the gene products involved in morphological development) to a near-minimal set of determinants for liquid-tissue behaviour.

The cnidarians (e.g., hydra, jellyfish), a group of animals with more than 10,000 species, are the simplest forms in which a full planar polarisation pathway is present, and which have basal laminae (reviewed in Newman 2016b, 2019a). Planar cell polarisation enabled elongation of the body, and of appendages such as tentacles, while the basal lamina was essential for the formation of true tissues, which differ from the transient and provisional cell layers of sponges and placozoans. The cnidarians are termed "diploblasts" because of their having two apposed *epithelial* tissue layers (epithelia being tissues composed of cells directly attached to one another via CAMs) with an intervening extracellular matrix composed in part of apposed basal laminae. The enigmatic ctenophores (e.g., comb jellies) are similarly diploblastic but might in fact be a sister clade to all the metazoans rather than just to the cnidarians. Their many unique morphological features, apparent loss of ancestral genes and unusually rapid evolution of others, and functional substitutions (e.g., Wnt, though encoded by the ctenophoran genome, is not employed during embryogenesis), make them difficult to place phylogenetically (King and Rokas 2017).

The ability to form and reshape sharply defined layers of tissue allowed diploblasts to develop in a determinate fashion, generating body plans with more stereotypical architectures than those of the basal metazoans. Arnellos and Moreno (2016) propose that the capacity to generate epithelia, which distinguishes the eumetazoans from the basal metazoans, is the hallmark of organismal identity in the animals, enabling the demarcation of inside and outside (and thus organism and environment), and providing the necessary cellular conditions for the formation of a nervous system.

It is plausible that the tissues of diploblasts and the more complex triploblasts that evolved from them are a novel form of matter within, but morphogenetically beyond, the basic liquid-tissue state. (See, for example, the discussion of electrical field-based consolidation and integration of pattern and form below.) In this sense, the cell-based materials that develop into these organisms constitute a subkind of matter within the liquid tissues that form the animals. This is analogous to asserting that the metals constitute a subkind of matter within the natural kind that collectively consists of the chemical elements. Of course, philosophers who reject the ascription of natural kindness to the category of chemical elements, or to any individual element or subset of them, are likely also to reject these biological attributions.

The evolutionary emergence of about 30 metazoan phylotypes in addition to the basal metazoans and diploblasts was based on the innovation of a third tissue or "germ layer" interposed between the two epithelial ones. Unlike the transition between the unicellular holozoans and the basal metazoans + cnidarians (and possibly ctenophores), the three-layered configuration, or triploblasty, appears to have been achieved in several different ways, involving different types of extracellular matrix molecules and factors promoting formation of the dispersed-cell tissue state known as *mesenchyme* (reviewed in Newman 2016b, 2019a). Appearing tens of millions of years later in the fossil record than cnidarians, triploblasts are thought to have arisen from diploblastic ancestors. One of the two tissue layers of ancestral forms outpocketed, separated, or disaggregated into a third (mesodermal) germ layer. The disaggregating mode, termed *epithelial-mesenchymal transformation* (EMT), seen in the development of arthropods, chordates, and several other phyla, appears to have evolved via the addition of molecules to the extracellular matrix of the middle zone separating the two layers, where they promote cell dispersion and invasion. Some mesoderm-promoting secreted proteins (e.g., fibronectin and tenascin in chordates) are phylum-specific innovations, while others, like the thrombospondin type 1 repeat (TSR) superfamily of proteins, are common to most triploblasts and appear to have been carried over from unicellular holozoan ancestors. Other extracellular matrix proteins (e.g., collagen, dentin) are capable of mineralising under certain conditions, adding the solid state to the physical repertoire of metazoan tissues (reviewed in Newman 2016b).

Since the resulting morphologies may be analogous to one another rather than arising from shared causes, the triploblasts as a group probably do not represent a single natural kind, if the criterion is taken to be a uniquely characterised material. Instead, at least some of the traditionally identified "bilaterians" (a term with a different sense – based on symmetry – but referring to the same group of organisms as the term "triploblasts") may individually be natural subkinds of the animals.

Multilayering provided a platform for prolific diversification of body plans. In extant animals and presumably ancestral forms, it occurs in several different ways, each at least partly dependent on phase-separation behaviours of the liquid tissue state and its defining genetic toolkit, with the addition (as noted) of some phylum-specific extracellular matrix molecules. Despite their being induced and mediated by a variety of different molecules, there are just a few main mechanisms by which *gastrulation*, the embryonic process that brings about multilayering in triploblastic organisms, is achieved. They are often overlapping and jointly employed, however. In topologically solid embryos (i.e., without an early-forming lumen) (i) an internal mass of cells can acquire greater cohesivity than the surrounding layer and pull away from it ("delamination"); (ii) a less cohesive cell mass can come to envelop a more cohesive one ("epiboly") or (iii) curl in on itself over the latter ("involution"); in embryos with interior spaces (which can form spontaneously by cells with A/B polarity), (iv) inward folding of the surface layer ("invagination"), or (v) migration of a subset of surface cells ("ingression") can give rise to a new layer (Forgacs and Newman, 2005; Newman 2016b).

Each of these morphogenetic processes is inherent to the liquid nature of metazoan tissues as embellished by the conditional apicobasal and planar polarisation of their cellular subunits, and optionally with the EMT-inducing effects of novel matrix proteins. This makes it plausible that the body-plan–defining process of gastrulation had its origins in the physically based rearrangements inherent to such materials rather than in multiple cycles of natural selection (Newman 2016a). In addition to the morphological motifs of gastrulation, triploblastic/bilaterian animals exhibit other recurrent forms: segmentation (in insect bodies, vertebrate backbones, individual tetrapod digits), branching tubes (in insect trachea and vertebrate respiratory airways, vascular trees, glands), regular two-dimensional arrays (in insect bristles, hair and feather follicles, pigment spots and stripes). The next section describes how these structural motifs, like those of gastrulation, are attributable, either in their present implementation or in their origination, to physical effects inherent in the materials of which the respective organisms and organs are composed.

7.4 The detachment of animal identity from genetic programming

Acoelomates, such as flatworms, are the morphologically simplest triploblastic animals. These organisms have an enteric cavity open at both ends but lack a second body cavity, the coelom, which surrounds a separate digestive tube in coelomate species. Acoelomate animals have a small number of organ types – ovaries and testes, and ganglia, localised clusters of neurons. In coelomates, in contrast, organ complexity is dramatically increased, since additional interfaces are available for epithelial-mesenchymal interactions during development. The morphological phenomena of organogenesis in these animals (producing, for example, circulatory systems, lungs, salivary glands) are essentially identical to those of gastrulation, utilising the same liquid, deformable sheet, and disaggregative behaviours of the component tissues. These, in turn, are mediated by the products of the same morphogenetic toolkit genes in conjunction with some phylum-associated extracellular matrix and other molecules (Newman 2016a).

It can be concluded that both development and evolution of animal form, at both the body plan and organ levels, draw (and have drawn) on inherent physico-genetic properties of tissues played out in different geometrical and topological settings. Most genes of metazoan organisms are concerned with the conduct of subcellular activities, which evolved long before animals existed and continued to be involved in primarily single-cell functions after the emergence of multicellularity. The small number of genes (the morphogenetic toolkit) that are specifically involved in creating animal identity (i.e., multicellular holozoans with the liquid-tissue properties discussed above) and in their partitioning into superphyla, phyla, and other sub-groups appeared coincidentally with animal life (Newman 2019a). Their roles are to mobilise physical forces and effects that shape and pattern tissues into often elaborate combinations of stereotypical motifs. While these effects eventually came to be programmed and orchestrated during development by interactions among groups of cells, no amount of evolution of these spatiotemporal programmes can effect a departure from the "morphospace" (the array of structural possibilities) afforded to animal tissues.

Some of the genes that evolved prior to multicellularity are involved with the requirement of all cells to store chemical and mechanical energy. Others are employed in positive and negative feedback regulation of gene expression. With the rise of multicellularity, energy storage and molecular feedback mechanisms in the context of direct and indirect cell-cell communication constituted tissues as "excitable media," spatially extended dynamical systems capable of propagating chemical and mechanical activities (Forgacs and Newman 2005). (Most early-evolved genes did not change their functions in the transition to multicellularity, but rather, with the change of scale, mediated the acquisition of tissue-level properties; see Newman 2019a.) Excitability enables liquid tissues to generate ("self-organise" into) regular patterns and forms, exploiting the potential that exists within the limits of metazoan morphospace.

Among the structures organised by such excitable effects during body-plan development are *segmentation* – full or partial subdivision of the body axis in animal groups as diverse as annelid worms, molluscs, arthropods, and vertebrates – and *patterned skeletogenesis* in the paired appendages (fins or limbs) of jawed vertebrates. Both these developmental processes employ genes and their products not as elements of incrementally modifiable ontogenetic programmes implicit in the standard evolutionary narrative, but as components of dynamical systems.

Additional evidence that the constitution of phylotypic identity is "detached" (Lenny Moss's term (Moss 2006)) from a sequential programmatic readout of the genome comes from the potential of animals to form vegetatively. While nearly all animals develop in an apparently programmatic fashion from a single-celled zygote resulting from fertilisation, marine and freshwater sponges, and colonial cnidarians such as corals, also develop from released multicellular propagules into organisms indistinguishable from ones produced from a fertilised egg. Other cnidarians, including free-living hydra, reproduce both sexually and vegetatively by extending buds from the body stalk that detach as fully formed individuals (reviewed in Newman 2014).

Vegetative development can also occur in morphologically more complex animals. Some tunicates (triploblasts with a larval body plan like that of vertebrates) preferentially reproduce asexually. Polyembryony, the production of multiple offspring from the blastomeres (pre-gastrula cells) of a single divided embryo, is an example of vegetative reproduction that occurs in some mammals. But while the animals in these examples can develop without passing through a zygote stage, they can also develop sexually. It is therefore possible, in principle, that a programmed genetic readout in a recent progenitor that developed directly from a zygote is remembered and redeployed in the vegetative propagules, buds, or polyembryonic blastomeres. Therefore, the existence of vegetative development is not itself sufficient to disconfirm the existence of genetic programmes for the establishment of the phylotype.

The most decisive evidence that phylotypic identity does not derive from a programmatic readout of the zygotic genome comes from interfamilial (within a taxonomic order) and interordinal (within a taxonomic class) embryo chimeras. Embryonic cells from sheep (which have 54 chromosomes) and goats (with 60 chromosomes), members of distinct families of the mammalian order *Artiodactyla* which diverged five to eight million years ago, can be mixed together and develop into a composite organism with a variable phenotype with features of each species. Even more strikingly, mixtures of embryonic stem cells from medaka and blastomeres from zebrafish (members of two different orders of ray-finned fish that diverged from one another around 320 million years ago) developed into chimeric fish, with the medaka cells accommodating themselves to the timing of zebrafish development and incorporating into several organs of the resulting composite organism. In both cases, the chimeric animals developed into coherent metazoan organisms from non-zygotic cells of species whose own development is strictly fertilisation-dependent (reviewed in Newman 2014).

If the causal basis for the establishment of metazoan phylotypic identity is not the readout of a zygotic genetic programme, what is its basis? As described earlier in this chapter specific genes are indeed involved in the establishment of animal identity, but not as part of a programme. Rather, the key genes are those of the morphogenetic toolkit, and their role is to constitute the material properties of metazoan tissues, including their liquid-like nature, their capacity to elicit apicobasal and planar polarity in contiguous populations of their cellular subunits, their ability to produce basal laminae or undergo EMT. None of these functions was present in unicellular holozoan ancestors (Newman 2019b). Correspondingly, they only begin specifying phylotypic morphological motifs at the multicellular, "morphogenetic" state (the blastula, blastoderm, or inner cell mass stage of development, depending on the species) when they initiate gastrulation (Newman 2011; see the following section).

The establishment of phylotypic identity is thus an inherently multicellular, mesoscale physics-based set of processes. This explains how it is possible, contrary to widely accepted tenets of both developmental and evolutionary biology (see Linde-Medina 2010, Minelli 2011), that embryo chimeras can develop into phylum representatives that are indifferent to any identification with known species. The seeming paradox here is that species, according to the standard picture, are the most honed products of evolution and the leading edge of its continued occurrence.

Notwithstanding the properties that first set the animals apart from other organisms (and indeed other kinds of matter), biological systems continue to evolve. Animals, like other organisms, are under selective pressure to conserve their useful features and propagate their types, and mechanisms that have evolved for these purposes have sometimes provided more reliable ways to develop true to type than the inherent physico-genetic ones described above. As discussed in the next section, this move from generic to specific causation has intensified and consolidated differences between types and promoted individuation of species and their members.

7.5 The consolidation and intensification of animal identity

When developmental mechanisms are looked at in detail, it frequently appears that evolution has generated a "belt and suspenders" solution to the reliable formation of important structures. A case in point is the role of the proteins nanos and hunchback in the segmentation of the *Drosophila* embryo. Both play key, well-characterised roles in determining the borders of expression of other spatially expressed genes leading up to the interdigitating seven-stripe patterns of still other genes that precede overt segment formation. Nanos is stored in the egg during oogenesis and represses the activity of hunchback. In its absence, segmentation is fatally perturbed. Elimination of nanos, however, can be entirely compensated for by also eliminating the (otherwise essential) maternally derived complement of hunchback. Embryos lacking both these maternal gene products are viable and can survive as fertile adults, although without the "overdetermining" circuitry the developmental system is probably more fragile (Irish et al. 1989).

A recently characterised and unexpected mode of conservation and rectification of animal morphology is the establishment of an elaborate set of electrical cues by ion channels, pumps, and communicating junctions at the plasma membranes of the cells of developing and regenerating tissues. This leads to a bioelectric scaffolding that influences proliferation, differentiation, cell shape, and apoptosis of stem, progenitor, and somatic cells. Voltage gradients appear to encode aspects of species-specific morphology that act in concert with the morphogenetic toolkit and other gene regulatory networks to facilitate embryogenesis, or the regeneration of missing structures after trauma (Levin 2014). While bioelectricity is not a morphogenetic force independent of the described physico-genetic mechanisms, its role is not merely passive. Experimentally reshaping the voltage pattern of a regenerating Planaria of one species to that of a different one, for example, will cause the alternative morphotype to form.

Voltage gradients can form in all animals that have intercellular gap junctions or ion channels. These are mediated by gene products called innexins and their homologs the pannexins (except for echinoderms, such as sea urchins and starfish) in all diploblastic and triploblastic animals, and additionally by connexins, a novel protein family of the chordates. Sponges and the placozoans lack these channel proteins, while echinoderms have gap junctions defined by different protein families (Bloemendal and Kuck 2013). Bioelectrical integration of morphology thus appears to be a novel material property of eumetazoans that set them apart from basal metazoans as a natural subkind within the animals.

The major arena for stabilisation and reinforcement of developmental outcomes, however, is the egg. This is notwithstanding the evidence (discussed above) that an egg stage of development is not decisive for generating phylum-specific features. Eggs themselves exhibit broad morphological variability even within a phylum (compare the minuscule mammalian egg with the huge avian one) and are the loci of widely disparate pre-cleavage intracellular signalling processes even in related species (see below). Despite this, embryos of a given phylum pass through a "phylotypic stage" in which they are morphologically very similar, before they again diverge to take on their class-, order-, and species-specific identities. This phenomenon of comparative developmental biology is referred to as the "embryonic hourglass" (Newman 2011). It finds a natural interpretation in the view presented here, in which phylotypic features depend solely on the inherent morphogenetic properties conferred by phylum-related physics-mobilising genes, and these are only expressed starting at the multicellular stage of embryogenesis.

In principle (as indicated by vegetative propagation), all that is required for parcels of metazoan tissue to generate bodies of a given phylotype are the respective phylum-specific morphogenetic toolkit genes in the context of the shared properties of metazoan cells. Moreover (as indicated by chimerism), aggregates of such cells with otherwise variant genotypes can generate bodies that reside within the phylotype but in other respects are evolutionarily unprecedented. This raises the possibility that organisms with morphological phenotypes of modern phyla (or propensities to generate them, since the phyla's origination as clades was likely not coincident with the emergence of their all body-plan characters (Budd and Jensen 2000)) may have first arisen in clusters of non-clonal populations of cells. This implies, in turn, that programmatic development from a single cell, which is the most common form of generation of animal bodies, is a derived feature (Newman 2011).

Since animals almost always develop from fertilised eggs, or have this option, this inference is counterintuitive and needs to be explained. Organisms that develop from aggregated cells will typically be genetically heterogeneous, and will therefore incur cellular competition and potential conflict (Michod and Roze 2001, Grosberg and Strathmann 2007). Initiating development from a zygote is a way to ensure that the cells of the embryo are clonal. The animal egg may thus have been an evolutionary innovation – appearing subsequently to the major body-plan motifs – that was selected for by its capacity to originate genetically uniform phylotypic lineages and thus ensure their phylogenetic stability (sexuality and fertilisation would have evolved afterwards in this scenario) (Newman 2011, 2014).

The egg has two properties that distinguish it from later developing cells of the embryo and suit it to its proposed role. The first is its size, always larger than the cells derived from it. The tendency of large cells to divide incompletely promotes cleavage, rather than cytokinesis (complete separation of daughter cells) after DNA replication. The egg has a second capability that promotes clonality even in the absence of cleavage. This is the presence of an extracellular container: the egg shell in birds and other groups, the zona pellucida in mammals. Blastomeres that fully separate from each other (as occurs during early development in some mammals) remain associated and are inhibited from mixing with the cells of other embryos owing to this containment (Newman 2014).

Development from an enlarged cell affords additional opportunities for ensuring reproducibility of developmental outcome while simultaneously promoting subphylum diversification. This set of phenomena has been reviewed at length elsewhere and will just be summarised here. Patterning processes within a large founder cell of an embryo (pre- or post-fertilisation) leads to nonuniform cytoplasm and thus cellular progeny that have regionally specialised properties by the time morphogenesis begins at the multicellular liquid tissue stage. Even though these intra-egg patterns do not prefigure the anatomy of the developed animal, having a set of landmarks laid out ahead of time causes the morphogenetic processes to occur in a standardised context and thus contributes to the fidelity of species-specific development and thus the intensification of species identity (Newman 2011).

Egg prepatterns, though not occurring in all species, are important, often critically, to subsequent development in the species in which they are present. They can be highly variable, however, differing within taxonomic orders, e.g., between mice and hamsters, both rodents. Evolutionarily more distant species like medaka and zebrafish can exhibit highly disparate intra-egg patterning processes despite having similar adult morphologies, and, as we have seen, the capacity to chimerise at later embryonic stages. In nematode worms, a phylum with little morphological diversity, formation of the embryo's anteroposterior axis can depend absolutely on egg patterning in some species (in which the patterns are highly divergent) and not at all in others, with the developmental processes then going on to produce essentially the same anatomy. Among species that require egg patterning, for example, the sperm entry point at fertilisation determines the head end in some nematodes and the tail end in others (reviewed in Newman 2011).

The overdetermining, bioelectrical and egg-dependent mechanisms that consolidate and integrate the morphological phenotypes apparently evolved to protect established developmental outcomes from being derailed ("canalisation," in Waddington's term (Waddington 1942)). Along with stabilising the respective phylotypes, though, they also made them tolerant to additional genetic change. This autonomy provided an arena for within-phylotype variability and individuality. By "individuality" I mean unique (in contrast to generic, e.g., phylum, genus, species) biological (including genetic), identity. While the standard evolutionary narrative holds that variability undermines the type-identity of individuals insofar as it contributes to incipient speciation (e.g., Okasha 2002), in the view outlined here, intensified biological individuality is compatible with, and an inevitable consequence of, genetic and epigenetic variation.

7.6 The elaboration of individuality

The previous section described inherited variation within animal types (e.g., phyla, classes) that clearly promotes the reliable development of typical subgroup members under different circumstances. A great deal of variation with no obvious function is also tolerated within types. Some of this may be pleiotropic, with negative effects balanced by positive ones. Some may be "neutral" in the sense of never having encountered an environment in which it made a difference. In the conventional evolutionary narrative all of this would be included under the rubric of "standing variation" in the respective populations, ready to be mobilised for the evolution of new types of organisms when conditions change.

If some categories of organisms are natural kinds, however, it may take more to dislodge them from their type-identity than the variations described. In analogy to an example from the physical sciences (mentioned above) in which the designation of natural kind is less controversial, a chemical element can tolerate some changes (i.e., incorporation of neutrons in its nucleus) without changing its type-identity, but not others (i.e., incorporation of additional protons). Some elements are indeed more stable with neutrons present than without them, making the "variant" more survivable. This does not negate the decisive role of protons (in contrast to neutrons) in defining the element's intrinsic nature, or essence (Ellis 2001).

All animal species may have "neutron equivalents" (the egg-patterning processes in fish, described above, for example, or the hunchback-nanos interaction during *Drosophila* embryogenesis) that may be necessary for their survival, but not in defining their identity, particularly their identity as animals or their membership in a given phylum. It would take much more (e.g., the loss of one or more metazoan- or phylum-defining morphogenetic toolkit genes) to alter the type-identity of an organism of animal origin.

Tunicates are a famous example of animals that lose the characteristic morphological signatures of their (chordate) phylum as adults. But tunicates still pass through a larval stage that exhibits characteristic chordate motifs such as an elongated body and notochord. The alchemy of true phylotypic transmutation has yet to be documented.

The evolutionary "rewiring" of gene regulatory networks shows that genetic change in a phylum's member organisms can be dramatic without converting their anatomies into non-canonical forms or even blurring the phylotype's boundaries. This phenomenon, termed "developmental system drift" (True and Haag 2001, Haag 2014), has been studied in several metazoan phyla, with the finding that structures that were established early in evolution, such as the neural tube (Harrington et al. 2009) and somatic segments of vertebrates (Stern and Piatkowska 2015), or the body plan and vulva of nematodes (True and Haag 2001), can be induced and regulated by different sets of genes in different species. At the core of these divergent developmental mechanisms, however, are the products of the metazoan morphogenetic toolkit – classical cadherins, catenins, Wnt, type IV collagen, and few others.

While the stated conclusion that "gene regulatory networks are constantly being reconfigured even when phenotypes are not" (Haag 2014) describes one aspect of developmental system drift, the other side of it consists of changes over evolution in the physical mechanisms involved in the morphogenesis of a conserved form (Newman 2019c). Together, these phenomena reflect the *autonomisation* of phylotypic anatomy over the course of evolution (Müller and Newman 1999). As described above, animals have inherent forms which are due to their origins as extracellular matrix-fortified liquid-tissue parcels. The stereotypical generation (based on their unique material properties), and subsequent integration and autonomisation (based on evolved consolidating and rectifying mechanisms) of the phylotypes make them resistant to reshaping.[2]

This still leaves plenty of room for genetic variation within the confines of a phylum. This can take the form of novel alleles, entirely novel gene segments or whole genes, or gene deletions, uniquely associated with subphylum clades, including species and individuals. A classic intraphylum morphological elaboration is the tetrapod limb, termed an anatomical "archetype" by Charles Darwin's contemporary and rival Richard Owen (Owen 1849). Although it appears in only a subset of species of jawed vertebrates, its basic plan is incredibly stable over evolution, as even Darwin himself had to acknowledge ("similar bones in the same relative positions"; Darwin 1872, p. 382) despite his own theory pointing away from morphological stasis.

The tolerance of organisms to genetic variation without losing their phylotypic identities is also manifested at the species level, where a given gene might not have the same role in different conspecific individuals. In humans, most alleles that have classically been considered disabling, or even fatal, have surprisingly been found to be compatible with health in individuals from different reference populations from those in which they were first identified. On average, a human individual has 54 mutations that, based on long-standing population studies, are predicted to "sicken or even kill their bearer" but do not (Check Hayden 2016). A recent study of a cohort of Pakistani descent in London, for example, identified a woman with a nonfunctional PRDM9 gene, the product of which is essential in defining the typical recombination hotspots during meiosis. Surprisingly, she was fertile (unlike mice that lack this gene) and had a healthy child, but the meiotic crossovers inferred from the two genomes were atypical, suggesting compensation by an unknown, alternative mechanism, employing gene products that normally have other roles (Narasimhan et al. 2016). Much recent work of this sort suggests that each individual organism, despite being a member of a hierarchy of groups, is developmentally unprecedented, utilising the genes it shares with its conspecifics in its own distinctive way.

7.7 Conclusions

The physico-genetic view of animal origins asserts that large-scale, macroevolutionary distinctions among organismal types did not originate through microevolutionary changes, but in the novel morphogenetic capacities of new kinds of biomaterials (Newman 2016a, 2019a). It is clear that the ability to draw energy from the environment, to persist, and to grow appeared during the evolution of unicellular life, and the thousands of genes necessary for these processes were in place millions of years before the emergence of the metazoans. But the liquid tissues that constituted the most primitive of the animals and their diploblastic and triploblastic descendents were not simply clusters of preexisiting cells. Moreover, natural selection of allelic variants of the genes of unicellular ancestors had little role in the macroevolutionary steps that produced the animals. Their origination depended on the appearance of genes with no counterparts outside of the metazoans: classical cadherins, Wnt, Stb/Vang, type IV collagen, peroxidasin, fibronectin, and a handful of others. The respective protein products, in the context of multicellularity, mobilised mesoscale physical forces and effects that were previously irrelevant to individual cells or their aggregates (Newman 2016b).

In ontological terms, the new materials that came to constitute the developing tissues of animals were *substances*, in the sense of being particular kinds of matter, but unlike nonliving liquids and liquid crystals, they were also *processes* (Dupré 2018). Composed of cells, they must metabolise and biosynthesise, consume energy, and emit waste, in order to maintain their decisive material natures. Their capacity to manifest physically predictable behaviours is thus intrinsically tied to their living activities: they are "biogeneric" substances (Newman 2016a).

While it continues to be claimed that natural selection is "the only explanation we have for the appearance of design without a designer" (Levin et al. 2017), the theory of emergence of animal form presented here does not

Origins and evolution of animal identity 143

appeal to selection or increased fitness of marginally different variants. Insofar as motifs such as body cavities, segments, or appendages have proved adaptive, there is no reason (since an independent explanation for their generation is now available) to reject the notion that fitness for an environment can be "after the fact" (West-Eberhard 2005). Novel combinations of intrinsically generated motifs may not be optimal for survival in the venue where they originated, but living organisms typically exhibit ingenuity and are not locked into preordained niches (Lewontin and Levins 1997).

The notion of biological natural kinds implied by the physico-genetic picture entails the metaphysical stance that Devitt terms "intrinsic essentialism" (Devitt 2008). Biological essentialism is often seen as a remnant of pre-Darwinian typological thinking (Mayr 1982). According to the modern synthesis, all organismal types are potentially mutable and the boundaries between all categories are permeable. Since microevolution, the driving force of natural selection in the Darwinian theory, is a species-level phenomenon, it is unsurprising that the debate in the philosophy of biology over essentialism and natural kinds has centred on the ontological status of species (Okasha 2002, Wilson et al. 2007, Devitt 2010, Lewens 2012, Dupré 2014). It is equally unsurprising that there is little consensus on what differences between the organisms of sister species, or between individuals within a species, could stamp them as distinct types.

While similarly "intrinsic-essentialist," the view I have put forward differs from that of Devitt (2008, 2010) in that it rejects an exclusively gene-centric notion of intrinsic identity of organismal categories, focusing rather on the material properties of animal tissues (which have, of course, a genetic dimension) (Newman 2016a, 2019a). (Analogous arguments can be made with respect to multicellular plants; see Benítez et al. 2018.) Reframing morphological evolution outside the assumptions of neo-Darwinism means that the level at which essential properties enter the picture is not that of species. A focus on the essential properties of the kingdom, superphylum and phylum ranks of the taxonomic hierarchy is consistent with a shift in perspective from the microevolution-first one of Darwinism to a macroevolution-first one.

John Dupré (2002) has presented five criteria that need to be satisfied for something to be considered a natural kind in the classical sense: (1) membership of the kind is determined by possession of an essential property or properties; (2) members of a natural kind are the appropriate subjects of scientific laws; (3) the properties or behaviour of the members of a natural kind can be explained by identifying the kind to which it belongs, and referring to the kind's governing laws; (4) the conformity of the kind's members to its governing laws of nature is due to an essential property or properties; (5) if a thing belongs to more than one natural kind it must be the case that the kinds to which it belongs are part of a hierarchy in which lower-level members are wholly included in higher-level members. Dupré marshals this set of stipulations to argue that biological species, or relatively more inclusive taxa, are not natural kinds in this sense, an argument I find compelling.

But the higher-level category of metazoan organisms as a whole, and, for example, the subcategory of diploblasts, described above, with the essential properties being, respectively, the liquid-tissue state and the liquid-crystalline-tissue state, satisfy criteria 1–5 above, as do some triploblastic categories derivative of the diploblasts.

The collection of inherent morphological motifs associated with the properties and dispositions of liquid tissues are not themselves animal bodies. But by providing animal-specific building blocks and structural elements beyond mere genes or gene regulatory networks they both underlie and promote the evolution of characteristic bauplans. Within the confines of phylotypic identity, which would not be breached other than by loss of a phylum's defining material properties (dependent on the described toolkit genes and the physics they mobilise in the multicellular context), subtypes (not all of them natural kinds) can be elaborated, consolidated, and intensified, down to the level of the individual organism. This "intensification of uniqueness" (Newman 1995), rather than the open-ended production of overt difference predicted by the Darwinian model, may thus be the characteristic mode of animal evolution once it has moved past its initial constitution as a hierarchy of novel biological materials.

Lenny Moss and I have suggested that a flexible appropriation of Immanuel Kant's "methodology of teleology," updated in light of developments in the biological and physical sciences in the nearly two and a half centuries since the *Critique of Judgement* (1790) was published, can inform philosophical approaches to the nature of multicellular organisms (Moss and Newman 2015). In this perspective the Kantian "organised being" is an evolutionary product. Thus, there is hypothesised to have been a period during which the mesoscale physics and self-organisational dynamics of liquid and liquid-crystalline tissues inevitably caused primitive ancestral metazoans to become multilayered, segmented, and otherwise patterned. Following this, as a result of stabilising and rectifying mechanisms such as those afforded by the bioelectrical integration and egg-dependent clonality and pattern standardisation described above, the prototypical animal bodies evolved into modern, stable organismal "types" with canalised taxon-specific developmental (but not strictly genetic) programmes. The plastic and variably implemented nature of such programmes (in contrast to strictly specified computational ones) generate biologically unique individuals. While we have moved very far from Kant's own terms of reference, the coinage of "natural purpose" for entities that are both the causes and effects of themselves seems an appropriate characterisation of animal identity.

Acknowledgement

I thank Anne Sophie Meincke and John Dupré for their careful and critical reading of this chapter and for alerting me to several philosophical subtleties and distinctions. Some of the concepts presented here were developed in collaboration with Ramray Bhat, Lenny Moss, and Gerd Müller.

Notes

1 Individual cells are not liquids (notwithstanding the viscous nature of cytosol), nor are arbitrary clusters of unicellular organisms. Cells themselves have fibrous submembrane cortices and their shapes and movements are subject to mechanical forces rather than fluid flow. Clusters and aggregates of cells will only flow as liquids if there is no constraint on their entering or leaving the surface of the tissue mass to minimise its free energy. For this to occur, their cellular subunits must be able to move randomly without losing contact with one another. These requirements make the liquid tissue state exceptionally rare.
2 Not all the accepted (i.e., based on both genetic and body-plan affinity) phyla show the same degree of body-plan stereotypy. Nematodes, with very similar anatomies, represent one extreme, whereas molluscs (snails, clams, octopuses) represent another.

References

Arnellos, A. and Moreno, A. (2016) 'Integrating Constitution and Interaction in the Transition from Unicellular to Multicellular Organisms', in: Niklas, K. J. and Newman, S. A. (Eds.), *Multicellularity: Origins and Evolution*, Cambridge, MA: The MIT Press, pp. 249–275.

Benítez, M., Hernández-Hernández, V., Newman, S. A. and Niklas, K. J. (2018) 'Dynamical Patterning Modules, Biogeneric Materials, and the Evolution of Multicellular Plants', *Frontiers in Plant Science* 9, pp. 871.

Bloemendal, S. and Kuck, U. (2013) 'Cell-to-cell Communication in Plants, Animals, and Fungi: A Comparative Review', *Naturwissenschaften* 100(1), pp. 3–19.

Budd, G. E. and Jensen, S. (2000) 'A Critical Reappraisal of the Fossil Record of the Bilaterian Phyla', *Biological Reviews, Cambridge Philosophical Society* 75(2), pp. 253–295.

Check Hayden, E. (2016) 'A Radical Revision of Human Genetics', *Nature* 538(7624), pp. 154–157.

Conway Morris, S. (2006) 'Darwin's Dilemma: The Realities of the Cambrian 'Explosion'', *Philosophical Transactions of the Royal Society London, B: Biological Sciences* 361(1470), pp. 1069–1083.

Darwin, C. (1872) *The Origin of Species: By Means of Natural Selection, or the Preservation of Favoured Races in the Struggle for Life*, 6th ed., with additions and corrections (twelfth thousand). London: John Murray.

Degnan, B. M., Adamska, M., Richards, G. S., Larroux, C., Leininger, S., Bergrum, B., Calcino, A., Taylor, K., Nakanishi, N. and Degnan, S. M. (2015) 'Porifera', in: Wanninger, A. (Ed.), *Evolutionary Developmental Biology of Invertebrates*, Vienna: Springer-Verlag, pp. 65–106.

Devitt, M. (2008) 'Resurrecting Biological Essentialism', *Philosophy of Science* 75(3), pp. 344–382.

Devitt, M. (2010) 'Species Have (Partly) Intrinsic Essences', *Philosophy of Science* 77(5), pp. 648–661.

Dupré, J. (2002) 'Is 'Natural Kind' a Natural Kind Term?', *The Monist* 85(1), pp. 29–49.

Dupré, J. (2014) 'Animalism and the Persistence of Human Organisms', *The Southern Journal of Philosophy* 52, pp. 6–23.

Dupré, J. (2018) 'Processes, Organisms, Kinds and the Inevitability of Pluralism', in: Bueno, O., Chen, R.-L. and Fagan, M. (Eds.), *Individuation, Process, and Scientific Practices*, Oxford and New York: Oxford University Press, pp. 21–38.

Ellis, B. D. (2001) *Scientific Essentialism (Cambridge Studies in Philosophy)*, Cambridge and New York: Cambridge University Press.

Fidler, A. L., Vanacore, R. M., Chetyrkin, S. V., Pedchenko, V. K., Bhave, G., Yin, V. P., Stothers, C. L., Rose, K. L., McDonald, W. H., Clark, T. A., Borza, D. B., Steele, R. E., Ivy, M. T., Hudson, J. K. and Hudson, B. G. (2014) 'A Unique Covalent Bond in Basement Membrane is a Primordial Innovation for Tissue Evolution', *Proceedings of the National Academy of Sciences of the U S A* 111(1), pp. 331–336.

Forgacs, G. and Newman, S. A. (2005) *Biological Physics of the Developing Embryo*, Cambridge: Cambridge University Press.

Foty, R. A., Forgacs, G., Pfleger, C. M. and Steinberg, M. S. (1994) 'Liquid Properties of Embryonic Tissues: Measurement of Interfacial Tensions', *Physical Review Letters* 72, pp. 2298–2301.

Grosberg, R. K. and Strathmann, R. (2007) 'The Evolution of Multicellularity: A Minor Major Transition?', *Annual Review of Ecology, Evolution, and Systematics* 38, pp. 621–654.

Haag, E. S. (2014) 'The Same but Different: Worms Reveal the Pervasiveness of Developmental System Drift', *PLoS Genetics* 10(2), pp. e1004150.

Harrington, M. J., Hong, E. and Brewster, R. (2009) 'Comparative Analysis of Neurulation: First Impressions Do Not Count', *Molecular Reproduction and Development* 76(10), pp. 954–965.

Irish, V., Lehmann, R. and Akam, M. (1989) 'The Drosophila Posterior-group Gene Nanos Functions by Repressing Hunchback Activity', *Nature* 338(6217), pp. 646–648.

Kant, I. (1790; trans. 1966) *Critique of Judgement*, transl. J. H. Bernard, New York: Hafner.

Karner, C., Wharton, K. A. and Carroll, T. J. (2006a) 'Apical-basal Polarity, Wnt Signaling and Vertebrate Organogenesis', *Seminars in Cell and Developmental Biology* 17(2), pp. 214–222.

Karner, C., Wharton, K. A., Jr. and Carroll, T. J. (2006b) 'Planar Cell Polarity and Vertebrate Organogenesis', *Seminars in Cell and Developmental Biology* 17(2), pp. 194–203.

King, N. and Rokas, A. (2017) 'Embracing Uncertainty in Reconstructing Early Animal Evolution', *Current Biology* 27(19), pp. R1081–R1088.

Levin, M. (2014) 'Molecular Bioelectricity: How Endogenous Voltage Potentials Control Cell Behavior and Instruct Pattern Regulation in vivo', *Molecular Biology of the Cell,* 25(24), pp. 3835–3850.

Levin, S. R., Scott, T. W., Cooper, H. S. and West, S. A. (2017) 'Darwin's Aliens', *International Journal of Astrobiology*, pp. 1–9, doi:10.1017/S1473550417000362.

Lewens, T. (2012) 'Species, Essence and Explanation', *Studies in History and Philosophy of Science Part C: Studies in History and Philosophy of Biological and Biomedical Science,* 43(4), pp. 751–757.

Lewontin, R. and Levins, R. (1997) 'Organism and Environment', *Capitalism Nature Socialism* 8(2), pp. 95–98.

Linde-Medina, M. (2010) 'Natural Selection and Self-organization: a Deep Dichotomy in the Study of Organic Form', *Ludus Vitalis* XVIII(34), pp. 25–56.

Loh, K. M., van Amerongen, R. and Nusse, R. (2016) 'Generating Cellular Diversity and Spatial Form: Wnt Signaling and the Evolution of Multicellular Animals', *Developmental Cell* 38(6), pp. 643–655.

Mayr, E. (1982) *The Growth of Biological Thought: Diversity, Evolution, and Inheritance*, Cambridge, MA: Belknap Press.

Michod, R. E. and Roze, D. (2001) 'Cooperation and Conflict in the Evolution of Multicellularity', *Heredity* 86(Pt 1), pp. 1–7.

Minelli, A. (2011) 'Animal Development: An Open-Ended Segment of Life', *Biological Theory*, 6, pp. 4–15.

Moss, L. (2006) 'Redundancy, Plasticity, and Detachment: the Implications of Comparative Genomics for Evolutionary Thinking', *Philosophy of Science* 73(5), pp. 930–946.

Moss, L. and Newman, S. A. (2015) 'The Grassblade Beyond Newton: The Pragmatizing of Kant for Evolutionary-developmental Biology', *Lebenswelt* 7, pp. 94–111.

Müller, G. B. and Newman, S. A. (1999) 'Generation, Integration, Autonomy: Three steps in the Evolution of Homology', in: Bock, G. K. and Cardew, G. (Eds.), *Homology (Novartis Foundation Symposium 222)*, Chichester: Wiley, pp. 65–73.

Narasimhan, V. M., Hunt, K. A., Mason, D., Baker, C. L., Karczewski, K. J., Barnes, M. R., Barnett, A. H., Bates, C., Bellary, S., Bockett, N. A., Giorda, K., Griffiths, C. J., Hemingway, H., Jia, Z., Kelly, M. A., Khawaja, H. A., Lek, M., McCarthy, S., McEachan, R., O'Donnell-Luria, A., Paigen, K., Parisinos, C. A., Sheridan, E., Southgate, L., Tee, L., Thomas, M., Xue, Y., Schnall-Levin, M., Petkov, P. M., Tyler-Smith, C., Maher, E. R., Trembath, R. C., MacArthur, D. G., Wright, J., Durbin, R. and van Heel, D. A. (2016) 'Health and Population Effects of Rare Gene Knockouts in Adult Humans with Related Parents', *Science* 352(6284), pp. 474–477.

Newman, S. A. (1995) 'Carnal Boundaries: The Commingling of Flesh in Theory and Practice', in: Birke, L. and Hubbard, R. (Eds.), *Reinventing Biology: Respect for Life and the Creation of Knowledge*, Bloomington, IN: Indiana University Press, pp. 191–227.

Newman, S. A. and Bhat, R. (2009) 'Dynamical Patterning Modules: a "Pattern Language" for Development and Evolution of Multicellular Form', *The International Journal of Developmental Biology* 53(5–6), pp. 693–705.

Newman, S. A. (2011) 'Animal Egg as Evolutionary Innovation: A Solution to the "Embryonic Hourglass" Puzzle', *Journal of Experimental Zoology B: Molecular and Developmental Evolution* 316(7), pp. 467–483.

Newman, S. A. (2012) 'Physico-genetic Determinants in the Evolution of Development', *Science* 338(6104), pp. 217–219.

Newman, S. A. (2014) 'Why Are There Eggs?', *Biochemical and Biophysical Research Communications* 450(3), pp. 1225–1230.

Newman, S. A. (2016a) "'Biogeneric' Developmental Processes: Drivers of Major Transitions in Animal Evolution', *Philosophical Transactions of the Royal Society London, B: Biological Sciences* 371(1701): 20150443, doi/10.1098/rstb.2015.0443.

Newman, S. A. (2016b) 'Origination, Variation, and Conservation of Animal Body Plan Development', *Reviews in Cell Biology and Molecular Medicine* 2(3), pp. 130–162.

Newman, S. A. (2018) 'Inherency', in: Nuno de la Rosa, L. and Müller, G. B. (Eds.), *Evolutionary Developmental Biology*, Cham, Switzerland: Springer.

Newman, S. A. (2019a) 'Inherency of Form and Function in Animal Development and Evolution', *Frontiers in Physiology* 10, p. 702.

Newman, S. A. (2019b) 'Inherent Forms and the Evolution of Evolution', *Journal of Experimental Zoology B: Molecular and Devevelopmental Evolution* 332(8), pp. 331–338.

Newman, S. A. (2019c) 'Inherency and Homomorphy in the Evolution of Development', *Current Opinion in Genetics and Development* 57, pp. 1–8.

Nicol, A., Rappel, W., Levine, H. and Loomis, W. F. (1999) 'Cell-Sorting in Aggregates of Dictyostelium Discoideum', *Journal of Cell Science* 112(Pt 22), pp. 3923–3929.

Okasha, S. (2002) 'Darwinian Metaphysics: Species And The Question Of Essentialism', *Synthese* 131(2), pp. 191–213.

Owen, R. (1849) *On the Nature of Limbs*, London: J. Van Voorst.

Rokas, A., Kruger, D. and Carroll, S. B. (2005) 'Animal Evolution and the Molecular Signature of Radiations Compressed in Time', *Science* 310(5756), pp. 1933–1938.

Shen, B., Dong, L., Xiao, S. and Kowalewski, M. (2008) 'The Avalon Explosion: Evolution of Ediacara Morphospace', *Science* 319(5859), pp. 81–84.

Stern, C. D. and Piatkowska, A. M. (2015) 'Multiple Roles of Timing in Somite Formation', *Seminars in Cell and Developmental Biology* 42, pp. 134–139.

True, J. R. and Haag, E. S. (2001) 'Developmental System Drift and Flexibility in Evolutionary Trajectories', *Evolution and Development* 3(2), pp. 109–119.

Waddington, C. H. (1942) 'Canalization of Development and the Inheritance of Acquired Characters', *Nature* 150, pp. 563–565.

West-Eberhard, M. J. (2005) 'Phenotypic Accommodation: Adaptive Innovation due to Developmental Plasticity', *Journal of Experimental Zoology B: Molecular and Developmental Evolution* 304, pp. 610–618.

Wilson, R. A., Barker, M. J. and Brigandt, I. (2007) 'When Traditional Essentialism Fails: Biological Natural Kinds', *Philosophical Topics* 35(1/2), pp. 189–215.

Wimsatt, W. C. (2015) 'Entrenchment as a Theoretical Tool in Evolutionary Developmental Biology', in Love, A. C. (Ed.), *Conceptual Change in Biology: Scientific and Philosophical Perspectives on Evolution and Development*, Dordrecht: Springer Netherlands, pp. 365–402.

8 Processes within processes
A dynamic account of living beings and its implications for understanding the human individual

John Dupré

8.1 What is a living being?

The central thesis of this chapter is that a living being is not a kind of thing but a kind of process.[1] An organism, for example, is essentially dynamic. It is activity that is fundamental to its being what it is rather than any static characteristic. Organisms certainly do exhibit an extremely intricate structure. But this structure is not something intrinsically stable, like the structure of a car or a house, but something constantly maintained by the uptake and expenditure of energy from its environment. It is a temporary and costly island in the flow of matter, an eddy in the flux of material process. In fact, with appropriate attention to the very different time scales of stabilisation, I think that this is the right way to see both living and non-living matter. But in this chapter I shall be concerned only with the living.

An organism is, of course, enormously more complex than an eddy. One of the ways this great complexity is built up is by the assembly of a hierarchy of processes at a hierarchy of both spatial and temporal scales. Starting again with an organism, its time scale is its life cycle. An organism is a process with a characteristic trajectory, and its stabilisation is not homeostatic but, to borrow a valuable concept from Waddington, homeorhetic: it is maintained on the trajectory rather than in a specific state. Within the organism, however, are many homeostatic processes, for example the regulation of cell division and apoptosis that maintains a stable population of cells comprising an organ, and a great number of metabolic processes inside the cell. These latter processes happen at smaller spatial and temporal scales than that of the organism. In the case of metabolic processes this difference may amount to many orders of magnitude.

The organism, moreover, can also be viewed as part of a much longer process, the evolving lineage (Dupré 2017). While this is homeostatically stabilised, notably by (stabilising) selection, it lacks any homeorhetically controlled trajectory. Over time a lineage may move into quite different states, the process we refer to as evolution, but evolution does not have a predetermined destination. While it will be important to bear in mind this broader picture, in which life is a complex hierarchy of very different processes, the main focus of this chapter will be on the organism. (As will become evident, I shall not assume that it is clear what exactly an organism is.)

8.2 Processes and things

What is the difference between a process and a thing? My simple working definition is that whereas the default state of a thing is stasis, for a process to persist it must undergo change. Related to this is a second crucial point: a thing, or substance, as traditionally understood, is ontologically independent. While it may, of course, interact with other things, its default state of stasis requires nothing outside it. A process may also sustain its activity entirely from its internal resources, but this is not generally the case, and is less feasible the longer a process continues. The erosion of a coastline, the red spot on Jupiter, and an elephant are very different processes all of which persist by drawing energy from their environments.

Given the default of stasis in the metaphysics of things, it is unsurprising that a great deal of attention has been accorded to understanding why and how things change. And while understanding change is of course important for a process ontology, equally important is understanding why certain "things", that is to say the processes that we generally take to be things, remain the same to the extent that they do. The point can be strikingly illustrated by reflecting on the nature of medical science. Health is not, as is often assumed, the default condition of an organism until something comes along to deflect it, perhaps an injury or a pathogen. It is a state that is maintained by countless interconnected activities occurring throughout, and even beyond, the organism.

The question of persistence leads to one further difference between processes and things. Persisting is something a process does, and the process exists as long as it continues doing this. Its continuity is the connectedness, typically causal, of this activity. This kind of persisting doesn't necessarily provide clear criteria for when a process is judged to have become a different process—a tributary becomes a river, for instance—or to have divided into two processes: when, if ever, is a root sucker a new plant, for instance. For a thing, however, persisting is just the state of nothing happening. But then the question arises, how much change to a thing is consistent with its persistence? As I discuss in the next section, does an insect persist, for instance, when it changes from larva to pupa to adult? The traditional answer is that what must persist is the essence, some property or set of properties that makes the thing what it is (as Locke said, "the being of anything whereby it is what it is" (1975, III. iii. 15, p. 417)). Perhaps there are such properties, but widespread scepticism about biological essences suggests that this may point to a heavy burden for theories of organisms as things.

If everything is ultimately process, some processes are very slow. A rock, even though its structure is ultimately maintained by activities at the molecular or atomic level, may be stable for many millennia. But an organism is stable for generally fairly short time periods, and its stability is hard won by its activities in wresting energy from its environment, and its internal activities deploying that energy in the maintenance of its form. It is a very complex eddy in the flux of process, but an eddy nonetheless. As Nicholas Rescher sums it up, "as process ontologists see it, enduring things are never more than patterns of stability in a sea of process" (Rescher 2006, p. 14).

It is important to recognise that the effectiveness of the stabilisation processes that make life possible not only explains why it has seemed natural to treat living systems as things but also shows that so treating them for many theoretical purposes may be harmless and even useful. This is because the time scales over which processes are stabilised in a living system range over many orders of magnitude. If we are trying to understand a very short-lived process, entities that are more or less stable over the relevant time scale may safely be treated as things. If, for example, we are interested in modelling rapid metabolic processes inside a cell, it may well be appropriate to treat a catalytic enzyme as a stable thing relative to that process. But of course no one is surprised to learn that if we are interested, rather, in the maintenance of a functional concentration of the enzyme, we will need to see the enzyme as merely one stage in a process of transcription, translation, folding, and degradation. More generally, it is the possibility of treating stabilised processes as stable things that underlies the success of mechanistic modelling at many levels of the biological hierarchy. It is important, however, that this success does not lead us to infer that life actually is a hierarchy of things.

A further crucial point about living processes is that stabilisation of a biological entity can be effected both by internal processes and external processes; and typically it requires processes of both kinds. A very simple example may illustrate the point. Consider the mammalian heart. Numerous internal processes of cell replacement, tissue repair, and so on are required to maintain the function of the organ over the sometimes considerable lifespan of the organism of which it is part. But as we are all constantly reminded, the activity of the organism, notably physical exercise, is also important for this stabilisation. At a more basic level, the heart pumps blood through itself, which carries many molecules essential for its maintenance. But if the heart is not embedded in an organism with functioning lungs, the blood will not acquire, on its journey around the system, one of the most essential molecules, oxygen. Outside extraordinary conditions in the laboratory or clinic, a heart can only survive in reciprocal relations with a body of which it is part. Even the organism is stabilised, in very different ways, by its position in a lineage (Dupré 2017) and by the sometimes very complex social environment in which it exists.

This last point provides the final component of the general picture I want to endorse. Life consists of a hierarchy of processes, stabilised both by internal and by external processes, and stabilised at a hierarchy of different time scales. Energy flows through the whole, and is deployed to sustain the complex low entropy structure. To take one final example, the cell in a multicellular organism is a relatively stable structure. Vast numbers of metabolic processes within the cell contribute to the maintenance of this stability, but also necessary for this stability is the embedding of the cells within its wider context, typically a tissue or organ. Molecules, organelles, cells, tissues, organs, organisms, and perhaps higher levels still such as social groups, demes, ecosystems, etc. are all entities stabilised in this way. The obvious next question I need to address is why we should consider these stabilised entities to be processes in the same broad ontological category as the activities more generally acknowledged to be processual that stabilise them—metabolism, genome repair, stabilising selection, etc.—rather than something quite different, things, or substances.

8.3 Organisms are processes

I shall address the last question with specific reference to the organism. I am confident that a similar argument can be made at other levels of the hierarchy, but I cannot attempt to make them all here. Most metaphysicians, while perfectly happy to admit the importance of processes, maintain a dualistic ontology in which there are also things which undergo these processes. So an organism is a thing that acts, breathes, digests, and so on. Even philosophers who acknowledge the more fundamental status of processes in our ontology may recognise things, or substances, as parts of the world ultimately explicable by processes (e.g., Simons 2018). Why do I insist that the organism itself is a process?

Consideration of the organism must begin with the realisation that it is not by any means uncontroversial what an organism is. Wikipedia tells us that an organism is "any individual entity that exhibits the properties of life", which raises the equally difficult question what is an individual. (Is a cell an organism?) There is a widely assumed answer to the question what an organism is, namely that there are actually two kinds of organisms: first there are micro-organisms, single cells, and second there are what I have elsewhere called monogenomic differentiated cell lineages (Dupré 2011). The latter are what we normally think of as multicellular organisms, the cellular descendants of an original founding cell, a zygote.

The trouble with this is that actual multicellular organisms are not typically monogenomic collections of cells. They almost always include vast numbers of symbiotic microbes, many of which are essential for the proper functioning of the whole system[2] (see, e.g., Bordenstein and Theis 2015). One can, by fiat, insist that the organism consists of only the cells originating from the founding zygote, and this may be appropriate for some theoretical purposes. But this is not an autonomously stable whole, for the simple reason that its functional stability depends also, as just noted, on many of the symbionts. This symbiotic whole, which has become known, following Margulis (1991), as the holobiont, is in this respect more autonomous, but suffers from another defect in regard to the standard expectations of a thing, that its boundaries are fluid and changing. Some symbiotic microbes are, as just noted, essential for the thriving of the whole, others are neutral, and others actually or potentially harmless. The role of microbes can change from beneficial to harmful or vice versa. And microbes can be lost from or recruited to the system. In short, which elements of the system are parts of it and which are passengers or parasites is constantly changing and to some extent indeterminate. As I shall shortly explain, all this makes perfect sense from the perspective of mutually interacting and stabilising processes, but it is hard to reconcile with an ontology of autonomous objects with sharp boundaries.

A second reason to see organisms as processes is, very simply, the second law of thermodynamics. Organisms are systems far from thermal equilibrium, and maintain that condition by constantly taking energy and matter from their environments (Nicholson 2018). Maturana and Varela (Varela 1979, Maturana and Varela 1991; see also Meincke 2019a) have observed that the identity of an organism is something it actively creates and maintains, and this elementary point about thermodynamics reminds us that this is necessarily the case. However much, in view of the preceding point, we decide is part of the organism, the latter is only maintained as a whole by the intake of energy that drives its metabolism. In a healthy human, for example, many trillions of events must happen every second to maintain the functional arrangement of the organism's parts, and these must, in the end, be driven by a throughput of energy.

One last reason for seeing organisms as processes is that rather than maintaining a fixed set of properties throughout their existence, organisms have life cycles. Some organisms go through an astonishing array of different forms, but the familiar transitions of insects through egg, larva, pupa and adult, or imago are quite sufficient to make the point. While an organism exhibits short-term homeostasis, in the longer term its stability is rather homeorhetic. Of course things change their properties to some extent through time, and a central philosophical problem since antiquity has been that of determining what changes to a thing were compatible with its continued existence. Without considering in any detail specific attempts to answer this question, all such answers appeal to some continuity of properties, whether one or a number of essential properties or some general preponderance of properties. However, there seem to be no good candidate properties to explain our conviction that the egg and the butterfly are somehow the same organism.[3]

Processes, however, necessarily involving change, are far more fluid entities. Processes not only change their properties, but can split or merge, as the canonical example of a process, a river, nicely illustrates. Crucially, no one imagines that the question when one process begins and another ends should have a fully determinate answer. How many distinct streams are there in a river delta, for instance? Exactly similar questions arise for many plants that spread by vegetative processes such as root suckers and rooting branches. How many trees are there in a grove of quaking aspens (Bouchard 2008)? When I try to dig up a patch of nettles, am I constantly creating new plants when my spade severs the root connections between one part of the clump and another? Or was there never more than one plant? No one, I think should expect there to be objectively discoverable answers to such questions. This situation is wholly inconsistent with traditional notions of a thing, but wholly unproblematic when the organism is recognised as being a process.

I mean to take very seriously the suggestion that the boundaries of an organism are far from uniquely determined. There are many ways to divide the flux of living matter into distinct organisms, and to some degree how we do so is dependent on the purposes for which such division is intended. A simple illustration of this point comes directly from the observation above about symbiosis. The boundary between a typical multicellular organism and the outside world is generally populated by a vast array of microbial life, much of which is useful or vital to the larger organism. Recalling that the outer boundary of a human is generally taken to include not only the external skin but also the surface of the digestive tract, all of this surface is populated more or less densely with microbial life. Are these microbes part of the human organism? An argument that they are is that many of them, especially but not only those within the gut, are essential for the healthy development and functioning of the organism.[4] If we are interested in describing the organism as a functional whole, then at least those microbes essential for its well-being would naturally be included. However, for certain evolutionary inquiries, since the microbial symbionts mainly have a quite distinct evolutionary fate, it may be appropriate to consider the human organism as no more than the monogenomic differentiated cell lineage.

More complex symbiotic systems provide even more challenging such problems. Consider the leafcutter ant,[5] for example ants of the species *Atta cephalotes*,[6] a dominant species in many parts of South and Central America. To begin with, these ants can be divided into multiple castes: reproductives (queens and drones); soldiers; and workers of various different sizes with different functions, such as foraging, masticating, nursing, and various kinds of housework. Workers may move through different roles at different life-cycle stages, for example older workers being consigned to the dangerous tasks associated with waste management (Bot et al. 2001). It has often been debated whether these millions of individual ants are really individuals or parts of a much larger system that encompasses them all, the *superorganism* (Wilson and Sober 1989). But there is much more to the colony than this multitude of ants. Of special note is the specialised fungus, in a dedicated chamber, which digests the leaves provided by the ants, and excretes the food that the colony consumes. Actually, this is not quite correct. Bacteria are also essential contributors to the process of breaking down cellulose in the leaves (Aylward et al. 2012). And bacteria play a variety of other essential roles in the colony. For example, they fix nitrogen to facilitate growth in the fungus gardens (Pinto-Tomas et al. 2009), and bacteria on the carapace of the ants protect the fungus gardens from a pathological, invasive fungus that can be catastrophic for the colony (Currie et al. 1999). It is a safe bet that even viruses play a role in this system, though I was unable to locate any research specifically on this point.

A system of this complexity allows the distinction of multiple subsystems as individuals. In addition to the question whether an individual ant or the entire set of ants is an organism (perhaps the answer is both), we might ask whether the fungus colony, with or without the bacteria involved in its metabolic work, was an independent individual or part of the (super)organism. Similarly with the microbial communities in the fungus gardens, in the dump chambers where waste products are disposed of, or in the digestive tracts of the individual ants. A perspective that makes these questions at least innocuous is that of multiple interdependent and deeply intertwined processes—ants, fungi, bacteria—together providing a highly successful and stable whole, itself part of a very productive lineage. Distinguishing subprocesses within this system may be essential for many investigative purposes, but no unique way of doing so is mandated by nature. This is a view I have elsewhere referred to as "promiscuous individualism" (Dupré 2012, p. 241).

I should stress that I am not suggesting that such complex systems are uniformly friendly and cooperative. As I have suggested elsewhere, symbiotic interactions may lie anywhere on a spectrum from perfect mutualism to parasitism and, together with my co-author Maureen O'Malley, we suggested that it would be more helpful to use a generic term "collaboration" than try to distinguish symbiotic relations into distinct categories of mutualism, commensalism, and parasitism (Dupré and O'Malley 2009). In fact, the existence of selfish genes (truly selfish ones, such as those capable of meiotic drive (Burt and Trivers 2006), not all genes, as unfortunately suggested by Dawkins [1976]) shows that even the monogenomic differentiated cell lineage is not necessarily a wholly cooperative individual.[7] This point will be relevant to the analogy I wish to draw later in this chapter between biological individuals and supra-individual social entities.

8.4 Humans are organisms are processes

No one, I suppose, denies that humans are a kind of organism. For purposes of the present chapter, I will not engage with the debate over whether human persons are the same entities as human organisms, and I shall assume a version of the contemporary animalism that argues that they are (see also Dupré 2014).[8] Given this, and the argument of the preceding section, human persons are human organisms, and therefore they are a kind of process. In some respects, the preceding arguments can be applied quite routinely to the human case, but it also turns out that there are respects in which this application brings to light some rather different points.

First, as with any other organism, humans are systems far from thermodynamic equilibrium. The inevitable fate of a human, if nothing happens to prevent it, is death; life is maintained (for as long as it is) by a highly organised set of activities (processes) that maintain the disequilibrium state. Second, could a human zygote, a fetus, a child, an adult, etc., all be the same thing at different times? All the properties that have occasionally been suggested as special to humans—reason, language, bipedalism, and so on—are developmental outcomes. They are certainly not properties of zygotes or even infants. If these were essential properties of humans, it would seem that a human is a sequence of entities with distinct essential properties and thus, from the perspective of a thing ontology, that the stages of a human life form a sequence of distinct things. If one wants to maintain the intuition that there is some common thread to a human life, then this can only be as a process that persists through time.

As already discussed, the assumption that humans have the sharply defined boundaries generally taken to be characteristic of a thing is challenged by the existence of trillions of symbiotic bacteria, ranging from the pathological to many that are essential for the proper functioning of the human system. But the parallel with ant colonies raises the possibility of another application of promiscuous individualism: are humans parts of social individuals as much as they are individual organisms? I believe that they are.

One reason for thinking that ant colonies form a superorganism is the extreme interdependence of the parts of the colony, the individual insects. No ant will survive for very long apart from the colony. But the same is true of humans, another notoriously social organism. While Adam Smith famously pointed out the ways that the division of labour facilitated vastly enhanced efficiency of production, division of labour has always been characteristic of humans to a significant extent, between sexes and castes, for instance. Whatever the significance of this history, today humans are wholly dependent on countless other humans as the production of necessities—food, clothing, shelter, and so on—is distributed across an enormously complex and interconnected system of production. From this perspective, and contrary to the great emphasis that recent political and economic thought has placed on the isolated individual, humans are every bit as much parts as wholes.[9]

While much more could be said about this last point, for present purposes its significance is just to stress the potentially radical implications of recognising the processual nature of the human. As complex arrangements sustained by the subtle interplay between internal dynamics and external interactions, humans are parts of almost unimaginably complex systems; and the discrimination of these systems into component parts is something that can be done in many different ways—individuals, families, villages, firms, etc.—for many different theoretical purposes.

I noted at the outset of this chapter that seeing an entity as a (continuant) process[10] rather than a thing draws attention to the question of what it is that maintains whatever stability that a process possesses. Organisms in general, and humans in particular, have a good deal of stability, of course, and, as I have stressed, many processes both within and without the organism contribute to maintaining this. Of particular interest are the ways that an organism interacts with the environment to maintain its stability, and one class of such ways that has been of particular concern to recent theoretical biology is so-called niche construction.

Niche construction is the reshaping of the environment that many, perhaps to some extent all, organisms effect in ways that conduce to their survival. The phenomenon was classically described, though not under that name, in Darwin's work on earthworms (Darwin 1892). Earthworms, he observed, by transporting organic material such as leaf litter into the soil, increased the water retention of the soil, thereby satisfying a necessary condition for their survival. By happy chance (perhaps) this also improved the growing conditions for many plants. More obvious familiar examples are the constructions of dams by beavers or nests by birds though, as just suggested, it can be argued that all organisms produce changes in their environments, and in the long run selection will ensure that successful organisms are those that produce changes that are conducive to their survival.

The present point, at any rate, is that humans modify their environments on a scale that is, to the best of our knowledge, quite unprecedented in the history of life on Earth.[11] Humans live in vast colonies, constructed out of materials gathered from across the globe and processed in enormous communal work sites where many humans cooperate in the processing of thousands or millions of tonnes of material. Elsewhere the materials so produced are transformed into devices that provide transport, clothing, and much else. A large part of the planet, finally, has been transformed into near monocultures generating food for humans or food for extensive populations of animals in turn consumed by humans.

The reason for drawing attention to these familiar facts about division of labour and niche construction is to stress once again how bizarre is the common assumption that humans should be treated as largely autonomous individuals. While a few individual humans have acquired the skills to survive in environments not much modified by human activities, or at least not deliberately so modified, even if these skills were widely distributed, only a tiny fraction of the current population could survive in these ways. The persistence of the billions of humans that now exist is entirely dependent on the wholesale reshaping of the planet by countless cooperating humans.

I have endorsed the position of promiscuous individualism, which allows that many different individuals may, for pragmatic reasons, be distinguished within the mass of intertwined process in which they are embedded. So it would be perverse if, on the basis of the observation of just how complexly and necessarily intertwined human organisms are, I were to claim that the distinction of the human individual in the traditional way was somehow illegitimate. The interesting question, therefore, is what are relevant purposes for individuating humans in this way, and what are less well-motivated contexts. This is, I take it, a set of questions in political and economic theory that could be the subject of many volumes. Here I will only give a small flavour of the issue, but enough, I hope, to suggest that reflection on biological reality may have significant value in curbing the worst excesses of contemporary ideological individualism.

First of all, humans are individuated for organisational purposes, for the distribution of responsibilities and rights essential for the functioning of the complicated structures that support the very large populations of humans. Whereas the much simpler division of labour in an ant colony requires no more than differentiation of castes—it is plausible that one forager, say, is pretty much fully interchangeable with another—human workers have much more specific and differentiated occupations. A human is not merely a retail worker, but one who works at a particular store on the early morning shift. Having retail workers show up at any convenient store at any arbitrary time would probably not be a successful way of managing the goods distribution function. Other aspects of human societies require, for instance, that the same individual who acquires a debt or orders a product is the one who must pay the debt or expect to receive the product. This is managed by uniquely individuating humans and through practices of re-identifying them.

There is, of course, a more private, as we say "personal" or "personalised", dimension to all this. First of all, individual humans do appear to have a special concern for what happens to themselves. Interestingly, nonetheless, we do not universally applaud this concern. Selfishness is considered a moral flaw, though a complicated one as we generally acknowledge that to some degree it is not only expected, but perhaps even deserves our approbation. More importantly, perhaps, selfishness is a highly variable feature of humans. In contemporary capitalist societies where selfishness is at least expected and sometimes even applauded, some individuals seem remarkably deficient in this character trait. Moreover, it is plausible, though not always easy to assess, that in many other historical, and perhaps even contemporary, societies selfishness was either unusual or perhaps even unimaginable.

Humans have strong bonds of affection to particular other humans, especially family. Perhaps, as many evolutionary theorists argue, the bonds to family are a deep part of our biology. There is no doubt that such bonds are part of our natural history, though there is a good deal of disagreement concerning how this aspect of our biology is sustained (see, e.g., Emlen 1995, Davis and Daly 1997). Still more controversial in terms of evolutionary origin, but no less indisputable, is the capacity, albeit highly variable, of humans to form a network of bonds beyond the immediate family (for a review, see Clutton-Brock 2009). At any rate, the network of personal relations that most people value is based on individuality: relations are with specific, not easily replaceable, individuals. Most of us would be unhappy if unpredictable strangers regularly showed up for family dinner rather than the familiar loved ones—though it is not an impossible ideal—indeed one identified somewhat ironically with some interpretations of Christianity that we might come to welcome such strangers as warmly as our relatives.

I will conclude with some brief reflections on the implications of a process ontology for two very different questions about human life, the search for immortality and the metaphysics of pregnancy. These reflections are not intended to present decisive arguments. It will be sufficient if they illustrate that adopting a processual understanding of human life provides strikingly novel perspectives on central philosophical questions about the human condition.

8.5 Immortality

If this topic needs an apology in an essay on human biology, I might remind the reader that the second longest chapter of Richard Dawkins famous book, *The Selfish Gene* (1976), is entitled "Immortal Coils", and argues for the immortality of genes. (The longest discusses the battle of the sexes.) Tempting though it is to have some fun with the appeal to immortality by such a notorious atheist, fairness to Dawkins requires me to note that the immortality in question is the immortality (or considerable longevity as we should surely more accurately describe it) of gene processes, processes in which the material genes are constantly replaced, rather than gene individuals.[12] And this provides a valuable reminder that the persistence of processes differs in many significant respects from the persistence of things.

The traditional conception of human immortality supposes the human essence to reside in the soul, a substance wholly different from the human body that we know to be so sadly perishable. But given the extreme epistemological obstacles in learning about this hypothetical thing it will not concern us further here. I should also acknowledge that if the essence of a human resides in a (genuinely) immortal substance, then my process interpretation is entirely misguided. I confess that I am not much concerned by this concession.

At the opposite extreme, and in parallel with Dawkins's account of the life of a gene, there is a relatively uncontroversial sense in which one might achieve immortality through one's offspring. This is not so different from the immortality that an Aristotle, Shakespeare, or Einstein may attain through their works: the world in the distant future may contain an identifiable causal trace of one's existence. Although this is something some people may care deeply about, it is hardly to be confused with the idea that the very process which constitutes one's life might exist for a long period after one's death.

More exciting for those who find themselves dissatisfied with the brevity of the human lifespan are the pronouncements of some who believe that medical science will soon be able to overcome the diseases of old age, and to extend our lifespan by an order of magnitude (De Grey and Rae 2007). If one thinks of the human organism as an object or machine subject to deterioration, this is at least a reasonable aspiration. Protection from sources of damage and repair or replacement of seriously damaged components should, in principle, enable a machine to be maintained indefinitely. However, recognising the human organism as a kind of process makes this story considerably more problematic.

First, we may wonder whether the motivation for seeking many centuries of life is confused. Processes do not persist by sustaining an unchanging and potentially immortal essence, but rather by the continuity and interconnectedness of the activities that maintain them. So does it make sense for me to be concerned about the existence in a thousand years of some person whose connection to me is merely historical, with whom I have little in common beyond mere shared participation in a continuous process? One may recall the impossibility for a child really to imagine that they will one day be an old person, and the consequent dubiousness of supposing the child to have any real concern with the coming to be of any such geriatric.

This leads immediately to more serious problems. As I have insisted, the human is fundamentally a life cycle. Stabilising one temporal stage of the life cycle (one hopes a relatively youthful one!) for a period many times the current lifespan of the entire cycle would be to create a fundamentally different kind of being. This will not, of course, concern the transhumanists who promote this vision. Some of them embrace the rather repulsive and parodically Cartesian notion that we might all continue our eternal lives as programmes in computers, certainly a more radical ontological shift (Sandberg and Bostrom 2008).[13] A more substantive point is that this ontological disjuncture between what we are and what it is supposed we might become suggests real doubts as to the possibility even in principle of the project.

As I have stressed from the outset, for a persistent process such as a human, a question that can always be raised is what maintains its more or less stable structure. As I also mentioned, the stability of an organism is, in Waddington's terms, homeorhetic rather than homeostatic. This is the kind of process illustrated by Waddington's famous diagrams showing a ball rolling down a valley. The slopes on either side of the valley represent the processes that keep the ball in the valley. Dysfunctions, for example cancer, could be represented by one of the branching valleys. And healthy or unhealthy modes of behaviour or dietary intakes could be thought of as changing the contours at the point where the valleys meet, in ways that affect the likelihood of the ball changing path.

In this model, then, it is easy to understand how appropriate interventions could maximise the chances of staying on the healthy developmental path for the full lifespan. But a ball cannot roll downhill for ever; and likewise it is not obvious that the developmental trajectory of a human can be extended forever. I do not want to argue here that this is demonstrably impossible. However, the misguided analogy with maintenance of a machine makes the plausibility of indefinite life extension much less problematic than would a more accurate picture of a process with a characteristic trajectory and a typical temporal duration.

8.6 Pregnancy

The ontological status of a pregnant animal is not one that has received much attention in the philosophical literature, and this is no accident. In a world of substances it is quite mysterious that an entirely new substance should originate within an existing one. Where did it come from and at what moment does it come into existence? If a substance can undergo changes at all (a proposition that has been effectively criticised by Meincke 2018, 2019b, 2019c), a change in number of individuals from one to two or more seems an especially problematic change. It was perhaps more reasonable than some feminists have been willing to admit to have assumed, with Aristotle, that the answer was to be found in the sperm, since this was at least arguably a new substantial individual entering the female's system, and hence a metaphysically credible origin for a new individual.

Even aside from the question of origin there is a problem that arises from the possibility of twinning. Suppose that somehow or other an individual human comes into being at conception. A few days later the embryo divides and now there are two humans. Since twinning is generally assumed to be a stochastic event that was not determined at the time of the original fertilisation, it appears that at that time it was indeterminate whether one human or two had come into being. This has led some to the bizarre conclusion that the human life does not begin until two weeks after conception, since this is the latest stage at which twinning is possible.[14] It seems more than a little strange that one's existence might have been deferred by a mere unactualised possibility. This is generally discussed in terms of theological assumptions about ensoulment (e.g., Shoemaker 2005), but any account of how individual animal substances come into being will face similar difficulties.

To avoid these problems we need to go beyond what Elselijn Kingma has aptly described as the "bun in the oven" or, more prosaically, the fetal containment view of pregnancy (Kingma 2019). Kingma herself considers the fetus to be a part of the maternal organism (Kingma 2019 and this volume), which is probably the best solution available to a substance ontology. But such a view seems to overlook the process of individuation, in which the fetus gradually becomes more independent from the mother, a process that reaches a kind of completion at birth. (I say "kind of" as a neonate is hardly fully independent.) This gradual separation perhaps underlies a phenomenon discussed by Maja Sidzinska (2017), drawing especially on Tyler (2000), a tendency of pregnant women themselves to reject the dichotomy of one being or two.

A process ontology immediately defuses these problems. A quite different way of thinking of pregnancy is as a gradual bifurcation in a flow of living process.[15] For a bifurcating process there is no expectation that there should be a sharply defined point at which one becomes two. A helpful parallel example is provided by sympatric speciation. As this is sometimes imagined, the behaviour, including mate selection, within a population[16] gradually begins to settle into two distinct clusters, until eventually there emerge two reproductively isolated populations. There is no moment at which the one population becomes two, but at some point it is clear enough that this has happened. In a similar way, within the mass of changes occurring to a pregnant organism a subset of metabolic processes is gradually becoming localised, stabilised, and increasingly integrated. Such a perspective also provides a basis for the experience underlying the reported unwillingness of pregnant women to answer unequivocally the question of one or two.

8.7 Conclusion

In this chapter I have aimed to summarise the reasons for thinking that biological entities, and especially organisms, are much better and more accurately thought of as processes than as things and substances. If this is correct, and given that humans are organisms, it follows of course that we are processes rather than things. I have not attempted to address here the question whether human persons are identical to human animals, but appeal only to the growing literature on animalism in support of this assumption.

If human persons are indeed a kind of biological process, this has profound implications across the whole range of philosophical questions standardly raised about ourselves. It recasts ancient problems such as those of free will (Dupré 2013) and personal identity (Dupré 2014, Meincke 2018, 2019c, under review and in this volume). In this chapter I have discussed the implications for the questions concerning the possibility of indefinite life extension and for the correct metaphysical understanding of mammalian pregnancy. Perhaps most importantly, a process ontology casts a critical light on the assumption of hard-boundaried, self-interested individualism that remains so central to the main stream of social and political thought.[17] I hope that demonstrating such a wide range of potential implications will at least encourage those who would defend the more traditional substance ontology to reflect on the arguments that the latter is difficult or impossible to reconcile with our best scientific insight into the nature of life.

Acknowledgements

This chapter is based on work carried out with funding from the European Research Council under the European Union's Seventh Framework Programme (FP7/2007–2013)/ERC grant agreement no. 324186, which I gratefully acknowledge. The chapter has also benefitted greatly from detailed comments on earlier versions by Anne Sophie Meincke.

Notes

1 For further elaboration and defence of this view, see Dupré and Nicholson (2018).
2 It is important to note that single-celled organisms also generally exist in highly integrated multi-species communities, such as biofilms (Bapteste and Dupré 2013). So both traditional kinds of organisms are actually problematic.
3 One might imagine that some description of the genome could serve this purpose. Apart from the problem that many organisms may have the same genome (a clone), the genome of an individual organism is considerably more complex and mutable than is sometimes supposed (Dupré 2012, ch. 7).
4 For an overview, see Cho and Blaser (2012). For a review of the function of gut bacteria, see, e.g., Jandhyala et al. (2015). For a discussion of beneficial effects of skin bacteria, see Cogen et al. (2008). A recent report suggests that a specific skin bacterium may provide some protection against skin cancers (Nakatsuji et al. 2018).
5 "Leafcutter ant" is a term that refers to several dozen species from the two genera Atta and Acromyrmex.
6 See www.arkive.org/leaf-cutter-ant/atta-cephalotes/.
7 An interesting example relevant to the present case is the presence of the endo-symbiotic bacteria Wolbachia in apparently all species of leafcutter ants (Van Borm et al. 2001). These bacteria are mainly inherited vertically from mother to daughter, and so are very closely integrated evolutionarily with their host species. However, they are well known to be capable of manipulating sex ratios, and in this respect are likely to have interests that conflict with those of their hosts.
8 For more detailed discussions of animalism, see the chapters by Meincke and Snowdon in this volume.
9 It is deeply ironic, incidentally, that Adam Smith's name has become so widely associated with competitive individualism. While it is of course true that he accounted for the distribution of the various products of individual or groups of individuals in terms of individual self-interest, surely of greater significance was his account of the ways in which cooperation could so massively enhance the production of the goods that people desired. A further irony is his deep awareness, quite lost on many of his contemporary followers, of the high cost of specialised roles in the production process (Gagnier and Dupré 1995, Gagnier 2000).
10 The assumption that a process can be a continuant at all will raise the hackles of many metaphysicians. This is a major thesis of much of the work of myself and my co-editor, but this is not the place to argue for it. See Meincke (in this volume, note 25) and (2018, note 45); Dupré and Nicholson (2018).
11 Perhaps an exception is the production of an oxygen-rich atmosphere by countless billions of cyanobacteria. This happened, at any rate, a long time ago.
12 One might be tempted rather to contrast the immortality of gene tokens and gene types. However, a type, I take it, is an abstract entity, and if it has any temporality it would have to be immortality. Better would be the unceasing instantiation of the type; but at this point it is surely better to focus on the process by which the type continues to be instantiated.
13 Following Bostrom's (2003) notorious simulation argument, there has been considerable discussion of the idea that we probably are now simulations in computers.
14 For more detailed discussion of both twinning and the bizarre conclusion, see Meincke (this volume).
15 For a detailed defence, see Meincke (under review). Sidzinska (2017), too, largely endorses this approach, though with some significant reservations.

16 There is a terminological problem here. "Population" is generally understood atemporally, as a current census of the members of a species in some area. What is needed here is a word more like lineage, clearly referring to an evolving group of organisms, a process with constantly changing material parts. But there is no entirely suitable word for referring to the constituent parts of the process. I take it that this reflects the embeddedness of substance thinking.

17 The political dimensions of process metaphysics are discussed in more detail by Ferner (this volume).

References

Aylward, F. O., Burnum, K. E., Scott J. J., Suen G., Tringe, S. G., Adams, S. M., Barry, K. W., Nicora, C. D., Piehowski, P. D., Purvine, S. O., Starrett, G. J., Goodwin, L. A., Smith, R. D., Lipton, M. S. and Currie, C. R. (2012) 'Metagenomic and Metaproteomic Insights into Bacterial Communities in Leaf-Cutter Ant Fungus Gardens', *The ISME Journal* 6(9), pp. 1688–1701, doi:10.1038/ismej.2012.10.

Bapteste, E. and Dupré, J. (2013) 'Towards a Processual Microbial Ontology', *Biology and Philosophy* 28, pp. 379–404.

Bordenstein, S. R. and Theis, K. R. (2015) 'Host Biology in Light of the Microbiome: Ten Principles of Holobionts and Hologenomes', *PLoS Biology* 13(8): e1002226.

Bostrom, N. (2003) 'Are We Living in a Computer Simulation?', *The Philosophical Quarterly* 53(211), pp. 243–255.

Bot, A. N., Currie, C. R., Hart, A. G. and Boomsma, J. J. (2001) 'Waste Management in Leaf-Cutting Ants', *Ethology Ecology and Evolution* 13(3), pp. 225–237.

Bouchard, F. (2008) 'Causal Processes, Fitness, and the Differential Persistence of Lineages', *Philosophy of Science* 75, pp. 560–570.

Burt, A. and Trivers, R. L. (2006) *Genes in Conflict: The Biology of Selfish Genetic Elements*, Harvard: Belknap Press.

Cho, I. and Blaser, M. J. (2012) 'The Human Microbiome: at the Interface of Health and Disease', *Nature Reviews Genetics* 13(4), pp. 260–270, doi:10.1038/nrg3182.

Clutton-Brock, T. (2009) 'Cooperation Between Non-Kin in Animal Societies', *Nature* 462(7269), p. 51.

Cogen, A. L., Nizet, V. and Gallo, R. L. (2008) 'Skin Microbiota: A Source of Disease or Defence?', *The British Journal of Dermatology* 158(3), pp. 442–455.

Currie, C. R., Scott, J. A., Summerbell, R. C. and Malloch, D. (1999) 'Fungus-Growing Ants Use Antibiotic-Producing Bacteria to Control Garden Parasites', *Nature* 398 (6729), p. 701.

Davis, J. N. and Daly, M. (1997) 'Evolutionary Theory and the Human Family', *The Quarterly Review of Biology* 72(4), pp. 407–435.

Darwin, C. (1892) *The Formation of Vegetable Mould, Through the Action of Worms, with Observations on Their Habits*, New York: Appleton.

Dawkins, R. (1976) *The Selfish Gene*, Oxford: Oxford University Press.

De Grey, A. and Rae, M. (2007) *Ending Aging: The Rejuvenation Breakthroughs that Could Reverse Human Aging in Our Lifetime*, New York: St. Martin's Press.

Dupré, J. (2011) 'Emerging Sciences and New Conceptions of Disease: Or, Beyond the Monogenomic Differentiated Cell Lineage', *European Journal for Philosophy of Science* 1, pp. 119–132.

Dupré, J. (2012) *Processes of Life: Essays in the Philosophy of Biology*, Oxford: Oxford University Press.

Dupré, J. (2013) 'How Much of the Free Will Problem Does (The Right Kind of) Indeterminism Solve?', *Philosophical Inquiries* 1, pp. 79–92.
Dupré, J. (2014) 'Animalism and the Persistence of Human Organisms', *The Southern Journal of Philosophy* 52, Spindel Supplement, pp. 6–23.
Dupré, J. (2017) 'The Metaphysics of Evolution', *Interface Focus*, Published online, August 18, 2017, http://rsfs.royalsocietypublishing.org/content/7/5/20160148\.
Dupré, J. and Nicholson, D. J. (2018) 'A Manifesto for a Processual Philosophy of Biology', in: Nicholson, D. J. and Dupré, J. (Eds.), *Everything Flows: Towards a Processual Philosophy of Biology*, Oxford: Oxford University Press, pp. 3–45.
Dupré, J. and O'Malley, M. A. (2009) 'Varieties of Living Things: Life at the Intersection of Lineage and Metabolism', *Philosophy, Theory and Practice in Biology* 1(3), http://hdl.handle.net/2027/spo.6959004.0001.003.
Emlen, S. T. (1995) 'An Evolutionary Theory of the Family', *Proceedings of the National Academy of Sciences* 92(18), pp. 8092–8099.
Ferner, A. (this volume) 'Processual Individuals and Moral Responsibility', in: Meincke, A. S. and Dupré, J. (Eds.), *Biological Identity. Perspectives from Metaphysics and the Philosophy of Biology, (History and Philosophy of Biology)*, London: Routledge, pp. 214–232.
Gagnier, R. (2000) *The Insatiability of Human Wants: Economics and Aesthetics in Market Society*, Chicago, IL: University of Chicago Press.
Gagnier, R. and Dupré, J. (1995) 'On Work and Idleness', *Feminist Economics* 1(3), pp. 96–109.
Jandhyala, S. M., Talukdar, R., Subramanyam, C., Vuyyuru, H., Sasikala, M. and Reddy, D. N. (2015) 'Role of the Normal Gut Microbiota', *World Journal of Gastroenterology* 21, pp. 8787–8803.
Kingma, E. (2019) 'Were You a Part of Your Mother?', *Mind* 128(511), pp. 609–646, doi:10.1093/mind/fzy087.
Kingma, E. (this volume) 'Pregnancy and Biological Identity', in: Meincke, A. S. and Dupré, J. (Eds.), *Biological Identity. Perspectives from Metaphysics and the Philosophy of Biology, (History and Philosophy of Biology)*, London: Routledge, pp. 200–213.
Locke, J. (1975) *An Essay Concerning Human Understanding* (1689), Nidditch, P. H. (Ed.), Oxford: Oxford University Press.
Margulis, L. (1991) 'Symbiogenesis and Symbionticism', in: Margulis, L. and Fester, R. (Eds.), *Symbiosis as a Source of Evolutionary Innovation*, Cambridge, MA: MIT Press, pp. 1–14.
Maturana, H. R. and Varela, F. J. (1991) *Autopoiesis and Cognition: The Realization of the Living*, Vol. 42, Berlin: Springer Science and Business Media.
Meincke, A. S. (2018) 'Persons as Biological Processes. A Bio-Processual Way-Out of the Personal Identity Dilemma', in: Nicholson, D. J. and Dupré, J. (Eds.), *Everything Flows: Towards a Processual Philosophy of Biology*, Oxford: Oxford University Press, pp. 357–378.
Meincke, A. S. (2019a) 'Autopoiesis, Biological Autonomy and the Process View of Life', *European Journal for Philosophy of Science* 9:5, doi:10.1007/ s13194-018-0228-2.
Meincke, A. S. (2019b) 'The Disappearance of Change. Towards a Process Account of Persistence', *International Journal of Philosophical Studies* 27, pp. 12–30, doi:10.1080/09672559.2018.1548634.

Meincke, A. S. (2019c) 'Human Persons – A Process View', in: Noller, J. (Ed.), *Was sind und wie existieren Personen?* Münster: Mentis, pp. 57–80.

Meincke, A. S. (under review) 'One or Two? A Process View of Pregnancy'.

Meincke, A. S. (this volume) 'Processual Animalism: Towards a Scientifically Informed Theory of Personal Identity', in: Meincke, A. S. and Dupré, J. (Eds.), *Biological Identity. Perspectives from Metaphysics and the Philosophy of Biology, (History and Philosophy of Biology)*, London: Routledge, pp. 251–278.

Nakatsuji, T., Chen, T. H., Butcher, A. M., Trzoss, L. L., Nam, S.-J., Shirakawa, K. T., Zhou, W., Oh, J., Otto, M., Fenical, W. and Gallo, R. L. (2018) 'A Commensal Strain of *Staphylococcus epidermidis* Protects Against Skin Neoplasia', *Science Advances* 4(2), doi:10.1126/sciadv.aao4502.

Nicholson, D. J. (2018) 'Reconceptualizing the Organism: From Complex Machine to Flowing Stream', in: Nicholson, D. J. and Dupré, J. (Eds.), *Everything Flows: Towards a Processual Philosophy of Biology*, Oxford: Oxford University Press, pp. 139–166.

Nicholson, D. J. and Dupré, J. (Eds.) (2018) *Everything Flows: Towards a Processual Philosophy of Biology*, Oxford: Oxford University Press.

Pinto-Tomas, A., Anderson, M., Suen, G., Stevenson, D., Chu, F., Cleland, W., Weimer, P. and Currie, C. (2009) 'Symbiotic Nitrogen Fixation in the Fungus Gardens of Leaf-Cutter Ants', *Science* 326, pp. 1120–1123.

Rescher, N. (2006) *Process Ontological Deliberations*, Frankfurt: Ontos.

Sandberg, A. and Bostrom, N. (2008) 'Whole Brain Emulation: A Roadmap', *Technical Report #2008-3*, Future of Humanity Institute, Oxford University, www.fhi.ox.ac.uk/reports/2008-3.pdf.

Shoemaker, D. W. (2005) 'Embryos, Souls and the Fourth Dimension', *Social Theory and Practice* 31, pp. 51–75.

Sidzinska, M. (2017) 'Not One, Not Two: Toward an Ontology of Pregnancy', *Feminist Philosophy Quarterly* 3(4), Article 2.

Simons, P. (2018) 'Processes and Precipitates', in: Nicholson, D. J. and Dupré, J. (Eds.), *Everything Flows: Towards a Processual Philosophy of Biology*, Oxford: Oxford University Press, pp. 49–60.

Snowdon, P. (this volume) 'The Nature of Persons and the Nature of Animals', in: Meincke, A. S. and Dupré, J. (Eds.), *Biological Identity. Perspectives from Metaphysics and the Philosophy of Biology, (History and Philosophy of Biology)*, London: Routledge, pp. 233–250.

Tyler, I. (2000) 'Reframing Pregnant Embodiment', in: Ahmed, S., Kilby, J., Lury, C., McNeil, M. and Skeggs, B. (Eds.), *Transformations: Thinking Through Feminism*, London: Routledge, pp. 288–301.

Varela, F. J. (1979) *Principles of Biological Autonomy*, New York: Elsevier.

Van Borm, S., Wenseleers, T., Billen, J. and Boomsma, J. J. (2001) 'Wolbachia in Leafcutter Ants: A Widespread Symbiont that May Induce Male Killing or Incompatible Matings', *Journal of Evolutionary Biology* 14, pp. 805–814.

Wilson, D. S. and Sober, E. (1989) 'Reviving the Superorganism', *Journal of Theoretical Biology* 136, pp. 337–356.

9 Activity, process, continuant, substance, organism

David Wiggins

9.1

In a manifesto entitled "A Process Ontology for Biology" and issuing from the Centre for the Study of Life Sciences, University of Exeter, John Dupré has proposed the question whether *things* or *processes* provide a better framework for interpreting science. He says that this question should be "a central concern for everyone interested in the metaphysics of science".[1] He is ready to think that it is a real option to abandon the substance-theorist's preoccupation with "the changes that occur to an entity and the conditions under which an entity can remain the same thing through change"[2] and to embrace instead a process-theorist's concern with "how a combination of processes can maintain the appearance of stability and persistence in an entity that is fundamentally only a temporary eddy in a flux of change".

Dupré is not alone in his readiness to embrace an ontology of processes (see, e.g., Guay and Pradeu 2016a).[3] But my own response to the question he has raised is not to disagree over the importance of processes. It is rather to contend that, if the idea of persistence is to have even a toe-hold within a scientific world-view that extends to the realm of organisms – if that worldview is ever to afford the option to make the barest cross-reference between some *x* or other at this time and some *y* at another time –, then there is need for an ontology not simply of *event*, *process* and *activity* but also and equally of *material things*, things that persist however temporarily from one time to the next. For present purposes it will suffice to call these latter, with W. E. Johnson (1924), continuants.[4] Continuants exist in time, have material parts and pass through phases. But such phases are not the material parts of the continuant. The phases are parts of the continuant's span of existence. Contrast processes. The phases of a particular historically dateable process *are* its parts.

There are further differences between processes and things. A process can be rapid or regular or staccato, or steady. It can even be cyclical and lifelong. Consider the Krebs cycle. Talk of organisms certainly demands talk of the processes by which they are maintained. But how is talk of organisms to be replaced altogether by talk of processes which submit to attributions such as rapid, regular, staccato, steady or cyclical? Organisms themselves cannot submit to these attributions. Meanwhile an organism can be the proud possessor of eight fingers and two thumbs. Can a process? In pressing these points, I shall appear to hark back to the archaic style of disputation proprietary to "linguistic philosophy". Yet, archaic or not, difficulties of this sort are suggestive of real distinctions – distinctions that are crucial perhaps for the philosophy of individuation. Does it not help towards the understanding of what an item is to ask what one can truly say about it?

9.2

Such arguments move much too swiftly, I fear, to carry full conviction among those who need convincing. Dupré himself is more interested in the radical redescription of biological reality, or so I surmise, than in the emendation of existing accounts of it. So let me begin afresh upon the effort of persuasion and invite the reader to try to imagine a world of pure process – of process without anything else. Let us try, for instance, to imagine a world consisting only of weather – a world where hurricane struggles with tornado for supremacy and powerful winds constantly oppose or cut across one another or combine to overwhelm all the other forces of the heavens. Such a world might seem to approximate to a world of pure process. But in a vacuum there is no weather. If there is to be weather, there must be not only process but also air or earth or water or ... some material principle which is other than process. (Could matter itself amount to no more than a process?) And, once there is any material principle at all, the collision of one process with another cannot help under some circumstance but make some quantities of matter collide and occasionally concresce with other matter. Not all the results of such concrescence need be momentary. To judge by the report of Diogenes Laertes IX.7 [Diels-Kranz A1], the thought is at least as old as Heraclitus: "the totality of things is harmoniously joined together through *enantiodromia* [the running of this against that]". In a world properly of process *and* matter, moreover – in a world such as can be the object of biological science – there wants at least one other thing, namely the possibility to refer twice to one and the same material concretion.

Here, in the world where we are, that condition is satisfied. Indeed our world-view has long since committed us to the existence not only of atoms and the rest, but also of re-identifiable organisms and microorganisms. At the subatomic and subsubatomic level all sorts of problems arise about the identification and reidentification of individual entities. Such problems have been thought to threaten the whole ontology that we try to apply there. But only a rather special kind of fanatic would claim that problems of this kind must undermine the possibility of genuine identity and difference of continuants at the macroscopic level. The subatomic level is not the level at which we have to account for the identification and reidentification of most of that which we know about from biology or enquire how living things relate to the processes that combine to enable or constitute their continuing existence.

9.3

In the effort to master these questions, some theorists of process without substance are apt to invoke the idea of *genidentity*. In their article "To be continued: the genidentity of physical and biological processes", Thomas Pradeu and Alexandre Guay write

> What does the concept of genidentity say? In a nutshell, [genidentity] says that the identity through time of an entity X is nothing more than the continuous connection of the states through which X goes. For example a "chair" should be understood in a purely historical way, as a connection of spatiotemporal states from its making to its destruction. The genidentity view is thus utterly *anti*-substantialist in so far as it suggests that the identity of X through time does not in any way presuppose the existence of a permanent "core" or "substrate" of X. It also leads one to replace the question "what *is* X, fundamentally?" by the question "How should I *follow* X through time?" [...] In this context, the notion of individual becomes derivative [...] It will not be unreasonable to [contend] that processes are ontologically prior and individuals should be conceived as the specific temporary coalescences of processes [...]
> (Guay and Pradeu 2016b, p. 317f.)[5]

In response to this I protest first that, according at least to my own avowedly "substantialist" account of what it takes to find a thing X and then find X again later, it is only required that in tracking X one should attend to the activity of X – attend that is to X's way of being and behaving.[6] Such a principle (I insist) need not invoke any "permanent core" or "substrate". The operation of a principle of activity for X's kind of thing will involve matter, but it is not excluded that that matter be exchanged constantly. Everything depends here on what kind of thing X is. Let me add that, among the proper parts or constituents of continuants of a given kind, nothing excludes the possibility that there be further continuants. Indeed, in the case of a continuant with the principle of activity of something alive, it may be discovered that it is essential to the life and survival of that continuant that it have within it certain other continuants, microbes, symbionts etc. This I learn from Guay and Pradeu themselves. But in this connection everything depends on the empirically discoverable demands of the particular principle of the activity that sustains the stability and persistence of the sort of organism in question.

The second point I put to Guay and Pradeu relates to what they mean by "temporary coalescences of processes". Do they mean the coalescence of *matter* with *matter*, a coalescence brought about by distinguishable processes? Or do they mean the process that results from the confluence or concurrence of the various processes which sustain the something or other that is some organism? On a literal reading of what they say, it seems they can only mean the second. They can only mean that the organism is *itself* a process. Only this literal reading distinguishes their position from the "substantialist" position from which they seek to distance themselves.[7] But are they content for this to be what they mean? The discomforts of taking such a position (see again Section 9.1, *ad finem*), would be out of all proportion with any real difficulties in the philosophy of continuants.

9.4

Where the idea of genidentity is concerned, I am drawn to a rival account:

> [An] examination of the concepts and principles of relativity [...] shows that they rest squarely on the ontology of things and events. A *world-line* is a sum of events all of which involve a single *material* body; any two events on the same world line are *genidentical*. That which cannot be accelerated up to or beyond the speed of light is something with a non-zero mass. But only a continuant can have mass. In like fashion, the measuring rods and clocks of special relativity, which travel from place to place, are as assuredly continuants as the emission and absorption of light signals are events. Nor does relativity entail that large continuants have temporal as well as spatial parts [...] [We] suggest that that rejection of the old (substantialist) ontology be postponed until such time as the promised better alternative is in a much more liveable state.
>
> (Simons 1987, p. 127)

Genidentity thus explained depends on the idea of a world-line and that idea depends on the idea of a material body. It is a relation between the events that involve a material body. On these terms – genidentity being glossed as Simons glosses it – it is simply impossible to reconstrue genidentity as a link between physical states whose concatenations can stand in for material bodies. We needed material bodies from the outset in order to say what a world-line was.

9.5

Guay and Pradeu may try to show that there is another way to say what a world-line is. But this is the moment for me to turn to their admirable suggestion that the philosophy of biology needs to focus upon the question "How should I follow X through time?" (Guay and Pradeu 2016b, p. 318). No proposal could be more welcome. I have long promoted the very same question – not because I have wanted to show the notion of individual continuant to be derivative in the way that Guay and Pradeu propose, but because I have wanted not to strain and strain in scholastic fashion after the idea of the singular essence of a thing. There had to be another way. The thought was that, in order to understand identity and individuation,

Activity, process, continuant... 171

one needs to study the way in which we learn to track continuants of any particular sort – at the same time, in the course of doing this, exploiting and extending our knowledge of how and why these things, of this or that sort, behave as they do.

Applying the question "How should I follow X through time?" to the context of biological science, Guay and Pradeu make a whole wealth of suggestions about how the competent inquirer must proceed: by consulting considerations of "causally significant process"; by reference to considerations relevant to "internal organization as measured in terms of intensity of interactions"; and by consulting considerations relating to "well-specified metabolic interactions" that contribute to the "cohesiveness" and unity of a whole organism, as well as to "higher level interactions", some of them on the part of the immune system, which themselves "exert control over metabolic interactions at a lower level".[8]

9.6

I have three kinds of doubt whether these proposals – interesting and enlightening though they are, and illuminating as they are of the ways in which scientists have tried to understand the secret life-cycles of strange creatures – can in the end help to vindicate any theoretical preference for a simplified ontology of process over the larger ontology that I advocate.

The first doubt concerns how well the thoughts, ideas and scientific practices of the investigators who have created the branches of biology that Guay and Pradeu appeal to can cohere and consist with the claim that the concept of individual organism is "derivative". Could some revised research practice of these investigators amount to their dispensing with all implicit reliance on the determinables *continuant* and *organism*? Could these investigators really think of the organisms they study as simply concatenations of states? If they were asked what concatenates the states in question, they would surely refer to their however provisional account of the life-cycle of a given organism. There is more to this preference than meets the eye. A concatenation, being defined by its membership, has its members necessarily. To imagine the smallest difference in a concatenation is to imagine a different (non-identical) entity. Is this not a special disadvantage? We need a "connection" between states – a connection that is more specific moreover than undifferentiated succession or causation. We need the provisional account of the life-cycle of an organism.

My second doubt concerns cohesiveness, unity, wholeness ... I submit that, au fond, such ideas depend for their application on a context that is framed by the kinds or kinds of kind in question. They need to be glossed in context as "one f", "a whole f", "something cohesive in the manner of an f", "unitary f" ... and in the presence of some adumbration at least of the concept f. Is it all right for the concatenation-of-states construal of continuants to wait for its implementation upon the prior operation of criteria informed by thoughts about the nature of the bearers of those states, namely (as I should say) the organisms themselves?[9]

The third doubt relates to something we all three agree about, namely the role of principles in the reidentification/tracking/following of a particular continuant (as I should say) or concatenation-of- states (as they would say). It relates also to something we do not agree about. I say that these principles depend not on the idea of a concatenation of states but on the idea of some continuant or other.

I begin upon this by remarking that simple logic requires that such tracking principles should respect both the reflexivity of identity and the indiscernibility of identicals. These imply the symmetry or reversibility and the transitivity of the relation which must hold between X and the this-or-that to which we trace it. Suppose that some putative principle **P** carries us from X to Y and carries us from Y to Z. Then **P** gives us a path back and forth between X and Z. Suppose however that **P** also gives a path between X via Y to Z' but **P** offers no path between Z and Z'. Then we shall have a contradiction. Z and Z' are distinct – there is no identificatory direct route between them – and yet also identical. For, via Y, **P** carries us back from Z to X and carries us back from Z' to X. So $Z = Z'$. But on the basis of **P** we had *also* supposed that $Z \neq Z'$.

Any tracking principle **P** that gives such a result in any of its applications will need to be reconsidered. I am uncertain what Guay and Pradeu will say that that involves. But anyone who takes the idea of a continuant in the way I do as primitive – and not as reducible to the account of a mere concatenation of states – will say that any workable tracking principle must arise from some however provisional conception of the particular kind of thing that is to be followed or tracked. In the face of contradiction it is this conception, the conception that animates principle **P**, which needs to be reshaped. Suppose, for instance, that in the sort of case we began with the conception was the conception of *human being*; and suppose that, as it stood, this conception allowed us to think of a human being as starting its existence as a zygote. The trouble would be that, as is well known, the human zygote may divide at any moment before the twelfth day after conception and give rise to two separate embryos (twins). It follows that the principle corresponding to the conception of human beings that we began with cannot stand. It is a mistake to think of a human being's existence as starting before the formation of the embryo (see Wiggins 2012, section 10). The conception we began with needs correction. The earliest moment a human being can begin is with the embryo. I am not sure how that point will come out on the concatenation of states conception.

9.7

Here, arising from the last point but moving on to something else, someone may offer an interesting objection. "*Genetically* speaking surely Z and Z' really are the same. And, in that case, is it completely clear that **P** must offer no direct path between Z and Z'?" Such an objection is highly instructive. Anyone who is a sortalist about identity in the same way as I am will insist that, if **P** is meant to track *human beings*, then **P** cannot stand. On the other hand, if **P** is a principle deriving from the non-singular idea of some [*particular*] *human genotype* or *lineage*, then there ought indeed to have been a path back and forth between Z and Z′.

The point I want to make now is that, where identity is concerned, everything depends on what *category* of thing and what *kind* of thing one is to single out from the rest of reality. Is the item in question concrete or abstract, is it a thing or a nature, a particular animal or an animal species, something singular or something plural, a member or a class, a plant or a plant-colony, a clone (specimen) or a clone (group), a token or a type, a continuant or an aggregate …? And is it a substance or a process?[10]

9.8

Having now, in this way, more or less reinvented the sortalist conception of identity, let me try to apply it to some part of the area where philosophers of biology such as I have already named experience the doubts that prompt them to try to dispense (or dispense initially) with continuants proper or cause them to long for a purer ontology of processes. Let me apply the sortalist approach to some of the remarkable creatures that Jack Wilson describes in *Biological Individuality: The Identity and Persistence of Living Entities* (Wilson 1999). I am not entirely sure how process-theorists such as Dupré or Pradeu and Guay will prefer to describe such organisms. But what I hope to show is that these creatures need not especially daunt a substance-theorist who embraces the pluralist ontology that I have sought to advocate.

(1) "A colonial siphonophore […] begins as a zygote. The zygote divides and forms a larva. The larva's ectoderm thickens and buds off zooids […] [which] remain attached together […] New zooids are budded off from one of the two growth zones located at the end of the nectophore region. Each colony is composed of a variety of zooids that closely resemble the parts of a normal jellyfish. The top of the colony is a gas-filled float. Below the float are the nectophones that move the colony by pumping water … Their action is coordinated […] The colony can swim and feed like a single organism […] Is a siphonophore colony an individual or is each single zooid an individual?" (Wilson 1999, p. 7)

(2) "At one point in the life-cycle of a certain species of cellular slime-moulds, a number of independent, amoeba-like single cells aggregate together into a grex. The grex is a cylindrical mass of these cells that behaves much like a slug. It has a front and a back, responds as a unit to light, and can move as a cohesive body. The cells that compose a grex are not always genetically identical or even related. They begin their lives as free-living single-cell organisms. The grex has some properties of an individual and behaves very much like one." (Wilson 1999, p. 8)

(3) "Blackberry plants reproduce both by sexual means resulting in seeds and also through vegetative growth. Some stands of blackberries are hundreds of years old and trace their origin back to a single sexually

produced seed. The seed grows into a plant, which send out runners. Some of the runners and roots remain connected underground and others have become detached. What should we count when we count blackberry plants?" (Wilson 1999, p. 8)

9.9

I begin with the problem (3). If there is a problem here, it is nothing special to biology. Consider the concept *crown*. It is clear enough how a thing has to be in order to count as a crown, and clear enough what it takes for crown C_1 to be the same as crown C_2. But there is no universally applicable definite way of counting crowns. The Pope's crown is made of crowns. When the Pope wears his crown there is no unique or definite answer to the question how many crowns he has on his head.[11] If we want to count under a concept *f* then either we must choose a concept that does not permit division of what falls under it or else *f* must be further qualified. Is our interest in counting particular genetically uniform stands (colonies) of blackberry, or in counting individual blackberry plants whether or not connected below ground to other plants, or what...? There is nothing reasonable in the idea that reality allows only one choice – one ontology and one ideology, one domain of individuals and one domain of properties of those individuals. Compare the distinction Hilary Putnam proposes between ordinary realism and "metaphysical" realism. See Chapter 2 (especially at p. 62) of his *Philosophy in an Age of Science* (Putnam 2012). A reasonable inquirer has to be prepared to attend to these things or to those different things. However palpable the things we refer to may be – no matter how strong their claim to be "there anyway" – they may require the one who attends to them (either optically or in thought or in both ways at once) to look for *this* sort of thing or for *that* sort of thing, whether singular or non-singular. Reality itself need not dictate what we are to heed. Still less will it forbid us to heed one kind of thing and *then* another kind.

9.10

I revert now to case (1). The idea of an individual or individual organism is the idea of a determinable. To ask whether something X is an individual is to ask whether there is some fully determinate kind *f* such that X is an individual *f*. In the case of the siphonophore colony it is indeed a particular kind of (quasi-jellyfish) creature with a specific principle of activity. There is nothing wrong with *siphonophore* as a specific kind. Let us forget the obsolescent idea that a substance is something that lacks substantial parts and is viable without parts that are substances. Nothing in the idea of a continuant demands this. Among the constituents of a siphonophore are numerous sortally further specified continuants, each of them with its very own principle of activity, co-ordinated and subordinated in important respects – but why not? – to the activity principle of the whole siphonophore.

I venture to think that there is nothing to forbid a similar treatment of the remarkable creature, the *grex*, which is the second puzzle-case.

9.11

Over the millennia, the philosophy of substance has created all sorts of mysteries and obscurities of its own. My claim is that, slimmed down in the form of a logically informed philosophy of continuants, taking each continuant not as indivisible but as possessed of its own determinate principle of activity, the philosophy of substance is ready and equipped to cohere and combine with any equally clear and coherent philosophy of process, activity and event. There must be room within any such philosophy for the idea of a continuant.

At the outset (see Section 9.2), I allowed that Dupré may be more interested in the possibility of a *radical redescription* of biological phenomena than in any scheme for the translation of existing descriptions into a language of pure process. So let me acknowledge that the very most I can achieve by the arguments I have advanced here is to suggest that radical redescription is not so urgent or so necessary as it has appeared.

Notes

1 See https://thebjps.typepad.com/my-blog/2014/08/a-process-ontology-for-biology-john-dupré.html.
2 Speaking for substance-theorists, let me insist that the words "remain the same thing" be replaced by the word "persist", unless our preoccupation is to be misdescribed.
3 But these authors state their aims differently: "[We] suggest a shift from an ontology of *substances, invariance* and laws to an ontology centred on *processes* and change." (*Editorial note*: Wiggins quotes from a previous version of Guay and Pradeu's chapter. In the published version of their chapter, the corresponding passage, with slightly different wording, occurs on p. 139.)
4 Johnson defines a continuant to be "that which continues to exist throughout some limited or unlimited period of time, during which its inner states or outer connections with other continuants may be altering or may be continuing unaltered" (Johnson 1924, pp. xx–xxi). It should go without saying that on these terms *continuant* is a determinable notion – the most that can be available in advance of empirical experience or inquiry.
5 *Editorial note*: Wiggins quotes from a previous version of Guay and Pradeu's chapter. In the published version of their chapter, the corresponding passages, partly reformulated and dispersed, occur on p. 317f.
6 See Wiggins (2012), sections 8–10.
7 In a further effort to reconcile Guay and Pradeu to "substantialism" in the form in which I have tried to present it, let me quote (yet again – see Wiggins 1980, p. vii) from a text I have long revered:

> The essence of a living thing is that it consists of atoms of the ordinary chemical elements we have listed, caught up into the living system and made part of it for a while. The living activity takes them up and organizes them in its characteristic way. The life of a man consists essentially in the activity he

imposes upon that stuff [...] It is only by virtue of this activity that the shape and organization of the whole is maintained.

(Young 1971, p. 86f.)

Another text by which I might seek to distance Guay and Pradeu from their reading of substantialism comes from Aristotle himself (*Metaphysics* 1050b2):

Substance or form is *energeia*

Perhaps this is to say, not without some grammatical obscurity, that substance or form is active being. What I should *like* Aristotle to be saying here is that for x to be a substance is for x to have a principle of activity (in the sense I give these words in the article referenced in note 6). But it will be for the scholars of the *Metaphysics* to unwind the syntactical and interpretive intricacies of Aristotle's sentence.

8 *Editorial note:* Wiggins refers here to a previous version of Guay and Pradeu's chapter.
9 Suppose the putative parts of a putative thing are all present but in the wrong array. Is the entity "cohesive" or "unitary"? If not, why not? Well, constituted so, the entity cannot participate in the mode of activity that is proprietary to it and definitive of its kind. Is it not in the light of *this* that "cohesiveness" has to be interpreted and determined?
10 A word more about the logical adequacy requirement upon principles such as **P**. It demands more than respect for symmetry, reflexivity, and transitivity. It demands that grounds for the identity of x and y be grounds for the indiscernibility of x and y. x and y must share all their properties. They must have the same life-history. If that seems implausible in a given case, then the fault (if there is one) lies with the conception of the *kind* that regulates the formulation of the principle **P**.
11 Compare Wiggins (1980), p. 72f., and Wiggins (2001), p. 74f.

References

Guay, A. and Pradeu, T. (Eds.) (2016a) *Individuals Across the Sciences*, New York: Oxford University Press.
Guay, A. and Pradeu, T. (2016b) 'To Be Continued: The Genidentity of Physical and Biological Processes', in: Guay, A., and Pradeu, T. (Eds.), *Individuals Across the Sciences*, New York: Oxford University Press, pp. 317–347.
Johnson, W. E. (1924) *Logic, Part III*, Cambridge: Cambridge University Press.
Putnam, H. (2012) *Philosophy in an Age of Science*, ed. M. de Caro and D. Macarthur, Cambridge, MA: Harvard University Press.
Simons, P. (1987) *Parts: A Study in Ontology*, Oxford: Oxford University Press.
Wiggins, D. (1980) *Sameness and Substance*, Oxford: Blackwell.
Wiggins, D. (2001) *Sameness and Substance Renewed*, Cambridge: Cambridge University Press.
Wiggins, D. (2012) 'Identity, Individuation and Substance', *European Journal of Philosophy* 20(1), pp. 1–25.
Wilson, J. (1999) *Biological Individuality: The Identity and Persistence of Living Entities*, Cambridge: Cambridge University Press.
Young, J. Z. (1971) *Introduction to the Sciences of Man*, Oxford: Oxford University Press.

10 Diachronic identity in complex life cycles
An organisational perspective

James DiFrisco and Matteo Mossio

10.1 Introduction

Questions about diachronic identity have sustained extensive philosophical reflection in general metaphysics and particularly in the context of personal identity. Less attention has been given to diachronic identity of organisms. This is in spite of a recent surge of work on biological individuality in philosophy of biology, which has largely focused on which sorts of parts should be included within the spatial boundaries of individuals, and not on which sorts of events should be included within the temporal boundaries of a life. However, temporal boundaries matter for some of the same reasons that spatial boundaries do.

One biological practice that requires criteria of individuation in space is counting individuals in order to measure population size and demographic change (Clarke 2011). For these purposes, it is often not enough to single out the appropriate individuals at specific times without also connecting them across time. One needs to be able to interrupt an observation period and later return to the system under study, to *re-identify* the same individuals as before, and distinguish them from the ones that are new. Otherwise, one risks double-counting certain individuals and missing others completely.

Similar considerations apply to measurements of reproductive fitness in terms of the number of offspring produced. Offspring number straightforwardly depends on how we individuate offspring. But it is not always obvious whether a biological system at a given time is the beginning of a new life or just a later developmental stage of the parent. This is especially the case in life cycles involving metamorphosis, metagenesis (alternation of generations), vegetative reproduction, fusion, fission, and symbiosis, as we will see. It is also important to be able to distinguish development, growth, and reproduction because these processes connect to different dynamics of life history evolution (Stearns 1992). A model that describes trade-offs and local optima for traits such as growth rate, time to reproduction, and mortality schedules will generate different predictions and different error rates depending on which segments of a life cycle it is applied to. For purposes of comparative biology, the formation of a reliable generalisation over life history patterns in different taxa demands that the patterns be genuinely comparable. Work on biological individuality can ideally provide a domain theory that identifies which intervals of organismic processes the models should apply to.

In attempting to provide diachronic criteria of identity we encounter a first difficulty that modes of development and reproduction differ markedly across different biological taxa. It is entirely possible that there are no shared criteria of identity that apply to the class of "organisms" as such, but only local criteria for different taxonomic groups. A second, related difficulty is that organisms can have very different properties at different stages of the same lifetime. In fact, many of the properties emphasised in proposed criteria of individuality are not present throughout the lifetime of the systems they are supposed to individuate (see Clarke 2011, DiFrisco 2019). Accounts of individuality face this difficulty when they are based on fitness (e.g., Folse and Roughgarden 2010), the capacity to undergo selection (Clarke 2013), and cooperation and conflict (Queller and Strassmann 2009). An organism during its post-reproductive lifespan has no fitness, its cooperation or conflict has no evolutionary effect, and it cannot undergo selection (excluding grandparenting effects). Yet post-reproductive organisms nonetheless exist, and they have effects on biological processes other than evolution. Non-evolutionary features of organisms such as histocompatibility and symbiotic interaction can also be time-variable in development. When the properties that are supposed to determine individuality change over developmental time, they need to be embedded in the context of a life cycle structure. Then we can still count something as an individual when those properties are absent.

Intra-lifetime variability sets up a general dilemma: if the criteria for continued persistence are too demanding, then developmental processes involving dramatic material and functional transformations like metamorphosis may count as a reproductive process involving the production of a new individual. If the criteria are too relaxed, then many reproductive processes will be re-cast as development of one and the same individual.

This chapter examines diachronic identity conditions for organisms specifically in light of the second problem of intra-lifetime variability. We argue that a suitable necessary condition for organismic persistence can be found in spatiotemporal continuity together with the presence of a causal structure known as *closure of constraints*. This dual condition is called *organisational continuity*. Organisational continuity is sufficient for diachronic identity only when there are no changes in the number of organised systems that are organisationally continuous with one another. Such changes can occur through events of fission or fusion, which call for special analysis. In explicating this qualification of sufficient conditions, we find that one also needs to take into account the way in which different temporal boundaries can be promoted by different biological perspectives—e.g., evolutionary, developmental, or physiological—with their distinctive modelling and generalisation strategies. We close by considering how the proposed organisational view can plausibly satisfy the above biological roles for diachronic identity conditions, and highlight open questions for future work.

10.2 Organisational continuity

Spatiotemporal continuity is widely considered by philosophers to be necessary for diachronic identity of objects. An object cannot pass in and out of existence, or disappear from one location and immediately re-appear in another, while remaining numerically the same object.[1] Spatiotemporal continuity alone does not entail contiguity of the spatial parts of an object—commonly called "unity" in philosophy and "cohesion" in perceptual psychology—but only that the spatial regions an object occupies are connected across time. For example, a biological population can be considered an object whose parts are not topologically connected, but it can nonetheless be tracked across time in virtue of the spatiotemporal continuity of its scattered parts. Unity is often considered to also be necessary for object identity, though we will remain neutral on this issue.

Mere spatiotemporal continuity however, without any specification of the relata that are supposed to stand in the relation of continuity, is not a sufficient condition for identity. This is because any object or event is spatially continuous with its immediate environment and temporally continuous with what precedes it and what succeeds it. Some background determination about the *sort* of item to be tracked—such as a horse, a computer, or a thunderstorm—is needed to get a grip on its spatial boundaries as well as the kinds of change it can undergo while remaining the same thing of that sort. For example, while relying on the criterion of spatiotemporal continuity, an observer may trace different paths through spacetime if she or he is tracking items in her perceptual field under the sortal *house* versus *set of bricks*. The application of a suitable sortal, together with the condition of spatiotemporal continuity, can be viewed as determining the identity of entities in space and time. This is the position defended in contemporary forms of substance ontology, from Strawson (1997), to Wiggins (2001, 2016), and Lowe (2009). Roughly, x and y are the same individual iff they are spatiotemporally continuous *and* they fall under the same sortal.

From this point of view, the problem of organism identity can be framed as a problem of finding an adequate definition for the sortal *organism* or for specific kinds of organisms. We propose to approach the issue of organism identity in a different way, by first shifting from the descriptive question to a more basic explanatory one. Rather than asking which concept of the organism determines the organism's persistence, we take a step back and ask which causal structure determines its survival. If we can understand the sorts of physical features and interactions that explain, from the bottom up, how organisms survive—taking "organisms" or "individuals" in an initially loose sense—then we gain insight into how organism identity ought ultimately to be conceptualised.

Biological organisms are physical systems that exhibit a type of macroscopic order based primarily on the energy provided by chemical reactions. In order to maintain their internal structure and energetic functioning against the global thermodynamic tendency towards entropic dissipation, organisms must continually occupy states far from chemical equilibrium by absorbing chemical or radiant energy (Ruiz-Mirazo and Moreno 2004). Unlike other macroscopic dissipative structures, however, organisms exhibit not only physical *order* but also functional *organisation*. We use here "organisation" in a theoretical specific sense, to refer to a specific causal regime in which groups of differentiated parts collectively maintain each other through their reciprocal interactions. The survival of organised systems as wholes depends on the overall activity of their parts. Hence, these systems can be said to be *self-maintaining* in a given environment (Moreno and Mossio 2015).

As thermodynamically open systems, organisms are traversed by flows of matter and energy in the form of complex chemical and physical transformations (processes or reactions). These transformations are controlled through the coordinated activity of system components, which are theoretically characterised as *constraints*. Constraints are material structures that asymmetrically act on transformations without being directly affected by them, and that remain conserved during the course of those transformations. For instance, the developed cardiovascular system of vertebrates constrains blood flow without being altered by it.[2] At smaller spatial scales, enzymes change the kinetics of chemical reactions without being consumed in the process (see Montévil and Mossio 2015 for further details).

In order to constrain a transformation, the constraint must be conserved at the time scale of the transformation it constrains. Yet on longer time scales, biological constraints also degrade and must be replaced or repaired. The cells that compose the vascular system must be nourished, and enzymes undergo degradation and denaturation over time and must be replaced. In general, the repair or replacement of a constraint occurs by means of another metabolic process or complex of processes, which is itself channelled under other constraints. For example, lipid membranes constrain the diffusion of enzymes and metabolites in such a way that the appropriate intracellular concentrations are maintained, but new lipids are also produced through the chemical reactions controlled by enzymes. Once we have reciprocally maintained effects of a group of constraints, the notion of organisation enters the picture. A system is "organised" in the relevant sense when the continued existence of a set of constraints depends on the effects that they *collectively* produce on the overall thermodynamic flow. If this occurs, the set of mutually dependent constraints can be said to realise *closure*. The mutual dependence between constraints, theoretically designated as "closure of constraints," is what allows organised systems to maintain themselves over time.[3] From the organisational perspective, we can view organisms or biological individuals as being organised systems (as a necessary condition), while recognising that not all conceptions of the organism will have this feature.

Closure of constraints plays a central role in explaining the survival of organisms, viewed as organised systems, as it explains how their parts are functionally arranged in such a way that the system can maintain itself dynamically over time. Accordingly, it seems reasonable to hold that, conceptually, closure of constraints is necessary for organism identity, since any organism that lost this property would quickly perish. Let us propose, then, that *a necessary condition for diachronic identity of organisms is that they continually exhibit closure of constraints*. Any temporal interval where this property of *organisational continuity* is not present then marks a temporal boundary of an organism's life, either due to death or non-existence before life.

What does it mean for a system to continually exhibit closure of constraints? A first interpretation might be that the particular material structures that act as constraints in the system—i.e., membranes, enzymes, vascular systems—are conserved over time. This would be a problematic requirement on organismic persistence, however. The material composition of organised systems changes over time—indeed, it *must* change if the systems are to survive, because functional constraints degrade and must be repaired or replaced.

A second interpretation might be that continually exhibiting closure of constraints means that a group of constraints, functionally individuated by their causal roles in self-maintenance, must persist despite continual material turnover of components and realising structures. As argued elsewhere (Mossio et al. 2009), constraints subject to closure can be said to ground biological functions. Diachronic identity would therefore be a matter of conservation of *function* rather than of material structure. The trouble with this solution is that material turnover is not the only kind of constitutive change that organisms undergo. In organisms that develop, i.e., single- and multi-celled eukaryotes, the constraints involved in self-maintenance can themselves arise, transform, and disappear within a lifetime—even if they are defined in abstract functional terms. A dramatic example of developmental transformation is metamorphosis (Figure 10.1).

Metamorphosis is a life history pattern characterised by successive life stages that differ markedly in their physiology, morphology, ecological niche, and reproductive capacity (Bishop et al. 2006). Although the evolutionary origins of metamorphosis are not well understood, one of the hypothesised adaptive effects of metamorphosis as a life history strategy is to permit the same organism to possess dramatically different functional specialisations at different stages of its life. In metamorphic life cycles, functions or constraints often fail to be conserved over entire lifetimes.

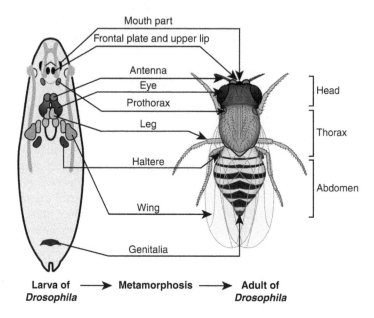

Figure 10.1 Material and functional change during metamorphosis in *Drosophila*. Many adult body parts are formed out of the set-aside cells in corresponding imaginal discs in the larva. Image reprinted from *Current Biology*, 20(10), Silvia Aldaz and Luis M. Escudero, "Imaginal Discs" PR429-R431, Copyright (2010), with permission from Elsevier. www.sciencedirect.com/journal/current-biology.

Moreover, intra-lifetime functional variation is not restricted to functions that are unnecessary for metabolic self-maintenance, such as camouflage or reproduction. The adult forms of many butterflies, moths, and mayflies lack mouthparts for eating, yet mouthparts—mechanical constraints on the flow of food energy—are necessary for self-maintenance in the larval forms.

Of course, many functional constraints *are* conserved over the course of development, particularly those having to do with core metabolism. One might try to reconfigure the criterion of functional conservation in light of developmental change by claiming that only those functions that are necessary for self-maintenance *throughout the entire lifetime* need to be conserved. This amendment, however, is self-defeating. If we say that the boundaries of a lifetime are determined by closure of constraints, but exhibiting closure of constraints means conserving functions that are necessary throughout the entire lifetime, then we already have to know what the temporal boundaries are in order to apply our proposed necessary condition for temporal boundaries. This solution is therefore circular or uninformative. Moreover, it is objectionable for departing from the explanatory role of closure of constraints. The set of constraints that are conserved throughout a lifetime will often be a *subset* of the set of constraints that realise closure at different stages. This conserved subset is not identified in terms of its special role in explaining survival, but by the fact of being conserved. But the main reason for proposing closure of constraints as a necessary condition for organism identity in the first place was that it is necessary for explaining survival. A subset of constraints that does not realise closure at the specific stage when it exists is not the relevant set of constraints for explaining survival. For example, the set of constraints in the mayfly larva that includes those that realise closure *minus* the mouthparts is not the relevant *explanans* for the survival of the larva. The relevant *explanans* is the *whole* set of constraints that realises closure at that stage. Subsets of constraints that happen to be conserved throughout a lifetime should not be construed as essential for tracking organism identity.

In order to continually exhibit closure of constraints, or to possess organisational continuity, an ontogenetic trajectory does not have to conserve anything in particular, neither structures nor functions. Nor do the different stages of life need to resemble each other. Being the same organised system over time, we argue, is based on a specific causal dependence relationship. If a system continually realises closure of constraints, then at any given moment the organisation of its parts is causally dependent on the effects of functional constraints that have operated at a previous moment. The organism at t_2 does not need to have the same set of constraints with the same inter-relationships as the organism at t_1, but some of the constraints at t_2 must be causally dependent on those at t_1. The *particular* set of constraints and their inter-relations—which we refer to as a *regime of closure*—can change over time as long as later regimes causally depend on earlier regimes. Hence, what must remain the same over time is closure construed as a general theoretical *principle* and not as a specific regime of closure (on this specific point see Mossio et al. 2016).

Note that organisational continuity implies spatiotemporal continuity, so by affirming the former as a necessary condition we include the latter. This implication stems from general physical relationships between causation, space, and time. In order for causal influence to propagate from x to y (without intermediaries), they cannot be separated by a temporal gap during which neither x nor y exists. Similarly, for the physical forces involved in biological interactions (primarily the electromagnetic force), interaction requires spatial proximity. This point might be taken to imply that organisational continuity, too, requires spatial proximity, but the implication is not obvious, and under-determined by current biological theories of organisation. The causal influence implied by constraint relationships must propagate between the parts, under specific spatial and topological conditions that depend on how a given system realises organisational closure, and on the level of organisation being considered. For instance, while it seems to be generally accepted that unicellular organisms cannot realise closure unless their parts are contained within an unbroken spatial boundary, this requirement might be relaxed in the case of ecosystems, viewed as candidate higher-level organised systems (see Nunes-Neto et al. 2014). Nevertheless, it seems reasonable to conjecture that it will generally be difficult to implement the productive relations of dependence involved in closure of constraints in systems whose functional parts do not remain spatially close enough to be able to causally interact with each other.

10.3 Organisational continuity: between genidentity and substantialism

The idea of conceptualising diachronic identity in terms of causal continuity, as opposed to material or functional conservation or resemblance, is not new. It was expressed in its modern form in Kurt Lewin's notion of *genidentity*, which has been promoted by some philosophers of biology as a criterion of diachronic identity for biological individuals (Hull 1978, Guay and Pradeu 2016).

Organisational continuity is a specific type of causal continuity and is thus more restrictive than genidentity. One issue with unrestricted genidentity is that many items are related by causal dependence that we would not want to consider parts or stages of the same individual. For example, organisms are causally continuous with their waste products as well as any artefacts they produce, but there seems to be no biologically interesting organism concept that includes these things as parts to be tracked through time. Organisms are also causally continuous with their parent(s) and the corpses they leave behind. Guay and Pradeu (2016) propose that temporal boundaries can be drawn where there are relatively fast rates of change, particularly change in "internal organisation," so that a single lifetime can comprise dramatic change as long as it is smooth and progressive. Although this rule would restrict genidentity in many cases, it is also likely to exclude metamorphic life cycles, as these can involve quite abrupt changes in internal organisation. It also seems that our ability to measure rates of change is epistemically dependent on our ability to identify and track what it is that is changing, rather than the other way around. Hull's (1978) writing on biological genidentity relies more narrowly on the idea that changes in "organisation" mark the temporal boundaries of individuals, though he does not explicate the notion of organisation in much detail. Our account of organisational continuity can be viewed as filling in the idea that biological identity is based on causal continuity (genidentity) and organisation. An important difference between the accounts, however, is that Hull (1978) and Guay and Pradeu (2016) use genidentity as a general model of persistence for *all* biological entities, including viruses, parts of organisms, and species. In fact, genidentity can be used as a criterion of identity for any kind of entity (e.g., molecules, stones, galaxies), which means that, as such, it does not restrictively apply to the biological domain. By contrast, organisational continuity applies only to systems realising closure of constraints, which include, at most, unicellular and multicellular organisms, and possibly symbioses and ecosystems.

Let us now examine the connection between organisational continuity and substance ontology and sortals raised earlier. As Hull (1978) and Guay and Pradeu (2016) point out, the theoretical decision to ground persistence on causal continuity rather than conservation or similarity runs counter to "substantialist" habits of thought, according to which being the same individual over time is a matter of retaining the same "substance," conceptualised either as matter or as form (see Simondon 1995). The most sophisticated contemporary defenders of substance ontology, such as Wiggins (2001) and Lowe (2009), are careful to drop the requirement of material conservation. In Wiggins's framework, the identity of a substance is grounded in its form, dynamically conceptualised as an Aristotelian *energeia* or activity (Wiggins 2001, p. 80). Every natural entity possesses a *principle of activity* or functioning that at once determines its conditions of identity as well as its membership in a natural kind (Wiggins 2001, p. 72). As Wiggins writes:

> [...] a particular continuant x belongs to a natural kind, or is a natural thing, if and only if x has a principle of activity founded in lawlike dispositions and propensities that form the basis for extension-involving sortal identification(s) which will answer truly the question "what is x?"
> (Wiggins 2001, p. 89)

To be the same individual over time is thus to conserve the same principle of activity, which implies spatiotemporal continuity and falling under the same sortal.

How does our proposal differ from Wiggins's activity-based substantialism? Initially, it would seem that the principle of closure could be interpreted as a kind of principle of activity for organisms. However, problems arise from the supposition that a principle of activity is not only (1) necessarily conserved across a lifetime but also (2) *sufficient* to determine the spatiotemporal extension or boundaries of an organism. These two roles pull the notion of a principle of activity in different directions, the first towards a coarser grain of individuation and the second towards a finer grain. There are often radically different modes of activity or functioning in different stages of metamorphosis, from physiological activity to morphological change to behaviour. In such cases, in order to capture the activity that is common to all developmental stages, the activity will have to be defined as a quite general one, such as "self-maintaining," "living," or "developing." But it is hard to see how generic activities like these could be sufficiently "extension-involving" to determine the temporal boundaries of organisms, to distinguish parents and offspring, or even to distinguish between organisms belonging to different taxa.[4]

In contrast, the conservation of a principle of *closure* over the course of ontogenesis does not entail that each stage displays the same characteristic functioning or activity. The principle of closure is something more abstract; this is why we treat it as a necessary condition and not as providing "thick" sufficient conditions for individuation. Sometimes a principle of closure will correspond to a unitary *regime* of closure that picks out temporal boundaries without difficulty. This is relatively more likely to be the case in organisms that do not undergo massive functional changes during ontogenesis, e.g., non-eukaryotes and some eukaryotes, and in direct developers. But in our proposal the concept of a regime of closure is not attributed an individuating role as a condition of persistence.

These features of our account entail a significant departure from core principles of the substantialist model advanced by Wiggins and assumed by many others. Against substance-based sortalism, there are unlikely to be sortals for organisms that, by specifying a distinct principle of activity, determine the spatial and temporal extent of a life cycle at one stroke. Instead, theoretically useful sortals for organisms are more likely to resemble sortals for *events*, events being inherently temporally extended and typically time-heterogeneous. Event sortals do not single out events by specifying a property or activity that remains the same throughout the event, but rather a characteristic sequence or series of stages or temporal parts, each with their own distinctive qualities, activities, and, in the present case, regimes of closure.

Ontologically, the key dynamic category that binds together stages into the same life cycle is not *activities*, but *developments*. Activities are homeomerous or "like-parted," in that what is happening in any sub-interval of the duration of an activity is itself the same kind of activity. Every sub-interval of the burning of a candle is also a burning. Developments are heteromerous or "non-like-parted" in the sense that what is present in any sub-interval of the development is not the same kind of development (on this distinction see Seibt 2008). The larval stage of holometabolous development is not itself a metamorphosis. This implies that the sortals capable of determining boundaries for developments will be complex, specifying a normal or typical series of stages as well as a relation that makes them stages of the same process (see DiFrisco 2017, 2018). In fact, this is what we see when biology texts introduce life cycles so that they can be identified by field workers (Gilbert 2010): not a conserved property or activity, but a normal sequence of stages of a specific sort of causal process.

The idea of organisational continuity is therefore more germane to a process model than a substance model of organism identity. To put this idea to use, however, we need more than just a necessary condition for organism persistence. The next section examines sufficient conditions and introduces distinctions between developmental and reproductive processes.

10.4 Development versus reproduction

One of the primary ways in which reflection on organism identity matters is in connection with the biologically important distinction between development and reproduction. The distinction is particularly difficult in complex life cycles of non-vertebrate multicellular entities such as plants, fungi, colonies of various kinds, and various invertebrates, where we see processes like metamorphosis, metagenesis, apomixis, symbiosis, fission, and fusion (see Fusco and Minelli 2019, chapters 1, 2, for many examples of borderline cases).

As we saw earlier, amid this tangle of generative biological processes, principled attempts to distinguish between development and reproduction face a general dilemma. On the one hand, if criteria of diachronic identity are too strict, they will not be able to include dramatic transformations such as metamorphosis within the same lifetime. On the other hand, if they are too permissive, they will often be unable to close the temporal boundaries of a lifetime, and reproductive events will be tracked as developments of one and the same organism. Organisational continuity is a quite permissive criterion of identity, given that it does not require conservation or resemblance. Accordingly, it can be expected to face difficulties marking reproductive events—or more precisely, fission and fusion.

To a first approximation, one might hold that organisational continuity provides both necessary *and* sufficient criteria for diachronic identity. The idea would be that any temporal interval where continuity is broken marks a temporal boundary of an organism's life, either due to death or birth, which appears to be straightforward from an organisational perspective. An organism dies when it loses the property of closure of constraints, as this is responsible for its capacity to maintain itself far from equilibrium, and is born when a closed regime appears. Any corpse or parts that remain after death, as well as any preceding material constituents will be spatiotemporally continuous but not organisationally continuous with the organism.[5]

To see why this solution generally does not work, recall that organisational continuity between two successive stages is present as long as both stages instantiate the principle of closure of constraints, the stages are spatiotemporally continuous with one another, and the later stage is causally dependent on constraints acting in the previous stage. The problem is that these conditions are satisfied not only between stages of metamorphosis, but also when the earlier stage is a parent and the later stage is an offspring. Parents and offspring are organised systems, they are spatiotemporally continuous with one another, and the offspring is dependent on constraints exerted by the parent, at least in the sense that the very generation of the former depends on constraints exerted by the latter. Hence, parents and offspring are organisationally continuous with one another.[6] If organisational continuity were not only necessary but also always sufficient for diachronic identity, then parents and offspring would count as continuations of the same organism.

This has the unwelcome consequence that whenever a parent continues to exist after reproducing, we would have a spatially scattered system comprising both parents and offspring as parts. This system lacks the right

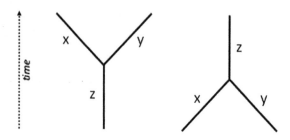

Figure 10.2 Fission (left) and fusion (right).

causal connections between parts to count as an individual organism in its own right, so we have to choose which system to track as a continuation of the pre-reproductive parental stages. Each tracking path satisfies organisational continuity equally well, however. A similar problem follows from treating fusion solely in terms of organisational continuity (see Figure 10.1). Two fusing systems may each individually be organisationally continuous with the fused system, but they cannot both be the same as the fused system, because before fusion they do not together comprise an individual organism (Figure 10.2).

To address this problem, we can modify our criterion by proposing that organisational continuity is sufficient for diachronic identity *unless there is a change in the local number of organised systems that are organisationally continuous with one another*, either via an event of multiplication (fission) or reduction (fusion). When a spatial separation of this sort appears or disappears, we have a temporal boundary and a change of identity.[7]

The issue of spatial separation raises complex problems that go beyond the scope of this chapter, as it would in some sense require a full-blown organisational account of biological individuality. Nonetheless, we can say the following. In general, spatial separation occurs when the involved systems do not realise a global closure with mutual dependence. In borderline cases where it is unclear whether there is one organised system or two, the critical consideration is whether the putatively distinct systems are functionally interdependent in virtue of constraint relationships. If organised systems share functional constraints, then the repair and replacement of the constraints depends ultimately on the action of other constraints in each system. In this case, they are not actually separately closed, and the wider system that includes them both is the one that realises closure of constraints. By contrast, mere causal dependence without constraint relationships does not make the involved systems parts of the same organised whole.

We can clarify and sharpen this characterisation of temporal boundaries by seeing how it deals with examples of fission, fusion, and life cycles that combine both.

10.4.1 Fission

In general, fission occurs when two or more spatially separated organised systems result from a preceding single one, with which they are all organisationally continuous.

Fission allows drawing a temporal boundary in a straightforward way: the preceding system ceases to exist and two new individuals appear. However, this interpretation is not the only possible one, and we need some way of determining whether one of the systems which result from a fission event is a continuation of the life of the preceding system, i.e., whether a parent persists after reproduction.

To tackle this issue, we propose that parent and offspring can be discriminated by the presence of transitory *asymmetrical dependence* relationships. In reproduction, the materials that will constitute the offspring begin as parts of the parent(s). This is expressed in Griesemer's (2000) condition of "material overlap." Eventually what begins as a part of the parent breaks the symmetry of dependence: it depends on the parent but the parent does not depend on it. When there is a transitory relation of asymmetric causal dependence during the process of multiplication of organised systems, the dependent system can be identified as the offspring, whereas the nondependent organised system is the parent. If an asymmetry is not present but there is still multiplication of organised systems—e.g., in binary fission— then neither of the resultant systems is a continuation of the parent and both are offspring (new individuals). In most organisms the asymmetrical parental dependence is quite short-lived on the scale of the organism's lifetime, and tends to cease either when there is a physical separation between bodies or the formation of a metabolic boundary such as an egg. Note the loose connection to spatial contiguity implicit in this characterisation of multiplication. In order for causal influence to propagate from parent to offspring, they must be at least partially contiguous. But they can remain contiguous without the parent continuing to exert constraints on the constitutive self-maintaining processes of the offspring, such as when eggs are carried around by the parent, or when plants like Aspen trees are connected by underground runners.

The simplest and most ancestral form of reproduction of organised systems is cell division.[8] When cell division is "symmetric" in the sense that cell biologists use the term—i.e., the resultant two cells have roughly the same kinds of components and the same size—it is thought that the parent cell ceases to exist upon division and gives rise to two offspring. Cell division can also be "asymmetric" in the sense that the parental components are unevenly distributed to the daughter cells, which are often different in size. When asymmetric division occurs in the cells of multicellular organisms, such as in stem cells, typically the parent cell is again treated as ceasing to exist upon division. But when it occurs in unicellular organisms, such as budding in yeast, it is more common to treat one of the resulting cells as a continuation of the parent.[9] In budding yeast (Saccharomycetales), the parent cell can be distinguished by several features—for example, it is usually larger, bears scars on the membrane where the bud formed, and retains the aging factors such as carbonylated proteins and DNA circles (Mortimer and Johnston 1959, Shcheprova et al. 2008). During the generation of a bud, the parent cell dedicates biosynthesis to the bud, whereas the bud dedicates biosynthesis to its own growth (Shcheprova et al. 2008, p. 728). Arguably this is a form of asymmetrical dependence in our (functional) sense, which marks budding as a reproductive process with an identifiable parent that survives it.[10]

It is quite likely, however, that there are similar asymmetrical dependencies that can be found in asymmetrical cell division *within* multicellular organisms, and so additional factors must be invoked to explain their different biological treatment. The most obvious difference is that although somatic cells and free-living cells are both organised systems, only the latter are able to develop and reproduce without being normally subordinated to the reproduction of a higher-level entity. Yeast can have complex life cycles with alternating asexual and sexual reproduction. Unlike somatic cells, they produce not only diploid daughter cells but also haploid "gametes" which fuse with gametes from another yeast to form another genetically unique diploid yeast cell. This process is distinguished from gametogenesis and syngamy in multicellular organisms because the fusion yields a yeast cell that is itself able to reproduce asexually and sexually, without intermediary stages of cell division and differentiation.

Attention to these differences illuminates why it is productive to view the yeast cell, but not a somatic cell, as an organism with a life history that persists through multiple reproductive events. Drawing the temporal boundaries in this way yields a developmental unit that is an informative base for generalisation and comparison with other developmental units. For example, yeast cells undergo aging and experience trade-offs between growth rate and reproductive rate due to antagonistic pleiotropy, like many developing organisms (Christie et al. 2018). If we stipulated that, like in somatic cells, the parent cell ceased to exist after each budding event, then the life cycle of yeast would become less comparable to these other life cycles in terms of aging and life history strategy, as well as in terms of ecological and evolutionary dynamics of alternating asexual and sexual reproduction. A similar pattern of reasoning could be applied to multicellular life cycles that include stages of asexual reproduction, such as fission by fragmentation in sea sponges, budding in hydrozoa, and cyclical parthenogenesis more generally. The result of this reflection is that in some contexts, the presence of asymmetric dependence between organisms is not sufficient to assign the identities of parent and offspring. We also need to take into account how drawing the temporal boundaries in one way rather than another can provide a more informative and inferentially rich classification scheme for that individual, giving access to generalisations and enabling the application of explanatory models of its developmental processes.

10.4.2 Fusion

Having looked at temporal boundaries in a few salient cases of fission, now we can turn to cases of fusion (Hull 1978, Guay and Pradeu 2016). In contrast to fission, fusion occurs when organisationally continuous systems merge in space in such a way that the persistence of one or both systems is affected.[11] An individual x and an individual y merge to form an individual z (see Figure 10.2). Fusion is not reproduction, but one kind of fusion—syngamy, or fusion of gametes—is part of the process of sexual reproduction, and all kinds of fusion pose issues of temporal boundaries. The question with fission was whether one of the two individuals that are produced is a continuation of the earlier individual (the parent), and with fusion it is the same except reversed in time. In fusion, one of two outcomes can occur:

1. x is the same individual as z and y becomes a part of z during fusion, or y is the same individual as z and x becomes a part of z during fusion; or
2. neither x nor y is the same individual as z, and x and y cease to exist upon fusion.

The first case corresponds to parental persistence after fission, and the second corresponds to parental cessation upon fission.

An outcome of spatial merging that is *not* fusion would be:

3. neither x nor y is the same individual as z, and x and y persist through merging.

Because this does not involve a disruption of the identities of the fusing entities, we will call this "integration" of x and y.

Continuing the parallelism with fission, what happens to biological individuals when they fuse is determined most basically by whether there is an asymmetrical dependence relation between the fusing systems once they have come together in space. When there is, we have outcome (1): the dependent system becomes *part* of the non-dependent system, and the non-dependent system persists through fusion. In Ceratioid anglerfish, for example, when a male finds a female he will bite into her skin and fuse their circulatory systems indefinitely. Males parasitise the metabolism of the females, which are much larger, in exchange for providing a constant source of sperm. Because the male becomes dependent on the self-maintaining processes of the female once they have merged, whereas the female is not similarly dependent on the male, we have an asymmetrical dependence relationship and thus (1). The female persists through fusion and the male becomes a part of her while (presumably) losing its individual closure of constraints. In fact, this assessment agrees with the naïve judgement of early investigators, who mistook the fused males as "lumps" on the female's body and searched in vain to find male exemplars of the group.

An example of fusion in which neither fusing partner persists through fusion (2)—instead becoming parts of a new organised system—would be syngamy, or fusion of gametes (Hull 1978, p. 346). In syngamy, each of x, y, and z is an organised system, and the gametes x and y do not maintain their respective closure once fusion occurs because their functional components get mixed together. This is what distinguishes syngamy from obligate fusion of anglerfish as a case of outcome (1). During fusion, male gametes do not retain sufficient cohesion as cells or even as unified parts of the female gamete to stand in a stable asymmetric dependence relationship[12]. The female gamete is, moreover, dependent on the ingression of the male gamete to initiate an individual process of development, giving an additional reason to consider them parts of a single individual.[13]

What distinguishes between outcomes (2) and (3) is whether x and y cease to be organised systems in their own right during merging. When x and y retain their closure of constraints, then (3) they persist through the merger. Outcome (2) implies that the fused individual (z) is an organised system. But outcome (3) is consistent with the possibility that the merged individual (z) is an organised system (which might actually be taken as a necessary requirement for being an individual from the organisational perspective), or that it is not.

"Integration" can be characterised as a biological relationship in which spatially contiguous partners influence each other's functioning in a way that benefits either some partners (asymmetrical dependence, as in parasitism or commensalism) or all partners (symmetrical dependence, as in mutualism). When the dependence is symmetrical, integration generates a new encompassing organised system that realises closure. An example of (3) that creates a new organised system would be slime moulds. Slime moulds are groups of free-living amoebas that facultatively aggregate and disaggregate depending on the presence of food. The aggregated amoebas each contribute to the self-maintenance of the whole and even undergo temporary differentiation and functional specialisation.

10.4.3 Sexual reproduction

With the preceding characterisations of fission, fusion, and integration in hand, we can now briefly consider life cycles that involve both. Sexual reproduction of multicellular organisms involves a complex succession of events: fission of gametes from parents or parental cells is followed by fusion of gametes (syngamy) which, in many organisms, is shortly followed by integration of the zygote with the mother.

Applying our criteria yields two interpretations of this sort of sexual reproduction (Figure 10.3). Under the first (A), the zygote is part of the mother due to its continuing asymmetrical dependence. The production of a new individual only definitively occurs when this dependence is broken due to the fission of mother and offspring, for example at birth.

A second interpretation (B) would count the zygote as a distinct individual from the mother on the grounds that its appearance constitutes the start of a developmental process. The asymmetrical dependence is an intermediate phase in the development of functional and spatial separation of the same entity, which allows tracking the zygote as the same individual as the offspring at birth. Under this interpretation, gametes form a new individual by syngamy; this zygote persists through its integration with the mother, and is the same individual as the eventual infant.

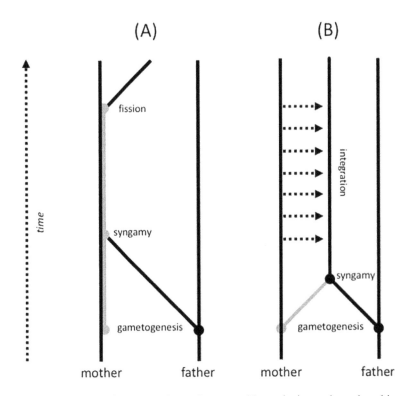

Figure 10.3 Differing interpretations of temporal boundaries and parthood in sexual reproduction in placental mammals. (Grey segments denote parts of the female parent, and circles mark fission or fusion events. Time intervals are not to scale.)

The temporary asymmetric dependence of metabolism between parents and offspring should not preclude our ascribing distinct individuality to the offspring in all contexts and for all purposes of inquiry. In line with interpretation (A), the idea that the developing embryo is *part* of the mother (in placental mammals) has been promoted in recent discussion of the "metaphysics of pregnancy" (Kingma 2019). However, from an evolutionary perspective, a zygote is already a new evolutionary individual due to its genetic uniqueness and homogeneity (Clarke 2013). This new individual plausibly persists through its integration with the mother during implantation in the uterine wall, though it is not yet a physically separate organised system throughout its existence. The distinction between mother and offspring is even more evident from a developmental perspective. The embryo is undergoing completely different ontogenetic processes and stages than those of the mother. For purposes of applying models of human development, then, it is more productive to think of the mother and embryo as distinct individuals. In the terminology of DiFrisco (2019), mother and embryo are parts of the same physiological and ecological individual, but they are distinct evolutionary and developmental individuals. Which classification scheme we choose in different classes of cases should be determined by our theoretical framework and its investigative and explanatory aims.

10.5 Conclusions

In this chapter, we have seen how diachronic identity for organisms can be plausibly grounded in a special kind of causal continuity. In order to capture sameness over time despite radical developmental changes such as metamorphoses, continuity should not be based on conservation of matter, function, form, or activity, but rather on a causal dependence relation called "organisational continuity." For two stages of a developmental process to be organisationally continuous with one another, they must be spatiotemporally continuous, must realise closure of constraints (by whatever particular regime of closure), and the later stage must be dependent on constraints exerted in the earlier state. This permissive conception of organism identity entailed a departure from the substantialist model of persistence due to Wiggins (2001) and others, while also enriching and specifying existing accounts of genidentity due to Hull (1978), Guay and Pradeu (2016).

Organisational continuity is consistent with dramatic developmental change, but it does not differentiate development and reproduction. This is because parent and offspring are organisationally continuous with one another just as the successive stages of parental development are. To distinguish parent and offspring, then, it is necessary to adopt working notions of fission and fusion. We argued that the reproduction of organised systems entails their spatial multiplication, and that the presence of a temporary asymmetrical dependence relationship between systems largely determines whether one of them is a continuation of the life of the parent. With fusion, similarly, the outcome is determined by the presence of asymmetrical dependence as well as by whether the fusing systems retain their separate closures.

With this account in hand, we can now assess it in light of the problems of individuality introduced at the start. On the problem of distinguishing development and reproduction, our account weaves between overly permissive and overly restrictive criteria of diachronic identity while making sense of a wide range of generative phenomena in biology. This provides a productive starting point for thinking about organism identity, but we also saw how, with the examples of reproduction in yeast and placental mammals, alternative temporal boundaries can be promoted from within different perspectives that bring different modelling and generalisation strategies. Further work will be needed to sort through this variation in a more fine-grained way, to determine exactly which empirical situations demand which kinds of temporal boundaries.

Other theoretical aims foregrounded by existing accounts of biological individuality can, in most cases, be fruitfully embedded within an organisational account of organism identity. Part of the reason these other criteria can be superimposed on an organisational account is that the latter starts from the biologically basic process of self-maintenance rather than more complex developmental, ecological, or evolutionary processes. There is a clear sense in which, in order to be able to participate in those dynamic processes, a physical system must first be able to maintain itself far from equilibrium with its environment. In many cases, criteria focused on these other processes will individuate subsets and supersets of organised systems across time, and may also single out different temporal segments of biological process, but closure nonetheless explains what keeps it all going physically. An open task for future work will be to account for how these elements ultimately hang together.

The picture of organism identity outlined here has an interesting implication for classificatory practice in biology. Substantialist views lend themselves to a "phenetic" approach in which the identity of organisms and classes of organisms is determined by qualitative similarity, resemblance, or sameness of properties or activities. This type of approach has historically been associated with classification errors in which stages of the same polymorphic life cycle are assigned to different taxa because of their dissimilarity (Hull 1978, Nyhart and Lidgard 2017). The propensity to make this error can be corrected by shifting from a similarity-based view to a causal continuity-based view of organism identity. Instead of looking for similarity between different stages of development, one looks for the right sort of causal connections between stages of the same four-dimensional developmental process.

Acknowledgments

The authors thank John Dupré, Anne Sophie Meincke, Alessandro Minelli, and Alvaro Moreno for insightful comments. James DiFrisco thanks the Research Fund – Flanders (FWO) and the Konrad Lorenz Institute for Evolution and Cognition Research for financial support.

Notes

1. The fact that so many authors have expressed the shared intuition that spatiotemporal continuity is necessary for object identity is plausibly a reflection of deep-seated psychological mechanisms that allow us to perceive objects as persistent, an insight explored by Scholl (2007). Thanks for Riana J. Betzler for pointing us to this work.
2. During morphogenesis, however, the structure of the vascular system is influenced by the dynamics of blood flow (hemodynamics), as Alessandro Minelli points out (personal communication). See Yashiro et al. (2007) and Kowalski et al. (2013).
3. Note that closure of constraints is importantly different from *causal* closure, a notion that is often invoked in discussions of physicalism or materialism. For more on the connection between physicalism and organisation, see Mossio et al. (2013) and DiFrisco (2017).
4. Curiously, Wiggins maintains that "live things exemplify most perfectly and completely a category of substance that is extension-involving, imports the idea of characteristic activity, and is unproblematic for individuation" (2001, p. 90). This may be true if the task is one of distinguishing organisms from artefacts and other non-living things, but problems of individuality in contemporary biology are more demanding.
5. In practice, it may be that we cannot specify an exact moment of death or birth, but only a temporal interval during which they have occurred. However, the ability to pinpoint an exact time of death does not seem to matter much from a biological perspective.
6. Recently, one of us has argued that organisational continuity among generations is actually a fundamental requirement for biological heredity, understood from an organisational perspective (Mossio and Pontarotti 2019).
7. Note that we are not attempting to define reproduction directly, but to identify temporal boundaries of lifetimes. One such temporal boundary occurs with the spatial multiplication of organisms, which is not quite the same as reproduction. To distinguish reproduction from mere production of a new organised system, it may be necessary that the offspring *resemble* the parent or belong to the same kind (Godfrey-Smith 2009, p. 78). Certainly some form of parent-offspring resemblance, probabilistic dependence, or covariance is necessary in order for reproduction to lead to evolution by natural selection, as Lewontin (1970) pointed out. That being understood, we sometimes still use the term "reproduction" for its familiarity.
8. Many authors think of *replication* as the most fundamental form of reproduction, the paradigm case being replication of nucleic acids. Our characterisation is not intended to cover reproduction or replication in non-organised systems such as nucleic acids, genes, or viruses.
9. Some yeasts, namely "fission yeasts" (Schizosaccharomycetes), reproduce asexually with symmetric cell divisions, so what follows does not apply to them.
10. Recent evidence suggests that some paradigmatic cases of symmetrical cell fission (e.g., *E. coli*) may actually involve asymmetries in the retention of ageing components (see Nyström 2007). Thanks to John Dupré for pointing this out.

11 Another phenomenon sometimes included under the heading of fusion is *conjugation*, a merger between cells in which genetic material is exchanged. Conjugation is a form of sexuality without reproduction (see Fusco and Minelli 2019). Since conjugation does not affect the persistence of organised systems, we do not include it as a case of fusion.
12 It should be noted that, in this framework, fusion is understood as a functional issue, and refers to the merging of constraints between two individuals. For this reason, the situation in which an organism eats another organism does not count as fusion, because the eating organism consumes the eaten organism and no functional integration occurs. The theoretical distinction between constraints and processes outlined above is at work here.
13 Isogamy, an ancestral state of sexual reproduction in which there are no male-female differentiated gametes, fits the interpretation of syngamy as (2) even more naturally. Other examples of fusions (2) would occur during the split-embryo development of certain freshwater fish (Cynolebias) (Minelli 2011, p. 11) and the formation of a syncytium.

References

Bishop, C. D., Erezyilmaz, D. F., Flatt, T., Georgiou, C. D., Hadfield, M. G., Heyland, A., Hodin, J., Jacobs, M. W., Maslakova, S. A., Pires, A., Reitzel, A. M., Santagata, S., Tanaka, K. and Youson, J. H. (2006) 'What Is Metamorphosis?', *Integrative and Comparative Biology* 46(6), pp. 655–661.
Christie, M. R., McNickle, G. G., French, R. A. and Blouin. M. S. (2018) 'Life History Variation Is Maintained by Fitness Tradeoffs and Negative Frequency-dependent Selection', *Proceedings of the National Academy of Sciences* 115(17), pp. 4441–4446, doi:10.1073/pnas.1801779115.
Clarke, E. (2011) 'The Problem of Biological Individuality', *Biological Theory* 5, pp. 312–325.
Clarke, E. (2013) 'The Multiple Realizability of Biological Individuals', *The Journal of Philosophy* 110, pp. 413–435.
DiFrisco, J. (2017) 'Token Physicalism and Functional Individuation', *European Journal for Philosophy of Science* 8(3), pp. 309–329, doi:10.1007/s13194-017-0188-y.
DiFrisco, J. (2018) 'Biological Processes: Criteria of Identity and Persistence', in: Nicholson, D. and Dupré, J. (Eds.), *Everything Flows: Towards a Processual Philosophy of Biology*, Oxford: Oxford University Press, pp. 76–95.
DiFrisco, J. (2019) 'Kinds of Biological Individuals: Sortals, Projectibility and Selection', *The British Journal for the Philosophy of Science* 70(3), pp. 845–875, doi:10.1093/bjps/axy006.
Folse, H. J. III and Roughgarden, J. (2010) 'What Is an Individual Organism? A Multilevel Selection Perspective', *Quarterly Review of Biology* 85, pp. 447–472.
Fusco, G. and Minelli, A. (2019) *The Biology of Reproduction*, Cambridge: Cambridge University Press.
Gilbert, S. F. (2010) *Developmental Biology*, 9th ed., Sunderland, MA: Sinauer Associates, Inc.
Godfrey-Smith, P. (2009) *Darwinian Populations and Natural Selection*, Oxford: Oxford University Press.
Griesemer, J. (2000) 'Development, Culture, and the Units of Inheritance', *Philosophy of Science* 67, S348–S368.

Guay, A. and Pradeu, T. (2016) 'To Be Continued: The Genidentity of Physical and Biological Processes', in: Guay, A. and Pradeu, T. (Eds.), *Individuals Across the Sciences*, New York: Oxford University Press, pp. 317–47.

Hull, D. L. (1978) 'A Matter of Individuality', *Philosophy of Science* 45(3), pp. 335–360.

Kingma, E. (2019) 'Were You a Part of Your Mother?', *Mind* 128(511), pp. 609–646, doi:10.1093/mind/fzy087.

Kowalski, W. J., Dur, O., Wang, Y., Patrick, M. J., Tinney, J. P., Keller, B. B., et al. (2013) 'Critical Transitions in Early Embryonic Aortic Arch Patterning and Hemodynamics', *PLoS ONE* 8(3): e60271, doi:10.1371/journal.pone.0060271.

Lowe, E. J. (2009) *More Kinds of Being: A Further Study of Individuation, Identity, and the Logic of Sortal Terms*, Oxford: Wiley-Blackwell.

Lewontin, R. C. (1970) 'The Units of Selection', *Annual Review of Ecology and Systematics*, 1, pp. 1–18.Minelli, A. (2011) 'Animal Development, an Open-Ended Segment of Life', *Biological Theory* 6, pp. 4–15.

Montévil, M. and Mossio, M. (2015) 'Biological Organisation as Closure of Constraints', *Journal of Theoretical Biology* 372, pp. 179–191.

Moreno, A. and Mossio, M. (2015) *Biological Autonomy: A Philosophical and Theoretical Enquiry*, Dordrecht: Springer.

Mortimer, R. K. and Johnston, J. R. (1959) 'Life Span of Individual Yeast Cells', *Nature* 183, pp. 1751–1752.

Mossio, M., Bich, L. and Moreno, A. (2013) 'Emergence, Closure and Inter-level Causation in Biological Systems', *Erkenntnis* 78(2), pp. 153–178.

Mossio, M., Montévil, M. and Longo, G. (2016) 'Theoretical Principles for Biology: Organization', *Progress in Biophysics and Molecular Biology* 122(1), pp. 24–35.

Mossio, M. and Pontarotti, G. (2019) 'Conserving Functions Across Generations: Heredity in Light of Biological Organization', *British Journal for the Philosophy of Science*, doi:10.1093/bjps/axz031.

Mossio, M., Saborido, C. and Moreno, A. (2009) 'An Organizational Account of Biological Functions', *The British Journal for the Philosophy of Science* 60, pp. 813–841.

Nunes Neto, N., Moreno, A. and El Hani, C. (2014) 'Function in Ecology: an Organizational Approach', *Biology and Philosophy* 29(1), pp. 123–141.

Nyhart, L. K. and Lidgard, S. (2017) 'Alternation of Generations and Individuality', in: Lidgard S. and Nyhart, L. K. (Eds.), *Biological Individuality: Integrating Scientific, Philosophical, and Historical Perspectives*, Chicago, IL: University of Chicago Press, pp. 129–157.

Nyström, T. (2007) 'A Bacterial Kind of Aging', *PLoS Genetics* 3(12): e224.

Queller, D. C. and Strassmann, J. E. (2009) 'Beyond Society: The Evolution of Organismality', *Philosophical Transactions of the Royal Society B* 364, pp. 3143–3155.

Ruiz-Mirazo, K. and Moreno, A. (2004) 'Basic Autonomy as a Fundamental Step in the Synthesis of Life', *Artificial Life* 10, pp. 235–260.

Scholl, B. J. (2007) 'Object Persistence in Philosophy and Psychology', *Mind and Language* 22(5), pp. 563–591.

Seibt, J. (2008) 'Beyond Endurance and Perdurance: Recurrent Dynamics', in: Kanzian, C. (Ed.), *Persistence*, Frankfurt: Ontos Verlag, pp. 133–164.

Shcheprova, Z., Baldi, S. S., Frei, B., Gonnet, G. and Barral, Y. (2008) 'A Mechanism for Asymmetric Segregation of Age during Yeast Budding', *Nature* 454(7), pp. 728–735.

Simondon, G. (1995) *L'individu et sa Genèse Physico-biologique*, Grenoble: Jérôme Millon.
Stearns, S. C. (1992) *The Evolution of Life Histories*, Oxford: Oxford University Press.
Strawson, P. F. (1997) *Entity and Identity and Other Essays*, New York: Oxford University Press.
Wiggins, D. (2001) *Sameness and Substance Renewed*, Cambridge: Cambridge University Press.
Wiggins, D. (2016) *Continuants: Their Activity, their Being, and their Identity*, Oxford: Oxford University Press.
Yashiro, K., Shiratori, H. and Hamada, H. (2007) 'Haemodynamics Determined by a Genetic Programme Govern Asymmetric Development of the Aortic Arch', *Nature* 450, pp. 285–288.

11 Pregnancy and biological identity

Elselijn Kingma

11.1 Introduction

Many philosophers have written on personal identity: the question of what makes us the same person or entity over time. Some will argue that questions about personal identity in humans are indistinguishable from, or at least overlap with, questions about biological identity, because persons are human animals (Olson 1997, Snowdon 2014; see also the chapters by Snowdon and Meincke in this volume). But we can consider the question of biological identity irrespective of animalism: what makes us the same organism over time? And when does that organism begin and end?

In sexually reproducing animals, it may at first glance seem fairly obvious that a sexual recombination event marks the beginning of the organism.[1] Its end is marked by death. But that may be too quick; determining the earliest beginning of mammals is a problem plagued by metaphysical obstacles. Anscombe (1984) famously argued that the possibility of twinning stood in the way of the zygote's/early embryo's being an individual (see also Smith and Brogaard 2003, Oderberg 2008, Rohrbaugh 2014). If this is an obstacle for determining the earliest beginning of human animals, it is surely also an obstacle for other mammals; armadillos always produce monozygotic quadruplets. Nor are metaphysical troubles confined to twinning: marmosets regularly *fuse* zygotes/early embryos, to produce genetic mosaicism in their offspring. Marmosets, as a result, often have multiple genetic (as well as social) fathers.[2]

That said, these metaphysical concerns tend to be confined to the very early stages of embryonic development, before the embryonic disc – the earliest beginning of the actual body – is formed. Once this disc forms and, through gastrulation, begins to take on a three-dimensional structure, it seems fairly reasonable to think that an identifiable organism – a biological identity – has begun to develop, which can be traced through time: through pregnancy, birth, adolescence, and adulthood, until death breaks it apart.

Or so many would think. In this chapter I want to argue that things are more complicated; the metaphysical problems don't stop at the point where fusion and/or fission become impossible. At least in placental mammals[3] there is a further metaphysical complication: candidate organisms in the embryonic and fetal stages develop inside another organism – their gestator. This should focus our attention on a widely overlooked question: the metaphysical relationship between the developing embryo/fetus and its pregnant maternal organism. Until recently, that question has received very little explicit attention. In Kingma (2019) I suggest an explanation: we harbour a widespread but implicit cultural understanding of pregnancy as *fetal containment*. According to fetal containment, one individual – the fetus – resides *merely inside* another individual – the pregnant organism. The fetus is not part of, or overlapping, the gestator. On this implicit and unquestioned understanding of pregnancy, there is no reason to consider maternal-fetal relationship either metaphysically interesting in its own right, let alone relevant to biological identity. But, I argue, this view is mistaken: the placental fetus is part of the pregnant organism. This *parthood view* of pregnancy then raises further questions for biological identity and individuality.

There are multiple ways in which one could argue in favour of the parthood view. In Section 11.2 I argue in favour of the parthood view by repudiating an argument by Barry Smith and Berit Brogaard (2003), which was supposed to support the containment view. I contend that, properly understood, their argument undermines containment and supports parthood instead. In Section 11.3 I summarise a different argument in favour of the parthood view: my (2019) application of criteria for biological individuality to pregnancy. In Section 11.4 I briefly discuss possible implications of the parthood view for biological identity and individuality.

But we need to do some housekeeping before we start. First, I emphasise that I focus on *organisms* in this chapter, rather than on persons, humans, or individuals. This allows us to consider answers without having to consider divisive questions about personal identity or ontology; allows for a unified treatment of mammalian reproduction; and is in keeping with this volume's focus on biological identity. Thus, although the embryological details in this chapter describe *Homo Sapiens*,[4] I invite the reader to constantly remind themselves that this chapter should apply to all placental mammals.

Second, and following Kingma (2018, 2019), I wish to introduce two bits of terminology: the *foster* and the *gravida*. The *gravida* is simply a shorter term for "pregnant organism", based on the Latin designation in medical casenotes. The *foster* is a Danish term that I stipulate denotes what the gravida is pregnant *with*, regardless of its developmental stage. Thus embryos and fetuses are both referred to as *foster*.[5]

11.2 Substance ontology and the parthood view: Smith and Brogaard[6]

What is the relationship between the foster and the gravida? Is the foster part of the gravida, or merely residing within the gravida, without being part of it? Barry Smith and Berit Brogaard (2003) argue that fosters are *not* part of the gravida in the context of an elaborate framework on biomedical ontology developed by Smith and colleagues.[7] Their argument is worth considering for two reasons. First, because theirs is one of the few explicit discussions of this topic and, other than my own work, the only detailed argument that considers maternal-fetal mereology. Second because – or so I will argue – their way of thinking speaks in favour of, rather than against fetal parthood. Note that in discussing this argument, I work with the substance metaphysics conception of the organism that Smith and Brogaard develop. I do not defend or endorse that conception, nor do I here criticise it. Regardless of whether it is correct, it provides a useful explorative start for discussing mereological questions about pregnancy.

Smith and Brogaard's argument is a small part of a larger argument that attempts to determine uncontrovertibly when a mammalian organism starts. They characterise organisms as topologically connected, persisting, physical *objects*, which exhibit certain necessary properties: an organism, they write, is a *substance* in the Aristotelian sense: a "three dimensional spatially extended entity, which exists *in toto* any time it exists at all" (Smith and Brogaard 2003, p. 47). Substances meet six conditions: they (1) can undergo changes; (2) must either stay numerically the same substance or cease to exist; (3) can have spatial parts some of which can be added, lost, and changed over time; (4) have a complete, connected external boundary which separates them from other substances; (5) are internally connected; and (6) are independent entities.[8] Organisms are those substances that meet four additional criteria, which identify them as unified causal systems that are relatively causally isolated from their surroundings: (7) it has an exterior membrane that (9) serves as a barrier; (8) it depends upon the maintenance of an internal climate that falls within a limited range of values;[9] and (10) it has mechanisms to reestablish and maintain that internal climate, and thus itself.

Smith and Brogaard maintain that everything that "satisfies conditions 1-10 [...], is of human descent and a product of normal fetal development, is a human being" (2003, p. 51). Because there is little in their account that is specific to humans, we can modify their statement to "everything that satisfies conditions 1–10 above and is of mammalian/*species x* descent and a product of species-normal development, is a mammal/member of *species x*".

With this account of the organism in place, Smith and Brogaard go on to explain when a human organism begins according to their framework. It is in this context that their argument against fetal parthood emerges.

11.2.1 Substance formation and start of the organism

When does a mammalian organism begin? Smith and Brogaard answer: when the process of *gastrulation* starts, 16 days after conception. Gastrulation marks a substantial change, according to Smith and Brogaard, because during that process the pre-embryo "ceases to be a cluster of homogenous cells and is transformed into a single heterogeneous entity – a whole multicellular individual living being which has a body axis and bilateral" (2003, p. 62). In other words, Smith and Brogaard maintain that human beings do not start *earlier* than gastrulation, because prior to that process, the cells in the blastocyst are not yet specialised in a way that mark them out as part of a larger contained whole. Thus they do not yet depend on each other in the right way to meet criteria 7–10: forming a unified causal system.[10]

What is relevant to this chapter, however, is Smith and Brogaard's argument that human beings cannot start *later* than gastrulation. For this they provide two reasons. The first is *intrinsic* to the foster: any further changes after gastrulation – such as neurulation, the acquisition of organs, hands, feet, etc. – they argue, are not substantial changes that change the foster into a new substance, but changes that the foster undergoes *as* a unified causal system and human being (criterion 1). The second, which is what concerns us, is *extrinsic* to the foster: Smith and Brogaard consider but emphatically reject the idea that the foster is part of the gravida. Birth, they claim, is *not* a substantial change, but *merely* a transition "from one environment into another", like an "astronaut leaving her spaceship" (2003, p. 65). What is their argument? And is it convincing?

11.2.2 The tenant-niche claim

Smith and Brogaard argue against the view that the foster is part of the gravida by invoking the concept of a *niche*. A *niche* "is a part of reality into which an object fits, and into and out of which the object can move" (2003, p. 70).[11] Smith and Brogaard maintain that a substance s can be inside another substance S, without being part of S, iff S contains a *niche* that contains s as an occupier or *tenant*. Gravidae, Smith and Brogaard argue, contain a niche of which the foster is a tenant. Therefore – appearances notwithstanding – fosters are not a proper part of the maternal organism, but instead substances in their own right.

Let's evaluate that. Can a foster move in and out of its niche, the way Smith and Brogaard claim a tenant supposedly can? What a preposterous idea. We cannot take "the bun out of the oven", check it, and stick it back in if it is not fully cooked. (If only!) Birth is irreversible: once a baby is out, it does not go back in – ever. In fact, nobody has even successfully transplanted a foster,[12] and even if we *did* gain that technology, that would not change things. Consider hearts and kidneys; although we have the technology to transplant these, that does not stop us from considering them parts rather than tenants. These organ transplants are possible, but they are exceedingly difficult and risky – and that is precisely *because* they remove and insert parts of organisms which involve severing major connections. The same will apply should we be able to transplant fosters.

It is clear that, on Smith and Brogaard's initial formulation of the tenant-niche relation, fosters and gravidae do not stand in a tenant-niche relationships. But, Smith and Brogaard (2003) do not discuss this initial formulation when discussing the foster. Instead, they posit three further characteristics for the tenant-niche relationship that they claim *do* apply to fosters. These are that niches and tenants (1) do not *overlap* or have parts in common, (2) do not share an external boundary, and (3) must be separated from each other by some liquid- or fluid-filled cavity. As an illustrative example, consider a fish inside an aquarium. Although the fish is *inside* the aquarium it is not *part of* the aquarium; rather it is a tenant in a niche. Not only because fish can be moved in and out of the aquarium, but also because (1) fish and aquarium do not share parts; (2) fish and aquarium do not share an external boundary; and (3) fish are in a (water-filled) cavity in the aquarium.

Based on these criteria, Smith and Brogaard (2003) give two arguments to convince us that fosters are tenants in a niche. First, and corresponding to the second criterion, they assert that a foster has its own, completely connected external boundary, marked by a physical discontinuity between foster and gravida; the foster is at no point topologically connected to the gravida. Second, and corresponding to the third criterion, they focus on the role of the amniotic cavity as "surrounding" the foster.

11.2.3 Testing the tenant-niche claim: boundaries

Let's assess whether, despite their not meeting the initial formulation of the tenant-niche relationship, fosters should be considered a tenant in a niche according to these new arguments and additional criteria. But before this assessment can be executed, we need to have two further bits of information. First, we must know what the *boundaries* of the foster are that the tenant-niche criteria refer to. Second, we must understand the concept of a *fiat boundary*.

What are the boundaries of the foster? The answer to this question is less than straightforward, and this is not something that Smith and Brogaard (2003) are explicit about. Rather than giving a definite answer, I will identify the three most plausible candidates for delineating a foster. Instead of picking one, I will argue for each of them that Smith and Brogaard's arguments and criteria don't apply, and thus that the foster fails to be a tenant in a niche. These three conceptions of the foster are:

1 "*Future Baby*" (FB), where the foster only comprises the parts that emerge as the future baby: the (future baby's) body, circumscribed by its skin and stopping at the umbilicus or some way along the umbilical cord;[13]
2 "*Baby with Placenta*" (BP), where the foster comprises the "future baby" plus the umbilical cord and placenta; and
3 the "*Chorionic Content*" (CC), where the foster comprises future baby, umbilical cord and placenta, as well as the chorionic and amniotic membranes and all their contents, including, e.g., amniotic fluid.

Before I can examine these three conceptions, it is helpful to understand Smith and Brogaard's concept of the "fiat boundary":

> Fiat boundaries are boundaries that correspond to no underlying physical discontinuities. Examples are found above all in the realm of arbitrarily demarcated geospatial entities such as postal districts, census tracts, or air traffic corridors.
>
> (2003, p. 72)[14]

As an example, the closed door to my office marks a real or *bona fide* boundary between my office and the hallway; when I open my door there is merely a *fiat* boundary between office and hall.

Smith and Brogaard's core arguments in favour of the foster's tenant-niche relationship to the gravida rely on the existence of actual, clear, and complete external boundaries. As they themselves state:

> [...] if the foster is connected to the mother – if, in other words, the boundary between the foster and the mother is a matter of fiat and not of bona fide boundaries – then the foster cannot stand to the mother in the niche-tenant relation [...].
>
> (2003, p. 73)

As the most expedient way of repudiating their claim about fosters being tenants, then, I will demonstrate that for each conception of the foster its supposed boundary very clearly is one that incorporates a section that is merely a *fiat boundary*.[15]

First, consider the *Future Baby* conception, where the foster has a boundary at the umbilicus or a bit further along the umbilical cord. During pregnancy, this clearly marks a *fiat boundary*; a physical discontinuity will only appear once the umbilical cord is severed after birth, but does not exist during the pregnancy. The umbilical cord is a clear example of topological connection, and moreover one that marks a functionally and metabolically essential connection. On the *Future Baby* view, the foster does *not* stand in a tenant-niche relationship to the gravida.

Second, consider the *Baby with Placenta* conception. Here the foster has a boundary at the maternal side of the placenta: somewhere inside the spongal maternal deciduous tissue – perhaps at the rough site of the placenta's future separation. This, again, is a *fiat boundary*; after birth, the placenta will detach and there will be a physical discontinuity between placenta and the (previously) pregnant organism. But before birth, there is no boundary here; the placenta is not a clearly defined mass with a smooth surface surrounded by a membrane the way that brains, kidneys, or lungs are. Instead, the placenta comprises tissues of maternal and fetal origins and grows direct into (or out of) the uterine wall. The interface between the placenta and the womb is best thought of as a zone, a zone that is not only marked by the intermingling of fetal and maternal tissue, but also traversed by not one but *many* functionally and metabolically essential arteries and veins over a large surface. In other words, the placenta has all the hallmarks of topological connection. A further testament to this is that the wound created upon placental detachment, despite its being a physiologically and biologically functional event, is of such severity that, even in our modern age of blood transfusions, it is one of the main causes of maternal death. On the *Baby with Placenta* view, the foster is not a tenant of the gravida.

Finally, on the *Chorionic Content* conception, the foster still has a *fiat boundary* on the maternal side of the placenta, just as it does on the *Baby with Placenta* view. For the placenta is part of – and in a sense even outside – the chorion. This view, then, inherits the problem of the *Baby with Placenta* view: on the *Chorionic Content* view, the foster is not a tenant of the gravida.

On none of the plausible conceptions of the foster does the foster stand in a tenant-niche relation to the gravida. On each of these views, the boundary of the foster involves a "fiat boundary" – a boundary that is not marked by a physical discontinuity, but recognised by us for other reasons. The existence of fiat boundaries means that the three additional criteria of standing in a tenant-niche relation are not met: foster and gravida *do* share overlapping parts (either at the level of umbilical cord, or at the level of the placenta/uterine wall); they share an external boundary at these locations; and they are not fully (but only partially) enveloped by a fluid-filled cavity. Nor do Smith and Brogaard's two arguments, which rely on a complete external boundary, marked by a physical discontinuity, and on the "surrounding" of the foster by a fluid-filled cavity, apply. The foster does *not* have its own, completely connected external boundary, which is marked by a physical discontinuity between foster and gravida; instead the foster is very clearly topologically connected to the gravida at the locus of umbilical cord or placenta, just like a tail is to a cat, your testes to your body, or your kidney to the rest of the organism. Second, the amniotic cavity only partially "surrounds" the foster, which remains topologically connected via a "stalk". Partial surrounding with stalk connection can*not* be the sign of tenant-niche status; such morphology is entirely common in mammals: hearts, lung, kidney, brain, pancreas, intestines, and so on are all suspended in fluid-filled cavities, but connected by a "stalk".

To conclude, either foster and gravida do not stand in a tenant-niche relation – meaning that instead fosters are parts of gravidae – or Smith and Brogaard must have had another conception of the foster in mind than the three I just outlined. Such a conception of the foster – that is free of fiat boundaries – would have to delineate the foster in such a way that no boundary is drawn anywhere between umbilicus and uterine wall. I, for the life of me, cannot imagine what such a conception would be – but I am open to suggestions.

11.2.4 Interim conclusion: the parthood view and substance ontology

I have repudiated Smith and Brogaard's argument *against* the parthood view: if we take an actual look at physiology, fosters, despite their assertions, do not meet any of their own criteria for being a tenant in a niche. Quite the opposite, such a look supports the conclusion that, by their own criteria, the foster is part of the gravida on Smith and Brogaard's substance-ontological account of the organism. On these grounds I offer the interim conclusion that fosters are part of their gestating organisms, until birth.

But that is not the only argument in favour of the parthood view. In other work (Kingma 2019) I argue that the foster is part of the gravida, not because of considerations specific to the substance ontology that Smith and Brogaard offer, but because of considerations that emerge from the literature on organisms/biological individuality. I summarise these arguments in the next section.

11.3 Organism and the parthood view: philosophy of biology

If we had an uncontested account of what is and is not part of the organism, then it would be easy to test whether fosters are or are not part of the gravida. But we do not have such an account; the literature on organism is a heavily contested terrain. There are many competing accounts on the table and a strong suggestion that a pluralist stance towards accounts of the organism may be appropriate.

Rather than focusing on one specific, inevitably contested, account of the organism, my (2019) therefore considers four distinct criteria that frequently recur in work on the delineation of organisms (Pepper and Herron 2008). Not only does this avoid committing to a specific account, it also avoids another problem: accounts of the organism tell us what entities are (and are not) organisms. But they are not – or at least not in first instance and not always – designed to tell us what is and isn't *part* of a given organism, or where, precisely, its boundaries are. That said, criteria for determining what organisms are, do as a matter of fact provide some insight into what is part of that organism (Kingma 2019). For these reasons I note:

> In focussing on these criteria, I do not commit to the view that any one (or any combination) of them provides either a necessary or a sufficient condition for being a part of an organism. I commit only to the claim that something meeting all of these criteria has a very strong initial case for being part of an organism.
>
> (Kingma 2019, p. 623)

11.3.1 Four criteria for organisms

The first criterion is "homeostasis and physiological autonomy". Organisms possess an internal environment that they actively maintain in a state of relative homeostasis, within a narrow range of parameters. This can be contrasted with their external environment, where much larger variations in conditions can be tolerated. It seems plausible that whatever is within the internal, homeostatic, environment is part of the organism; what lies outside it is not.

The foster appears to be within the internal environment of the gravida for two reasons. First, because of its precise location; the foster does not reside in the uterine cavity, as a tenant in a niche, but is implanted in the uterine wall. At least in its early stages of development it is completely covered by maternal deciduous tissue. Second, because the foster appears to be homeostatically regulated by, and within the context of, the rest of the gravida. Thus the entire entity can maintain its internal environment within the narrow range of parameters that are compatible with life. This, I argue in my (2019), is consistent with recognising that the foster has a certain degree of physiological autonomy. For the foster still relies on the rest of the gravida for many of its important physiological functions, including the regulation of the pregnancy. Moreover, the existence of parts with their own internal environments, and so with some degree of physiological autonomy, is a universal feature of multicellular organisation.

The second criterion is that of metabolic unity and functional integration. Just as the foster is homeostatically regulated as part of the gravida, I argue, it is metabolically integrated with other parts of the gravida, such that the gravida, including fetus, forms one single metabolic system. Not only does the foster depend upon the gravida for several of its metabolic activities, but the gravida actively *integrates* the foster into its metabolic system, making anatomical and metabolic adjustments that are both functional and adaptive. Pregnancy is a functional and key part of the female mammalian life cycle, and successful maintenance and conclusion of the pregnancy is a complex interplay between the foster and other regulatory parts of the gravida that remains poorly understood.

These considerations also speak in favour of functional integration. The foster and the rest of the gravida – or, if one wants, the gravida inclusive of the foster – works as one functional unit towards a common goal: successful reproduction. That does not mean that the evolutionary interests of the foster and the rest of the gravida completely coincide. Because these only share half their genome, genetic inclusive fitness may come apart and there is both potential for and a reality of evolutionary genetic conflict. A gravida may better serve the promotion of its inclusive fitness by investing comparatively less in its present foster, instead spreading out its resources and investment over a lifetime of reproduction, for example, either by investing more in older, born offspring, or by saving resources for future pregnancies. The foster, by contrast, may wish to claim comparatively more of the gravida's resources for itself.[16] But this does not affect the point. Although, as we will see, the potential for genetic conflict generates interesting questions for biological individuality, the criterion under consideration is not *complete coincidence of evolutionary interest*, but functional integration. And the high degree of integration of the foster within the rest of the gravida, and the functional, adaptive accommodation of the foster by the gravida speak in favour of the parthood view.

A third criterion for biological individuality is topological continuity. Here we can be quick; our discussion in Section 11.2 settles this criterion: there is no topological discontinuity between foster and the rest of the gravida. As we saw, placenta and umbilical cord grow directly out of the foster's abdomen and into/out of the maternal uterine tissue.

Fourth, there are immunological criteria that are frequently appealed to for delineating and identifying organisms (Tauber 1994, Pradeu 2010, 2012). These, too, speak in favour of the parthood view; mammals interactively tolerate fosters. This active immunological non-rejection signals acceptance by the gravida of the foster as one of its parts (Howes 2007, Pradeu 2012).

Thus, I argue in Kingma (2019) that these four recurrent criteria for the delineation of organisms deliver a verdict in favour of the parthood view. The foster is part of the gravida, at least up until birth.

11.4 The parthood view and biological identity

This chapter has presented multiple arguments in support of the parthood view: the view that the foster is part of the gravida. What does this mean for biological identity? For the beginning, end, and transtemporal continuity of the organism?

11.4.1 The beginning at birth view

One may think that the parthood view has direct implications for when mammalian organisms begin, because one holds the view that no organism could be part of another organism. On such a view, the parthood view entails that mammals cannot precede their birth. Formally:

(P1) An organism cannot be part of another organism.
(P2) Fosters are part of another organism (gravida). [parthood view]
=> (C1) Fosters cannot be organisms.

This conclusion – that mammalian organisms only begin at birth – is certainly one that Smith and Brogaard appear to be committed to if I am correct about fetal parthood. Their substance ontology explicitly commits to the view that substances – and hence organisms – cannot be part of other substances/organisms.[17] So, by their own criteria, fetal parthood commits them to a "beginning at birth" view. Contrary to their stated claims, then, birth *is* a substantial change: at birth fosters cease to be, and new substances – baby organisms – come into existence. Human beings do not begin 16 days after conception as they claim, but, usually, nearly eight-and-a-half months later: at birth (Kingma forthcoming).

But such a view is surely highly counterintuitive. Surely, on any plausible view of biological and numerical identity, newborn mammals were fosters only seconds before? Surely dogs gestate canine fosters that become puppies and later dogs themselves? Some, no doubt, will take this to be further evidence that substance ontologies are doomed and must be replaced with a process ontology (Meincke this volume and under review, Dupré this volume). Others may wish to make different adjustments; not everyone thinks a substance ontology must be committed to the view that substances can't be part of other substances (e.g., Hoffman and Rosenkrantz 1994).[18]

11.4.2 Organisms as parts

If one grants – contra Smith and Brogaard – that organisms can be parts of other organisms, then the parthood view does not threaten the intuitive view that mammalian organisms start at or not long after a sexual recombination event, and develop and persist throughout pregnancy, birth, and childhood. One can retain, on such a view, biological identity throughout pregnancy – as parts – and after pregnancy, as non-parts.[19] Still this view generates interesting questions.

First, the literature on biological individuality appears fairly comfortable with the idea that organisms are parts of other organisms. Considerations of our gut microbiome, for example, suggest that our microbiomes should be considered part of us (Dupré and O'Malley 2009, Hutter et al. 2015; for dissenting voices see Queller and Strassmann 2016, Skillings 2016). If we are organisms, on this view, then we have many organisms as our parts. But, as I discuss briefly in my (2019), this starts to raise questions that take us straight to the heart of contemporary debates over levels of selection, biological individuality, and the nature of organisms. For how do we determine the difference between a symbiotic collaboration between organisms and a parthood relation? And how do we settle this for pregnancy? And – finally – how is this affected by the fact that, in pregnancy, the organisms involved are of the same kind?

Second, mammalian pregnancy appears to raise questions that are related to debates about growth versus reproduction (Janzen 1977). These are questions about the delineation of individual organisms of the same species, including distinguishing parents and offspring. Usually these questions have been considered in the context of plants, fungi, and asexually reproducing animals. A good example is the layering of strawberry plants. But perhaps mammalian pregnancy can be usefully investigated here. For although there can be no doubt that pregnancy must culminate in reproduction, that does not settle at what point a part becomes independent; it may be that there is a period of growth before actual reproduction happens in the form of breaking off. Such a picture is certainly appropriate for several non-sexual reproducing life forms.

Finally, once we accept that mammals begin and develop as parts of other mammals – and more generally that organisms can be parts of other organisms – then a next question is whether this relation is restricted to pregnancy. What about breastfeeding? What about other high-investment parenting strategies? Can they, too, present metaphysical questions of the kind discussed here?[20]

All in all, whether substance or process, close considerations of pregnancy, and the maternal-fetal relationship raise interesting questions. These deserve far more attention than has been customary in discussions of biological identity and individuality so far.

Acknowledgements

I am to grateful to Anne Sophie Meincke, Suki Finn, and John Dupré for useful comments on this chapter, and to Cambridge University Press and the editors of the Royal Institute of Philosophy Supplement for letting me partially reproduce my (2018) here. This chapter is part of a project that has received funding from the European Research Council (ERC) under the European Union's Horizon 2020 research and innovation programme, under grant agreement number 679586.

Notes

1 Clarke (2010) cites several accounts of biological individuality that pinpoint either sexual recombination or a bottleneck stage as marking out a new organism. In mammals the two criteria coincide.
2 See also Meincke (this volume) for a discussion on twinning.
3 Placentals are the subset of mammals that have prolonged placental pregnancies, to be contrasted with marsupials (e.g., wombats) and monotremes (e.g., platypus). I use "mammal" to mean "placental mammal" throughout the chapter.
4 Although embryogenesis is similar and remarkably robust amongst placental mammals (indeed amongst all vertebrates), there are relevant differences. Humans mostly have singleton pregnancies, for example, whereas many other mammals do not. Placental physiology also shows considerable variation. If this affects my arguments then my claims are restricted to human mammals and other mammals with sufficiently similar placental physiology.
5 I do not want to suggest that "foster" is a morally or even metaphysically unified category; there are many relevant and interesting differences between zygotes, term fetuses, and the many stages in between. But for the purposes of this chapter, i.e., investigating the relationship between the foster and gravida *during* pregnancy, they can be collapsed into one category (though with a possible exception for pre-implantation – see Kingma (2019), pp. 639f.).
6 This section reproduces material that appeared earlier in Kingma (2018).
7 See listings on http://ontology.buffalo.edu/smith/.
8 In the sense that they do not require other entities as their "bearers" or "carriers", such as a smile which needs a face to bear it.
9 This criterion bears a strong resemblance to traditional ideas of homeostasis.
10 Oderberg (2008) argues, I think correctly, that differentiation and causal unity actually precede gastrulation.
11 See also Smith and Varzi (1999).
12 After implantation, that is. Before implantation this is a different matter, as IVF illustrates. If and how the arguments apply prior to implantation will have to be addressed another time.

13 Oderberg (2008) notes that in discussions of the metaphysical status of fetuses it is conventional to take the *Future Baby* view of the foster (he calls it the *fetus proper*). But, he says, there is nothing inconsistent about taking something like the *chorionic content* or *baby with placenta* view.
14 See also Smith and Varzi (2000).
15 For a more detailed version of this argument, see Kingma (forthcoming).
16 This is one explanation for why (pathological or obstetric) maternal-fetal conflict is so much more frequent than "maternal-kidney" conflict (see, for example, Haig 1993). Yet such maternal-fetal conflict is still pathological, promoting neither the foster's nor the gravida's fitness.
17 See, e.g., Smith and Brogaard (2003, p. 47), criterion 4: "Substances are distinguished [...] from the undetached parts of substances. The latter can become substances, but only through becoming detached [...]." This they reassert, e.g., (2003, p. 53): "We might attach a new tail to a tailless cat. Before the attachment, cat and tail are separate substances. As a result of the attachment, what had been a separate substance is now a part of the cat."
18 Kingma (2018) discusses questions that arise for the substance ontologist who wishes to accommodate both fetal parthood and the numerical identity of fetuses and babies.
19 For a critical discussion of this view, see Meincke (under review).
20 I am grateful to John Dupré and Samir Okasha for suggesting these questions to me.

References

Anscombe, G. E. M (1984) 'Were You a Zygote?', *Royal Institute of Philosophy Lecture Series* 18, pp. 111–115.

Clarke, E. (2010) 'The Problem of Biological Individuality', *Biological Theory* 5, pp. 312–325.

Dupré. J. (this volume) 'Processes Within Processes: A Dynamic Account of Living Beings', in: Meincke, A. S. and Dupré, J. (Eds.), *Biological Identity. Perspectives from Metaphysics and the Philosophy of Biology*, (*History and Philosophy of Biology*), London: Routledge, pp. 149–166.

Dupré, J. and O'Malley, M. (2009) 'Varieties of Living Things: Life at the Intersection of Lineage and Metabolism', *Philosophy, Theory and Practice in Biology* 1(3), http://hdl.handle.net/2027/spo.6959004.0001.003.

Haig, D. (1993) 'Genetic Conflicts in Human Pregnancy', *The Quarterly Review of Biology* 68, pp. 495–532.

Hoffman, G. and Rosenkrantz, G. S. (1994) *Substance Among Other Categories*. Cambridge: Cambridge University Press.

Howes, M. (2007) 'Maternal Agency and the Immunological Paradox of Pregnancy', in: Kincaid, H. and McKitrick, J. (Eds.), *Establishing Medical Reality: Essays in the Metaphysics and Epistemology of Biomedical Science*, Dordrecht: Springer, pp. 179–198.

Hutter, T., Gimbert, C., Bouchard, F. and Lapointe, F. (2015) 'Being Human is a Gut Feeling', *Microbiome* 3: 9, doi:10.1186/s40168-015-0076-7.

Janzen, D. (1977) 'What Are Dandelions and Aphids?', *The American Naturalist* 111, pp. 586–589.

Kingma, E. (2018) 'Lady Parts. The Metaphysics of Pregnancy', *Royal Institute of Philosophy Supplement* 82, pp. 165–187.

Kingma, E. (2019) 'Were You a Part of Your Mother?', *Mind* 128(511), pp. 609–646, doi:10.1093/mind/fzy087.

Kingma, E. (forthcoming) 'Nine Months', *The Journal of Medicine and Philosophy*.

Meincke, A. S. (this volume) 'Processual Animalism: Towards a Scientifically Informed Theory of Personal Identity', in: Meincke, A. S. and Dupré, J. (Eds.), *Biological Identity. Perspectives from Metaphysics and the Philosophy of Biology*, (*History and Philosophy of Biology*), London: Routledge, pp. 251–278.

Meincke, A. S. (under review) 'One or Two? A Process View of Pregnancy'.

Oderberg, D. S. (2008) 'The Metaphysical Status of the Embryo: Some Arguments Revisited', *Journal of Applied Philosophy* 25, pp. 263–276.

Olson, E. (1997) *The Human Animal: Personal Identity Without Psychology*, Oxford: Oxford University Press.

Pepper, J. W. and Herron, M. D. (2008) 'Does Biology Need an Organism Concept?', *Biological Review of the Cambridge Philosophical Society* 83, pp. 621–627.

Pradeu, T. (2010) 'What Is an Organism? An Immunological Answer', *History and Philosophy of the Life Sciences* 32, pp. 247–267.

Pradeu, T. (2012) *The Limits of the Self: Immunology and Biological Identity*, transl. E. Vitanza, Oxford: Oxford University Press.

Queller, D. C. and Strassmann, J. E. (2016) 'Problems of Multi-species Organisms: Endosymbionts to Holobionts', *Biology and Philosophy* 31(6), pp. 855–873.

Rohrbaugh, G. (2014) 'Anscombe, Zygotes and Coming-to-be', *Noûs* 48, pp. 699–717.

Skillings, D. (2016) 'Holobionts and the Ecology of Organisms – Multi-Species Communities or Integrated Individuals?', *Biology and Philosophy* 31, pp. 875–892.

Smith, B. and Brogaard, B. (2003) 'Sixteen Days', *Journal of Medicine and Philosophy* 28, pp. 45–78.

Smith, B. and Varzi, A. (1999) 'The Niche', *Noûs* 33, pp. 198–222.

Smith, B. and Varzi, A. (2000) 'Fiat and Bona Fide Boundaries', *Philosophy and Phenomenological Research* 60, pp. 401–420.

Snowdon, P. (2014) *Persons, Animals, Ourselves*, Oxford: Oxford University Press.

Tauber, A. I. (1994) *The Immune Self: Theory or Metaphor?* Cambridge: Cambridge University Press.

12 Processual individuals and moral responsibility

Adam Ferner

12.1 Introduction

The organism is a process, not a substance. This is the view endorsed by a growing number of analytic philosophers of biology (most prominent among these, of course, are John Dupré and his PROBIO group at the University of Exeter). And we – you and I – are organisms. Given that the identity relation is transitive, we seem to be led to the conclusion that we are processes too. This is a thought that finds explicit expression in the work of one of Dupré's colleagues, and the editor of this collection, Anne Sophie Meincke. "Human persons are biological higher-order processes..." (Meincke 2018, p. 454).

It is a curious claim and one that bears closer scrutiny. Is it metaphysically road-worthy? Does it make sense for me to think of myself as a process? The aim of the first section of this chapter is to examine the metaphysical proposal in greater depth and to re-examine and re-state some of the objections to it raised by David Wiggins. The prime focus here, however, is not abstract ontological musings. Following the work of Sally Haslanger, Judith Butler and, in certain moods, Dupré himself, the intention is to investigate the *political* repercussions of this processual view of biological individuals. More specifically, I will ask, and try to answer, how the posited processual individual fits – or rather fails to fit – into our models of moral responsibility.

This chapter is motivated by a concern with an idea of apolitical metaphysics. Analytic metaphysics aspires to describe reality as it really, truly, honestly, actually is. Which beings are the most fundamental? Which entities exist? *What is reality like?* Interesting as these questions are, they are not, by themselves, particularly important. It matters very little, I think, whether processes actually *are* more fundamental than substances, or if Pegasus subsists, or if the Queen perdures. I *do* care, however (and I think you should too), how our claims about what's true and what's not are mobilised in the ethical and political sphere. I care about the way metaphysics is entangled with ideological biases – and I care about the lives our metaphysicians live outside the seminar room.

The first half of this chapter is, I'm afraid, rather negative. But process philosophers, or those sympathetic to the process picture, should persevere – I promise the second half is much nicer. I begin by suggesting that processual individuals fail to mesh with our everyday notions of moral responsibility, but I end by saying that this might be a *good thing* and that the process view of organisms has a liberatory potential that the neo-Aristotelian substance view lacks. I provide no metaphysical support, nor transcendental defences for either position – but I do not, as I say, care very much about that side of things. I care about what, following Butler, we might call the "strategic power" of metaphysics, and the aim here is to draw out some of the politics latent in contemporary process philosophy, and the avenues for political thought that it may open.

12.2 Substance and process

"[T]he world – at least insofar as living beings are concerned - is made up not of substantial particles or things, as philosophers have overwhelmingly supposed, but of processes" (Dupré and Nicholson 2018, p. 3). This is how John Dupré and Daniel J. Nicholson start the introduction to their recent edited collection, *Everything Flows: Towards a Processual Philosophy of Biology* (Nicholson and Dupré 2018). The debate – about what living beings actually *are* – is often construed in these, oppositional terms, a choice between *things* (or *substances*) and *processes*. But what are these different entities?

They have, we are often told, different metaphysical characters. A substance, a *primary* substance as Aristotle would have had it, is something like this man, this woman, or that boy, or that cat, or this dog – some stable, determinable *thing* that persists, in its entirety through space and time. David Wiggins is one of the most well known and most subtle of defenders of today's neo-Aristotelian substance metaphysics (see also, e.g., Lowe 2002, Michael Thompson 2008 and Helen Steward 2015). In his essay – titled, helpfully, "Substance" (1995) – he describes how our daily interactions are premised on the primitive idea of a "persisting and somehow basic object of reference" (1995, p. 41). Our conceptual scheme contains

> the thought, which we do not know how to do without, that we can gradually amass and correct a larger and larger amount of information about one and the same thing, the same subject, and can come to understand better and better in this way how these properties intelligibly cohere or why they arise together.
>
> (Wiggins 1995, p. 216)

Here, in this proudly descriptivist statement (more on this later), Wiggins aims to revitalise Aristotle's ancient notion of substance, *ousia, on hupokeimenon*, etc.; a substance is an entity with a particular ontological character; it is informed by "a principle of activity" (it persists and behaves in a specific kind of way); it is neither in anything else nor predicable of anything else; it is conceptually *prior* to our concepts of colour, shape, etc. (again, more on this, below); it exhibits *internal cohesiveness* (it is *unified*). These are the metaphysical features of the entities that we trade with in our daily lives, be they cats or dogs or other people.[1] We could not get up in the mornings, or find our way to work, without the *substance* concept shaping our thoughts.

Broad brushstrokes, but that is roughly what Wiggins and the neo-Aristotelians think a substance is. And given Wiggins's views about personal identity (which I have written about in my 2016), he also believes that, as organisms, we – you and I – are substances too.

What about processes? Well, we are familiar with a whole host of processes in our everyday lives: the process of making a coffee in the morning, the process of necrosis or maturation (think of those sped-up films of rotting fruit), the process of thinking, jumping, taking a bite out of an apple. These phenomena have a very different metaphysical character to *things* or *substances*. We often think they happen *to* substances (you, for example, are currently undergoing a process of maturation). They are dynamic (constantly moving), temporary eddies in a flux of change (as characterised in e.g., Dupré 2014).

Roland Stout thinks that processes are a type of "occurrent" (Stout 2016). An occurrent is a being that *occurs*, or *happens* – of which the paradigm is an event, like a football match. Unlike a substance, such phenomena are extended *through* time. A football match has *temporal parts* (e.g., its kick-off) in the same way that we have spatial parts (limbs in different spatial locations). Unlike a continuant – or, at least, a continuant as classically conceived by the substantialists – a football match does not progress, temporally, in its entirety. Rather, it has different parts in different times: its start stands at 4pm and its end (the penalty shoot-out) occurs at 5.30pm. As Helen Steward makes clear in "What Is a Continuant" (2015), events do not seem to change – because they do not pass through time. Their temporal parts are fixed.

> The event itself does not change, any more than an apple changes which is redder on one side than on the other. It is merely that some of its parts – in this case, temporal, as opposed to spatial parts – possess properties which are different from those of certain other of its parts.
> (Steward 2015, p. 113)[2]

Consider the process by which some zygote developed into a human adult. According to one understanding that puts processes on a par with events, the process itself did not continue *in its entirety* through time. There was the "germination" part of the process (very near the start), then the "puberty" part and so on. These parts of the process are not all present at once in the way one's spatial parts are.

What else? As Wiggins points out, a process "can be rapid or regular or staccato, or steady. It can even be cyclical and lifelong" (2016, p. 270 = This volume, Chapter 9, p. 167). So there is some debate, hinted at here, about the extent to which processes – in contrast to events – can be said to change through time. One might think, as I believe Wiggins, Stout and Steward all do, that a process's parts might be said to be, e.g., faster or slower – and might then be said to be capable of losing properties (and thus capable of change) – but I think all would agree that, whatever the final analysis, such parts cannot be said to change through time in the same way as continuants classically conceived by the substantialist philosophers.[3,4]

These, then, are the two options pointed to by Nicholson and Dupré. We are substances. Or we are processes. In arguing about which is which and who is what, the substance and process metaphysicians tend to disagree on three main points. The first, which I will only mention into, is the degree to which they are actually disagreeing. Wiggins writes often about the importance of flux, and constant activity, for his view of *substance* – and he quotes, constantly and emphatically (e.g., Wiggins 1980, 2001, 2016), from J. Z. Young's *An Introduction to the Study of Man*:

> The essence of a living thing is that it consists of atoms of the ordinary chemical elements we have listed, caught up into the living system and made part of it for a while. The living activity takes them up and organizes them in its characteristic way.
>
> (Young 1971, p. 86f.)

Personally, I am sympathetic to Wiggins's view here – but I am content, for present purposes, to overlook his processual leanings (or the processualists' substantialist ones).

The second, more relevant point of conflict is over whether processes or substances are more *fundamental*. Dupré and Nicholson (2018), Johanna Seibt (2017) and others – broadly following in A. N. Whitehead's footsteps – claim that, in the hierarchy of being, *processes* are more fundamental than substances (and that we are more fundamentally processes than substances). Wiggins, by contrast, endorses a position similar to one once favoured by Dupré himself, and advocates a kind of metaphysical pluralism – which gives him the option of saying (as he does) that, at the end of the day, we are still substances (Wiggins 2016). His reasons for taking this line are of a piece with his descriptive metaphysics and are considered in the next section.

It is his descriptive approach – or "linguistic philosophy", as Wiggins sometimes calls it – that also lies behind the third point of conflict. The process philosophers say that we are processes. The descriptive metaphysicians, like Wiggins, say that we simply cannot think of ourselves as having a processual character – and our inability to think of ourselves thus is *metaphysically significant* and stands as a tangible obstacle to the process position. This is the focus of Sections 12.4 and 12.5.

12.3 Fundamentality

"[T]he minimal condition for a position to count as a form of process ontology is that processes must be, in some sense, *more fundamental than things*" (Dupré and Nicholson 2018, p. 4; emphases added).[5] That is Dupré and Nicholson again. But in which sense are processes more fundamental than things? How do they (and might we) analyse *fundamentality*?

They state their meta-metaphysical method at the start of their introduction to *Everything Flows*:

> Scientific and metaphysical conclusions do not differ in kind, or in the sorts of arguments that can be given for them, but in their degree of generality and abstraction. If it turns out that process is indeed the right concept to make sense of nature, then this is as good a reason as we can expect for taking nature to be ontologically composed of process.
> (Dupré and Nicholson 2018, p. 4)

This, I suppose, is why they describe their project as falling within the somewhat ludically titled "metaphysics of science" (ludic, because metaphysics is by definition the metaphysics of *everything*). Their meta-metaphysical approach is science-based (or, ungenerously, "scientistic").[6] They look to the kinds of posits demanded by biology (and physics and chemistry) and thereupon write their ontological call sheets. The *process* concept is, they say, much better than the *substance* concept when it comes to making sense of the natural world – and they substantiate this claim (persuasively, in my opinion) by looking at metabolic turnover, life cycles and ecological interdependence (Dupré and Nicholson 2018). Based on the primacy of process in these phenomena, they draw the above metaphysical conclusion. *Processes are more fundamental than substances.*

This is all well and good – as far as it goes. There are, however, some not inconsiderable objections to this claim about fundamentality that these metaphysicians of science have yet to respond to. Foremost among them is a thought aired by David Wiggins in his 2016 paper, "Activity, Process, Continuant, Substance, Organism" (reprinted in this volume, Chapter 9).

Methodologically, Wiggins falls within what Peter Strawson called the "descriptive" tradition.[7] In contrast to the "revisionary" attitudes of the science-led metaphysicians – revising their conceptions of reality in accordance with empirical findings – the descriptivist is happy to examine reality by looking at the way that we *experience* it. For Wiggins, like Strawson, we should describe the structure of the world *as it appears in our pre-theoretical conceptual framework* – i.e., the structure of our minds – and we gain access and insight into this framework by, most notably, studying our human language (hence Wiggins's "linguistic philosophy").[8]

The revisionist's response to this meta-metaphysical methodological move is, predictably, to say that any metaphysical picture grounded in the categories of the human mind will only ever be *local* (or, worse – gasp! – *idealist*);[9] they, in contrast, aspire to reach *beyond* the bounds of human experience towards something more objective – and science is, they think, a surer guide than the architecture of our limited mortal minds to identify the true warp and weft of the world.

The catechism has been well rehearsed, and the descriptivist will quickly retort that scientific practice is a *human* practice and as such, firmly embedded in the pre-theoretical framework that the revisionists hope to disavow. The scientists cannot even state their claims without relying on certain basic concepts, most notably, *substance* and *identity* (and Dupré, Nicholson and Seibt all, I think, agree that everyday language contains this commitment to substances and the identity relation).[10] This, roughly, is the thought that undergirds Wiggins's point about the processual picture of biological individuals:

> Could some revised research practice of these investigators amount to their dispensing with all implicit reliance on the determinables *continuant* and *organism*? Could these investigators really think of the organisms they study as simply concatenations of states?
> (Wiggins 2016, p. 275; emphases in original
> (= This volume, Chapter 9, p. 171))[11]

And again, in his paper "Identity, Individuation and Substance" (2012):

> At one and the same time, how can we deny ordinary substances their status as proper continuants, insist that ordinary substances are really *constructs*, yet lean shamelessly upon our ordinary understanding of substances when we come to specify that from which these constructs are to be seen as constructed or assembled?
> (Wiggins 2012, p. 12; emphasis in original)

The research that motivates Dupré and Nicholson's metaphysical proposal is conducted in laboratories in which all the materials, documentation and equipment are geared towards *substances* (the human scientists conducting these inquiries). To do away with, or to metaphysically discredit, the *substance* concept is thus to saw off the branch on which they are standing. Wiggins makes this point time and again – e.g., in reference to the pervasive notions of *cohesiveness*, and the concept of *genidentity* deployed by the processualists (Wiggins 2012, 2016) – and while there may well be a convincing reply to his concern, I have yet to hear the process philosophers offer it.

On my own reading of his work (Ferner 2016), Wiggins follows Strawson in thinking that *ontological dependence* – and consequently fundamentality – is correlative with *conceptual dependence*.[12] Our scientific concepts – of biological processes, etc. – are premised on other, more basic concepts – most notably, the *substance* concept. Despite what Dupré and Nicholson's empirical findings show, the human mind cannot begin to think of processes before we have thought of the substances in which they occur – and this, says the descriptivist, is metaphysically significant. This is the position the revisionary process philosophers need to consider and consider seriously. It is, as discussed below, susceptible to various objections – but whether these objections support the processual picture remains to be seen.

12.4 Moral processes

> A conscious being cannot think of itself – or of the persons, organisms or physical objects that it encounters – as having the shape of a complete succession of events and present only in part at the moment of action or deliberation. Nor can it think of itself as an event or an event in the making. To act and think as it does, a conscious being must think of its whole self as present at the moment of reflection, perception or action and poised to persist in that way in the future.
> (Wiggins 2012, p. 13f.)

When I am out and about, at banks, in bars, on the bus, I do not think of myself as *a process*. I can only really think of myself as a *thing* – as a *substance*. I think of myself as the kind of entity that continues in its entirety *through* space and time. Certainly, I know I am constantly losing and gaining atoms – I am in constant flux – but at the end of the day, I think of myself as a single, relatively stable individual who exists in its entirety at any one moment in time. This thought, as we have seen, gains a certain prominence in the kind of descriptive, "linguistic" philosophy that Wiggins and his ilk endorse.

Crucially in these commonplace situations, I do *not* think of myself as temporally extended. I do not think I have temporal parts. In fact, I am not sure I can even really comprehend (or state) that claim. In my daily life, I do not think there is a *now*-Adam, a *tomorrow*-Adam and a *next-week*-Adam, who are all parts of one great big processual Adam. I think I am Adam, me, in my entirety, right now, and that I will be around (hopefully) tomorrow, and next week too.

Maybe that is not how you experience reality? If so, well, I'm sorry – because it seems to me (and, again, I think Seibt and Dupré would agree)[13] that most cultural practices and societal norms are geared towards people who experience themselves as substantial continuants rather than processes. Whether this is a function of a pan-human pre-theoretical framework is, for the time being, moot; few metaphysicians would want to deny that society is structured for substances.

Now, process philosophers, if they are taking themselves and their project seriously, will have a problem with this societal bias. They will say that though there may be some societal bias that figures agents as substances, we have in reality a very different metaphysical character. I may talk about how, yesterday afternoon, I went to the cinema before enjoying a stroll along the Southbank, but this is, in the end, what we might call *ontological shorthand*.[14] It is useful, for conveying information, but ultimately misleading. What I *should have* said was that a neighbouring temporal part of this four-dimensional process (Adam) enjoyed a trip to the cinema before another temporal part of that process enjoyed a walk along the Southbank – or something of that sort.

I am caricaturing slightly, but there is an important issue here – because, in defending their metaphysics, some processualists may find themselves at odds with certain important, everyday institutions, besides cinemas, and including ethical and legal ones, whose moral claims have, I think, a much greater immediacy than the metaphysical ones assayed above. How, for example, are we to think that a human process is to be held responsible today for a crime committed yesterday? A process is spread over time – and temporal parts (putatively) do not continue. How are we to conceive of one temporal part being *responsible* for the actions committed by another temporal part? We may be able to point to a certain moment where some wrongdoing occurred, but the means of *addressing* (and sentencing) this wrongdoing are much less clear.

Moreover, I doubt that the processualists will take a deflationary tack here given how radical they take their thesis to be.[15] The effects of their processualism are far-reaching (consider, for instance, its implications for our comprehension of causation in Anjum and Mumford 2018) so it would be odd if the shift from a substance to a process ontology was seen to effect no change in our moral practices or analyses.

This has a certain intuitive force as well. It is, perhaps, a failure of imagination on my part, but I cannot understand how prison sentences, say, could be fairly applied to temporal parts. Surely there will be some temporal part of a process that exists within a prison's walls – but a different temporal part to the one that committed the crime? How does responsibility function within these parameters?

This is all rather gestural – but would it really be so surprising to find our common assessments of moral responsibility are, in fact, closely conceptually connected to a *substance* metaphysics? Not particularly. On the one hand, Wiggins (in a Strawsonian mood) may say that since our pre-theoretical framework is geared towards substances, our moral systems will likely follow (and, as we will see, this is fully coincident with his neo-Aristotelian ethics). On the other hand – and, I think more plausibly – we might shift to a historiographical register and look at how these ideas about responsibility and substance seem both to be embedded, and deeply intertwined, in a long and well-established intellectual tradition reaching back to, and probably beyond, Aristotle. The long-standing notion of individual responsibility is interwoven with the idea of an individual substance.

12.5 Moral responsibility

I am not a historian of philosophy (as will be painfully clear), but I believe there is some worth in at least nodding towards areas where our moral concerns about responsibility and agency overlap with metaphysical claims about, or implicit commitments to, *substance*.

One area is, of course, Aristotle and neo-Aristotelian ethical naturalism, as represented in the anglophone tradition by Martha Nussbaum (1988), Rosalind Hursthouse (1999) and Michael Thompson (2008). The view of these neo-Aristotelians, roughly stated, is that our moral thoughts are not insignificantly tied to us being the kinds of beings that we are. And for Thompson and the others we are *substances* – maybe not fleshed out in quite the same way that Aristotle (or Wiggins) fleshes out that concept, but certainly substantial *continuants*, and certainly not possessed of temporal parts. As Thompson puts it, his aim in *Life and Action* (2008) is

> to show that certain leading concepts in our various spheres – *life-form, action-in-progress, intention, wanting, practical disposition* and *social practice* – are all 'form concepts'. Anything that falls under any of them will exhibit some of the attributes Aristotle attaches to form/*eidos* in his general metaphysics and natural philosophy.
> (Thompson 2008, p. 11; emphases in original)

Metaphysics and morals are understood to be deeply entangled here – and the processualists would do well to take note. As the quotation from Thompson indicates – and as the title of his book attests – *life and action* are seen to be powerfully interconnected. Our ideas about agency correlate with our ideas about what it is to be a living being (and again: "an appeal to notions of life and organism and life-form would seem to be implicit in all departments of ethical thought" (2008, p. 27)). If Thompson is to be believed, it is unlikely that our thoughts about moral responsibility will be undisturbed by a shift from a substance view of organisms, to a processual one.

Let us repeat the question: if we are to hold each other responsible for our actions (which I think we should), if we are to apportion to each other praise and blame, is it necessary to think of ourselves as *substances*? If so, why? Is it because our minds are geared in this way, or because there are, historically, subterranean ethico-metaphysical connections we have yet to fully unearth?

In Aristotle's *Nichomachean Ethics*, we find one of the earliest renderings of a theory of moral responsibility (Eshleman 2016). There, he states that only certain kinds of being can bear responsibility: those who can consciously decide to act. For such an agent to be responsible for some action they must first have decided to do it freely (avoiding coercion) and they must know what it is that they are bringing about (they must be fully informed). They must be *authors* of their own actions, with complete command over themselves and their bodily movements. An individual can only be held responsible for the actions they have willingly committed.

There is something undeniably seductive about this account – and in some form or other it pervades almost all major legal institutions. (Consider, for instance, how strange it would seem to hold one person responsible for a crime performed by someone else) But how closely tied is it to the metaphysical concept of a *substance*? Is the view of organisms as *processes* antithetical to this picture? I have neither the space nor the brain power to investigate the precise metaphysical commitments, but it seems clear that *individual* responsibility does at least presuppose an *individual*, and that this individual must (if it is to be accurately tracked by the appropriate bodies) first of all be clearly defined – distinct, boundaried and stable – so that we can distinguish it from other individuals. Further, it must be able to claim *authorship* of its actions so its agency – which expresses itself diachronically – is the agency of a thing that persists (barring the occasional loss of body parts and so on) through space and time. There is some core, individual continuing *thing* that is subject to our moral appraisals.[16]

This is all somewhat schematic. Still, I think it shows that *some* work needs to be done to show how, for instance, one temporal part can be held responsible for the actions of another. There is a *small* onus on the processualist to demonstrate how our moral and legal practices – of imprisoning or otherwise punishing criminals – can be accommodated by a conception of ourselves as beings that do not actually move through time…[17]

12.6 Sovereign subjects

Up to this point, I have demanded rather a lot of my processual readers. Here, I hope, the Adam-of-the-second-half can make amends for the gentle joshing of the Adam-of-the-first-half. I do not know if, in the end, the processualists will be able to square their metaphysics with our understanding (neo-Aristotelian or otherwise) of moral responsibility, but their failure to do so may not be quite as troubling as I have suggested.

As I have said, there is *something* compelling about the Aristotelian account of individual responsibility. We tend only to hold people responsible for actions they have, themselves, willingly and knowingly committed. There is also *something* compelling about the thought that we are substantial beings – continuants – that move in our entirety through space and time.

But what are these mysterious *somethings*? What makes them so compelling?

The descriptivist would have us believe that the *substance* concept, under which we fall, is part of our pre-theoretical framework. They might say, moreover, that given that this concept structures our understanding of the world, and organises the way in which we navigate space, it will inevitably undergird our ethico-legal practices as well. And maybe the descriptivist is right. But there is reason to be cautious here. Following the work of postmodernist thinkers – like Judith Butler and N. Katherine Hayles – there are a number of analytic philosophers – including Sally Haslanger – who are growing both weary and wary of this kind of approach to metaphysical inquiry. In her book, *Resisting Reality* (2012), Haslanger construes the problem like this:

> [I]f metaphysicians wrongly assume that we have unmediated access to reality when in fact our access is culturally conditioned by background sexist and racist beliefs, and if metaphysics also functions to constrain our theorising within the limits it sets, then this poses a very serious problem for any effort to overcome oppressive attitudes and practices.
> (Haslanger 2012, p. 146)

Let us precisify slightly, by turning to E.A. Burtt and Tsu-Lin Mei's critique of the descriptive project.[18] Focussing specifically on Peter Strawson, both Mei and Burtt have shown how the methods of "linguistic" philosophy can reify, i.e. make real, temporary artefacts of local languages and practices. In Strawson's case, his metaphysics is grounded in a linguistic analysis of the grammar of *subject* and *predicate*. What Mei and Burtt show is that the subject/predicate distinction is importantly restricted to Latinate languages and Greek. That is, the grammatical distinctions, which Strawson takes to represent *a central strut of some universal human schema*, are notably lacking in e.g. Sino-Tibetan languages and Quechua. If Burtt and Mei are right (and I think they are), Strawson's project universalises specifically Western forms of thought (as present in Western grammars) – and the metaphysical picture he extrapolates is a reflection not of some pre-theoretical framework but of, unfortunately, a white, upper-middle-class Englishman's mind.

There is no such thing as apolitical metaphysics – and substance metaphysics is, I think, powerfully political, not least in its silence about this aspect of metaphysical discourse. This is not, by itself, an insurmountable obstacle for the substantialist's position (as Haslanger has shown, metaphysical inquiries can work, generatively, in tandem with political or ethical projects). However, one may (and I do) object to the way in which substance metaphysicians, like Wiggins, either consciously or unconsciously attempt to depoliticise their disciplinary space.[19] Moreover, one may (and I do) object to the conception of the human subject nurtured within this neo-Aristotelian framework – the stable, distinct, boundaried, rational substance that forms intentions and acts as an isolated persisting agent.

The gap between analytic philosophy of biology and postmodern analyses of subject-hood is vast – but let us try to jump it. Here is N. Katherine Hayles, in the introduction to her 1999 essay collection, *How We Became Posthuman*, suggesting some possible lines of critique for the substantial human subject:

> Feminist theorists have pointed out that [the liberal humanist subject] has historically been constructed as a white European male, presuming a universality that has worked to suppress and disenfranchise women's

voices; postcolonial theorists have taken issue not only with the universality of the (white male) liberal subject but also with the very idea of a unified, consistent identity, focusing instead on hybridity; and postmodern theorists such as Gilles Deleuze and Felix Guattari have linked it with capitalism, arguing for the liberatory potential of a dispersed subjectivity distributed among diverse desiring machines they call 'body without organs'.

(Hayles 1999, p. 4)

It does not, in fact, require too great a leap to see how these critiques might apply to the neo-Aristotelian substance conception of the human and the correlative models of moral responsibility, which revel in individual agency. We direct praise or blame towards an agent because of *their* (and nobody else's) actions – and this emphasis on the intentions and actions of the "unified, consistent identity" marginalises, indeed *conceals* various powerful and persuasive forms of injustice. Consider, for example, the injustices routinely perpetrated by *corporations*; big businesses commit all sorts of criminal acts – as conglomerates (or, as the philosophers of biology might prefer, as aggregates).[20] Think also of *historical injustices*: the agents directly involved in Britain's various and horrific colonial enterprises are all dead. Though British people continue to benefit from their activities, no one – according to Aristotle – is any longer responsible for those crimes.[21]

Moreover, N. Katherine Hayles (1999), Donna Haraway (1991) and C. B. Macpherson (1962) have each examined how this unified, conscious, nucleic, atomic subject stands front and centre in liberal and neo-liberal discourse (connecting with Deleuze and Guattari's position, above); for the political liberal, the human subject has to be single, complete and the ultimate proprietor of its own body and actions (it is in this that its liberty rests; it owns its body and is consequently free from the wills of others). The discrete, boundaried self is an essential posit for those who need a "natural" agent to prefigure (and thus stand as foundations) for market relations (whereby we can sell our labour – our body's labour – for wages).[22]

All of this is just to say that there is distinct *political pressure* for us to conceive of ourselves as genuinely unified, rational beings with distinct, discrete, impermeable boundaries. That is, as *substances*. And given the insights – and cautions – offered by Haslanger and others, this should, at the very least, give us pause for thought. Perhaps the process philosophers, with their radical re-description of reality, might be able to offer us a more politically productive alternative.

12.7 Processual futures

In Section 12.4, I alluded to a possible meta-metaphysical heuristic. Perhaps, I wondered, it is a failing of a metaphysical theory if it blocks the possibility of broader ascriptions of moral responsibility. Our ability to say that some action is right or wrong trumps any possible abstract solutions to metaphysical and/or scientific puzzles.

Maybe, as suggested in the previous section, there *are* reasons to move away from the neo-Aristotelian conception of human organisms, as substances. The kind of metaphysical picture that Wiggins endorses *may* be tied, problematically, to a neo-liberal, exclusively choice-focussed worldview that occludes serious and pervasive forms of injustice (not least, historical injustices, complicity and corporate wrongdoing).[23] Nonetheless, Wiggins and his friends are surely right to suggest that the substance view undergirds our (Western) everyday ways of thin(g)king; as such, an outright rejection of this picture will present obstacles to our ascriptions of right and wrong. Bluntly put, we may need to talk about substances if we are going to continue to address wrongdoing – and process philosophy presents an obstacle to this.

This kind of objection is not uncommon when it comes to metaphysical challenges to established norms. Indeed, we see a similar dialectic emerging from Judith Butler's famous (in some camps, infamous), postmodernist critique of the *pre-discursive self* (e.g., Butler 1990). Having leapt the gap between philosophy of biology and postmodernist critique once, we may benefit from a second attempt.

Like Haraway and Hayles, Butler claims that our "everyday" conception of self (which stands at the centre of Wiggins and Strawson's metaphysical project) is a political construct (Butler 1990). There is, she says, no "I" prior to discourse; there is no rational, sovereign subject that exists outside the socially embedded power structures – there is only

> a repeated stylization of the body, a set of repeated acts within a highly rigid regulatory frame that *congeal over time to produce the appearance of substance, of a natural sort of being*.
> (Butler 1990, p. 33; emphases added)

I am, as you might expect, sympathetic to this analysis, but it is hard not to be swayed by the objection raised by those like Seyla Benhabib (1995) and Sabina Lovibond (1989). Their complaint, which parallels the one articulated above, is that Butler's metaphysical claims undermine – damagingly – our notion of political agency.

> Along with [the] dissolution of the subject into yet "another position in language" disappear of course concepts of intentionality, accountability, self-reflexivity, and autonomy. The subject that is but another position in language can no longer master and create that distance between itself and the chain of signification in which it is immersed such that it can reflect upon them and creatively alter them.
> (Benhabib 1995, p. 20)

That is, a (metaphysical) critique of the sovereign subject will undermine moral autonomy. This applies directly to the processual subject as imagined by Dupré, Meincke, Seibt and others. Can the human organism, conceived by the processualists as thoroughly dynamic, constantly in flux, vaguely bordered, and possessing temporal parts, really be the kind of agent for whom autonomy choice and self-determination are possible? And if not, surely this stands as a failing of their position? Happily (depending on where you are standing), Butler – along with Ladelle McWhorter (1990) and Barbara Applebaum – has developed a productive response to this.

> [T]he question of whether or not a position is right, coherent, or interesting is, in this case, less informative than why it is we come to occupy and defend the territory that we do, what it promises us, from what it promises to protect us.
> (Butler 1995, p. 127f.)

That is Butler, in "For a Careful Reading". Unlike most mainstream analytic metaphysicians, her interests do not lie, primarily, in whether her metaphysics *actually, really, truly* maps reality (though she thinks it does). Her aim is not to create some final system, to be engraved in stone. Rather, her metaphysics is used as what McWhorter calls a "strategic tool". Her metaphysical claims about subject formation are tactical moves, or ways of "opening up possibilities that are closed off by our 'common' ways of thinking" (Applebaum 2010, p. 55). As Applebaum puts it,

> [h]er framework is like a pair of glasses that bring into sharper focus things that tend to remain blurred or invisible. Butler's work draws our attention to the need to always be open to questioning our certainties and to the significance of taking the position that nothing – neither our self nor our good intentions – is unaffected by power relations.
> (Applebaum 2010, p. 55)

My thought here is that Butler is not, in fact, uncommon in using metaphysics like this – but that she is uncommon in doing so *honestly* (in the analytic sphere at least).[24] We saw, albeit briefly, how Strawson's descriptivist project is involved in reifying a certain, local world-view; the Latinate subject/predicate structure was taken to be true of all human cultures, and consequently (and problematically) rendered universal. Butler's approach to metaphysics is a prophylactic against such a lack of self-criticality. As such, she points the way forward for process philosophers. She encourages a more engaged approach to metaphysics, one that is responsive to the systems – public, academic and theoretical – in which it is situated. Hers is a subtle approach to metaphysical declarations – a *maybe* at the end of every sentence – that renders every such declaration subject to interrogation. There is no stable resting place for Butler's metaphysician – and that, perhaps, might be both appealing and advisable for a group of philosophers who hold, like Heraclitus did, that everything flows.

From the discussion in Section 12.6, you will now (I hope) have an inkling of how a processual understanding of the human subject might encourage us to explore previously hidden avenues of thought. The human process is in constant flux, with no clear, definable core; it is always exchanging parts, so the distinction – concretised by Aristotle, between biological individuals – is brought into question. The processual human here is heavily embedded in society; it is just one midway process in a dynamic "hierarchy of processes", each of which has mutually dependent persistence conditions. It is the work of a much longer piece than this to explore how the shift away from a substance view might combat the concerns raised above, about the marginalisation of collective responsibility, and so on – the aim, here, has simply been to show that *there is this work to be done*.

12.8 Conclusion

I started this chapter by describing a concern with *apolitical* metaphysics. Metaphysicians, I said, should be conscious of their political commitments. At the very least they should consider the real-world effects of their views. What would it actually mean if we were to think of ourselves as *processual beings*?

My enthusiasm here may have given the impression that I believe the process philosophers to be blind to this concern. In actual fact, I think that philosophers of biology – and John Dupré is a good example – tend to be very considerate about how their metaphysical musings intersect and overlap with, not just science but also societal concerns more broadly. Anglophone philosophy of biology has been the site of a whole host of fascinating, important and politically engaged discussions, whether about the myth of "natural", biological "races", or critiques of human (and species), or sex essentialism, by authors like Judith Jack Halberstam (1998), Tommie Shelby (2005), Richard Lewontin (1972) and Naomi Zack (2011). As I say, some such work has been done by Dupré and his colleagues (e.g., Dupré 2001) – and the hope, in this chapter, is to provoke a similar debate about the liberatory aspect to processual conceptions of the biological individual.

All too often, analytic metaphysics is done behind closed doors, in seminar rooms, in offices, in laboratories sometimes; this creates the illusion that our theoretical work has little to no direct and immediate bearing on our day-to-day living – or vice versa. For all their failings, and for all their uncritical readings of "everyday life", descriptive metaphysicians have at least tried to resist this impulse; they have pushed our lived experience to the fore in their metaphysical practice. My aspirations for this chapter have been modest (and its achievements likewise), but I hope at least that it has left you with a sense that the positing of processual individuals has repercussions not just in biology or medicine, but in our daily lives and in how we interact with one another. These consequences are normally side-lined in philosophy of biology – but it would be advisable, I think, to pay them greater attention.

Acknowledgements

I am grateful to Anne Sophie Meincke, John Dupré, Thomas Pradeu, David Wiggins and Rowland Stout for helpful comments on this material (much of which was presented at the final conference of the PROBIO project, "A Process Ontology for Contemporary Biology", in March 2018 at The Royal Society of Great Britain, London, UK).

Notes

1 There are, of course, curious and important issues with (non-living) artefacts – but for present purposes, I leave these to one side (for more, see Ferner 2016, chapter 3).
2 And it is, I should say, clear that some process philosophers accept at least some of this characterisation of processes: "Processes are extended in time: they have temporal parts [...]" (Dupré and Nicholson 2018, p. 10) (cf. Meincke 2018 and 2019 for arguments for the view that temporal parts of processes are abstractions rather than real ontological entities).
3 It should be noted that one of the main aims of the process project in Exeter was to explore the idea that at least some processes *are* a kind of continuant (and to challenge the traditional equivalence – made by Wiggins and his neo-Aristotelian associates – of substances and continuants).
4 While writing this chapter, I have received some – very helpful – editorial encouragement to precisify the distinctions between "process" and "event", and to clarify phrases like "continuing in one's entirety through time". I have tried to make things as clear as I can, but the worry may remain that I have failed to properly represent the most persuasive form of the processualist's view. John Dupré and Anne Sophie Meincke have a very specific, and well-articulated view of what a process is (and how a process is a continuant and – in line with Steward and Stout – distinct from events) and it may be that my rendering does not do it justice. However, it is also worth mentioning that the relevant conflict is not borne directly out of the clash between precise conceptions of process, but rather between the methodological frameworks that sustain those conceptions. In anticipation of the discussion below, Dupré and Meincke are *revisionists*, Wiggins is a *descriptivist* and all that is necessary for the problematic to emerge is that the processualist's view conflicts with the everyday world-view that apparently sustains the neo-Aristotelian substance metaphysics.
5 We find the complementary claim in Guay and Pradeu 2016: "Soon it will not be unreasonable to sustain that processes are ontologically prior, and individuals should be conceived of as specific temporary coalescences of processes" (Guay and Pradeu 2016, p. 342).
6 In this connection, Wiggins describes his own position as "anti-scientistic" (Wiggins 2012, p. 14).
7 See Haack's paper for insight into Strawson's distinction, described at the start of his 1959.
8 See chapter 1 of Ferner (2016) for a more in-depth overview of Wiggins's position.

9 This is the charge often brought against the descriptive practice (discussed in Ferner 2016, chapter 1, sections 3 and 4).
10 For example: "This pervasive bias towards things is reflected in our everyday language [...]" (Dupré and Nicholson 2018); "an ontology of things, or Aristotelian substances, has dominated western philosophy since the Greeks" (Dupré and Nicholson 2018, p. 11); "It is arguable that our thinking is for deep reasons anchored to conceptions of objects describable in static terms [...]" (Dupré 2012, p. 8); "the standard theoretical tools of Western metaphysics are geared to the static view of reality" (Seibt 2017).
11 Some processualists may object to Wiggins's construal here – both in terms of the implied opposition between processes and continuants (see note 3), and with respect to the phrase "concatenation of states" (suggestive of atomistic thinking). Nevertheless, I think some form of the objection still stands; Wiggins will say that our understanding of *organism* is necessarily substantial and that the *substance* concept is indispensible to our understanding of living beings.
12 Haack, in her 1978, provides a concise and precise statement of the Strawsonian position:

> Ontological priority is defined in terms of our capacity to pick out and talk about things; Strawson suggests that identifiability of at least some individuals of a given kind is necessary for the inclusion of that kind in our ontology, and thus connects identifiability with ontological commitment, and identifiability-dependence with priority of ontological commitment.
> (Haack 1978, p. 362)

13 See note 10 emphasising the pre-eminence of the substance world-view.
14 Of the kind deployed by biological minimalists, like Peter van Inwagen (1990), in describing everyday objects like chairs (a shorthand for "atoms-arranged-chair-wise").
15 See, e.g., Dupré and Nicholson (2018, p. 10), and Seibt's entry in the *Stanford Encyclopedia of Philosophy* (2017).
16 This stands in stark contrast to the picture offered by Guay and Pradeu (2016): "The genidentity view is thus utterly *anti*-substantialist in so far as it suggests that the identity of X through time does not in any way presuppose the existence of a permanent 'core' or 'substrate' of X [...]" (quoted in Wiggins 2016, p. 272 (= This volume, Chapter 9, p. 169)) (*Editorial note*: The passage quoted by Wiggins appears in modified form on p. 318 of the published version Guay and Pradeu's chapter.).
17 There is a similar question to be raised for the four-dimensionalist philosopher – a close cousin of the processualist metaphysician (and the connections between these positions merit, I think, closer attention than they have received – and will receive here). Of course, not all forms of processualism are as closely aligned (Dupré, for instance, claims to be an ardent 3.5 dimensionalist and Meincke defends a version of process ontology that is meant to provide an alternative to both three-dimensionalist ("endurantist") and four-dimensionalist ("perdurantist") accounts of persistence, see Meincke 2018, 2019), but they are, I think, still interestingly connected in virtue of their scientific/scientistic methodological approach.
18 And for a further elaboration, see Ferner (2016, chapter 1, sections 3 and 4).
19 Of course, this by no means fully exculpates the revisionary process philosopher – since Wiggins's original complaint persists: scientific practice is a human practice.
20 For an analysis of this see, e.g., Haslanger (2012, p. 21).
21 For an important survey of these kinds of injustices, and how they are marginalised, see Applebaum's (2010).

22 For a more careful examination of these issues, see Hayles's introduction to her 1999 (e.g., p. 3).
23 I was interested to see the reference to Engel's *Dialectics of Nature*, and dialectical materialism (Dupré and Nicholson 2018, p. 6), in the introduction to *Everything Flows*, and wonder whether the thoughts outlined here have already occurred to Dupré and Nicholson.
24 Though we should point out, too, that things are changing, thanks to work being done by e.g., Sally Haslanger and Arianne Shahvisi.

References

Anjum, R. L. and Mumford, S. (2018) 'Dispositionalism: A Dynamic Theory of Causation', in: Nicholson, D. J. and Dupré, J. (Eds.), *Everything Flows: Towards a Processual Philosophy of Biology*, Oxford: Oxford University Press, pp. 61–75.

Applebaum, B. (2010) *Being White, Being Good*, New York: Lexington Books.

Benhabib, S. (1995) 'Feminism and Postmodernism: An Uneasy Alliance', in: Benhabib, S., Butler, J., Cornell, D. and Fraser, N. (Eds.), *Feminist Contentions*, London: Routledge, pp. 17–34.

Butler, J. (1990) *Gender Trouble*, London: Routledge.

Butler, J. (1995) 'For a Careful Reading', in: Behabib, S., Butler, J., Cornell, D. and Fraser, N. (Eds.), *Feminist Contentions*, London: Routledge, pp. 127–144.

Dupré, J. (2001) *Human Nature and the Limits of Science*, Oxford: Oxford University Press.

Dupré, J. (2012) *Processes of Life*, Oxford: Oxford University Press.

Dupré, J. (2014) 'A Process Ontology for Biology', *Auxiliary Hypotheses*, BJPS blog, https://thebjps.typepad.com/my-blog/2014/08/a-process-ontology-for-biology-john-dupr%C3%A9.html (August 21, 2014).

Dupré, J. and Nicholson, D. J. (2018) 'A Manifesto for a Processual Philosophy of Biology', in: Nicholson, D. J. and Dupré, J. (Eds.), *Everything Flows: Towards a Processual Philosophy of Biology*, Oxford: Oxford University Press, pp. 3–45.

Eshleman, A. (2016) 'Moral Responsibility', in: Zalta, E. (Ed.), *The Stanford Encyclopedia of Philosophy* (Winter 2016 Edition), https://plato.stanford.edu/archives/win2016/entries/moral-responsibility/.

Ferner, A. M. (2016) *Organisms and Personal Identity*, London: Routledge.

Guay, A. and Pradeu, T. (2016) 'To Be Continued. The Genidentity of Physical and Biological Processes', in: Guay, A. and Pradeu, T. (Eds.), *Individuals Across the Sciences*, Oxford: Oxford University Press, pp. 317–347.

Haack, S. (1978) 'Descriptive and Revisionary Metaphysics', *Philosophical Studies* 35, pp. 361–371.

Halberstam, J. J. (1998) *Female Masculinity*, Durham, NC: Duke University Press.

Haraway, D. (1991) 'A Cyborg Manifesto: Science, Technology, and Socialist-Feminism in the Late Twentieth Century', in: Haraway, D., *Simians, Cyborgs and Women*, London: Routledge, pp. 127–149.

Haslanger, S. (2012) *Resisting Reality: Social Construction and Social Critique*, Oxford: Oxford University Press.

Hayles, N. K. (1999) *How We Became Posthuman*, Chicago, IL: University of Chicago Press.

Hursthouse, R. (1999) *On Virtue Ethics*, Oxford: Oxford University Press.

Lewontin, R. L. (1972) 'The Apportionment of Human Diversity', *Evolutionary Biology* 6, pp. 381–398.

Lovibond, S. (1989), 'Feminism and Postmodernism', *New Left Review* I/178 (November–December), pp. 5–28; reprinted in: Lovibond, S. (2015), *Essays on Ethics and Feminism*, Oxford: Oxford University Press, pp. 18–44.

Lowe, E. J. (2002) *A Survey of Metaphysics*, Oxford: Oxford University Press.

Macpherson, C. B. (1962) *The Political Theory of Possessive Individualism*, Oxford: Oxford University Press.

McWhorter, L. (1990) 'Foucault's Analytics of Power', in: Dallery, A. B., Scott, C. E. and Roberts, P. H. (Eds.), *Crises in Continental Philosophy*, Albany, NY: State University of New York Press, pp. 119–126.

Meincke, A. S. (2018) 'Persons as Biological Processes: A Bio-Processual Way Out of the Personal Identity Dilemma', in: Nicholson, D. J. and Dupré, J. (Eds.), *Everything Flows: Towards a Processual Philosophy of Biology*, Oxford: Oxford University Press, pp. 357–378.

Meincke, A. S. (2019) 'The Disappearance of Change. Towards a Process Account of Persistence', *International Journal of Philosophical Studies* 27(1), pp. 12–30, doi:10.1080/09672559.2018.1548634.

Nicholson, D. J. and Dupré, J. (Eds.) (2018) *Everything Flows: Towards a Processual Philosophy of Biology*, Oxford: Oxford University Press.

Nussbaum, M. (1988) 'Non-relative Virtues: An Aristotelian Approach', *Midwest Studies in Philosophy*, XIII, pp. 32–53.

Seibt, J. (2017) 'Process Philosophy', in Zalta, E. N. (Ed.), *The Stanford Encyclopedia of Philosophy* (Winter 2017 Edition), https://plato.stanford.edu/archives/win2017/entries/process-philosophy/.

Shelby, T. (2005) *We Who Are Dark*, Cambridge, MA and London: The Belknap Press of Harvard University Press.

Steward, H. (2015) 'What is a Continuant?', *Aristotelian Society Supplementary Volume* 89(1), pp. 109–123.

Stout, R. (2016) 'The Category of Occurrent Continuants', *Mind* 125(497), pp. 41–62.

Strawson, P. F. (1959) *Individuals: An Essay in Descriptive Metaphysics*, London: Methuen.

Thompson, M. (2008) *Life and Action*, Cambridge, MA and London: Harvard University Press.

Van Inwagen, P. (1990) *Material Beings*, Ithaca, NY and London: Cornell University Press.

Wiggins, D. (1980) *Sameness and Substance*, Cambridge: Cambridge University Press.

Wiggins, D. (1995) 'Substance', in: Grayling, A. C. (Ed.), *Philosophy 1: A Guide Through the Subject*, Oxford: Oxford University Press, pp. 214–249.

Wiggins, D. (2001) *Sameness and Substance Renewed*, Cambridge: Cambridge University Press.

Wiggins, D. (2012) 'Identity, Individuation and Substance', *European Journal of Philosophy* 20(1), pp. 1–25.

Wiggins, D. (2016) 'Activity, Process, Continuant, Substance, Organism', *Philosophy* 91(2), pp. 269–280 (reprinted as Chapter 9 in this volume).

Young, J. Z. (1971) *An Introduction to the Study of Man*, Oxford: Clarendon Press.

Zack, N. (2011) *The Ethics and Mores of Race: Equality after the History of Philosophy*, Lanham, MD: Rowman and Littlefield.

13 The nature of persons and the nature of animals

Paul F. Snowdon

13.1 Introduction

We are in a period when the biological notion of an animal (or an organism more generally) has become prominent in philosophical discussions of our nature. Now, on one fairly orthodox conception of the notion of an animal (a conception that I assume in this discussion) it is a highly general natural kind notion, and so the theory of what animals are, what nature they have, will come ultimately from the empirical scientific study of animals, from, that is, biology, or the life sciences. That is why engagement between philosophers and biologists, and of course philosophers interested in biology, cannot but benefit our joint concerns. It might be felt that only philosophers who support animalism in what philosophers call "the theory of personal identity" have an interest in the nature of animals, because according to them that is *what we are*. That is, however, a mistake, since, of course, anyone wishing to argue *against* animalism has to endorse certain claims about animals to sustain their own negative claims, and so they cannot be uninterested in the question what animals are, and hence in what biology tells us.

I write here as a supporter of animalism, a view which employs the notion of an animal, but I do not come with anything approaching a full theory of what animals are or with what might be called a theory of animal persistence over time, commonly although unilluminatingly called "theories of animal identity". I approach such issues rather as a learner, hoping to build up, in the light of what is said by the properly informed, those whose meat and drink is engagement with and reflection on animals, more information and a better understanding of how to think about animals. It might seem odd to be an animalist without a full theory of what animals are, but in reality we can advance and defend theories which employ notions which are not attached to informative definitions or final theories. We can employ such notions while awaiting new theories and information about the things in question. What I want, rather, to contribute to the ongoing debate is an exposition of my own understanding of animalism, and the arguments surrounding it, which, although it is not by this stage in the debate at all revolutionary or novel, will, I hope, shed some light on the role of the notion of an animal in that theory, and which may be informative to readers not steeped in standard metaphysics. One question that I wish to engage with, and make, if I can, some progress with, is where in the ongoing *philosophical debates* the biologist can make a contribution that is of aid in the evaluation of philosophical arguments. These tasks will occupy the next two sections. In the final section I shall consider in a rather preliminary way how we should think about animals, and, in particular, whether thinking of them as processes represents an improvement in our thought.

13.2 Issues about the nature of persons

Philosophers tend to employ "ism"s to stand for theses or theories. "Animalism" is just one of many "ism" words in our subject. Now, it is my impression that in many subjects, say, biology or medicine, when technical terms are introduced and spread through the practitioners of the subject, somehow or other, the process brings it about that people acquiring the term understand it *in the same way*. It is also my impression that in philosophy the spread of technical terms is less controlled and that they often end up being understood in different ways, to the obvious detriment of discussion in the subject. In the light of this there is perhaps some value on an attempt to propose a way of understanding and hence anchoring the term "animalism".[1]

I want to suggest, then, that one way to understand "animalism" is that it is the name for the claim that we, the people here and others of our sort, are animals. We can replace, or expand, the verb "are" here by "are identical to" animals. So animalism is an *identity thesis*. Formulated in this way the claim implies, because identity is what is called a symmetrical relation, that there are certain animals (some of which are here) which are identical to us. That is an identity claim about some animals. Of course, there will be other animals here which are not identical to any of us – e.g., some spiders, flies, mice.[2] Having the notion of identity as central to the thesis enables the debate to employ insights given to us by generations of philosophical reflection on identity.

So understood it may sound unimpressively simple, but we can point out that it provides an answer to a long-lived question, a philosophical question if you will, that we face, namely, *what are we?* (Sometimes that is asked in the words: what is our place in the universe?) What kind of thing are we? It is not built into the thesis as explained here that it is a priori or conceptual, or can be proved by armchair reflection alone. It is merely advanced as being a *true* identity claim.[3]

I want to return to the proposed formulation. On the left-hand side is the indexical designator "we" or "we and things of our sort". "We" is a designator that does not have a precise rule for determining its reference. Thus if I say to someone "We shall visit you next week", I can be asked "who do you mean by 'we'?". By contrast there is a precise rule for determining the reference of "I" in normal contexts of speech – it is the speaker. If I say "I shall visit you next week", it would be weird for the listener to ask: "Who do you mean by 'I'?" The involvement of "we" in the formulation of the thesis therefore means that there is some lack of precision about the content of animalism. It can, though, be further pointed out about "we" that the determination of its reference in a context is not entirely rule-less. First, the speaker him- or herself must be part of the group that is being referred to. It would be incongruous for someone to say, "We shall arrive at six o'clock but I shall not". Second, the group picked out cannot include things which in their nature are incapable of consciousness. Thus if I and my clothes get wet on a walk I cannot report that by saying "We got wet". However, it would be perfectly acceptable if I go for a walk with my dog and he and I get wet to report that by saying "We got wet". This means that there is something misleading in our standard classification of "we" as a *personal* pronoun. The range of its reference is not confined to persons in the normal sense, whatever that is exactly. Even if these rules do apply to "we" it remains true that there is something imprecise about its reference, but my own inclination, having noted that, is to live with the imprecision (until one gets to a point where it matters and then deal with it). Two things can be added. (i) When confronted by arguments for and against animalism, one can consider them from the first-person point of view. Each of us can formulate the thesis as "I am identical to an animal" (which contains no referential looseness), and all the significant considerations can be formulated and assessed in this first-person formulation without imprecision. (ii) If the left-hand designator in the general thesis is to be left imprecise, it might be asked: why not substitute the general term "person"? Why not just say: persons are identical with animals? There are two reactions to that. First, can we say that "person" is precise and clear? Maybe that term is introduced into our language as standing for us and creatures like us. It is a myth to think that general nouns are precise. It is not implausible, it seems to me, to think that "person" is used in this imprecise way, in which case it can be substituted for "we" in the formulation, but without any gain in precision. But, second, if our use of the term "person" is linked to, or grounded in, that elucidation of it coming from Locke and the Lockean tradition, according to which "person" is defined in terms of the capacity for having certain advanced psychological states, such as memory and self-consciousness, then it would be a mistake to employ it to pick out what is being claimed by animalism to be animals. For one thing it is entirely open whether all extant advanced psychologically endowed entities are animals; some might be machines, some might be brains, and half-brains, angels (should there be such things), etc. It also certainly seems an open question whether Lockean-type persons have to be animals. Animalism is not a thesis about "persons" in this Lockean sense, or in any sense, although perhaps not specified in precisely Locke's way, which retains the idea that a central condition for being a person is the presence of advanced psychological capacities.

Next, I want to specify some assumptions which form the framework of the discussion of animalism as it is usually carried out, noting that there may be some who are critical of these assumptions. (i) There is wherever each of us is one animal occupying the same space that we occupy. One way to deny this is to deny there is such a thing as the animal here. Another way is to affirm that there is more than one animal here. However, neither of these opinions is usually or standardly affirmed. (ii) In the standard framework the animal in question is conceived of as an enduring thing which has a spatiotemporal location, and is composed of matter; it is solid and has a shape, and is not itself a process or an amalgam of processes. The processes related to the animal are rather things that are going on in it or with it, however important they may be for its continued survival. Moreover, this is how we think of at least most of the animals with which we are acquainted, such as dogs, cats and rabbits etc.... There are, though, clearly people who wish to challenge these assumptions, and I want later, in the fourth section, to say something about the issues raised by what seems to me to be the most important of such challenges. (iii) The third assumption in the standard framework that I want to pick out is that when we are thinking about ourselves, when each of us is asking – am I identical to an animal? – we succeed in picking out an object, the object that each of us is. I do not think that anyone has suggested that biological reflection gives a source for scepticism about this basic assumption but some participants in the debate do reject it. I have in mind some cognitive scientists and philosophers who say that the "self" is a myth, the reason that influences them being that it is an illusion to think of ourselves as strongly unified or highly rational things, and who think that this means that "I" does not pick out anything. People who say this should be asked why the supposed mythical status of the self means that "I" has no reference, and they should also be asked why the things that there undoubtedly are which do use "I" in their speech cannot be re-educated about the referential role of "I" and thereby acquire a device for reference to themselves, whatever they are, in which case we would be back to what most of us already assumed was the case anyway, that something is picked out by "I".[4]

Formulating animalism in this way as an identity thesis has an interesting consequence about its scope. It is standard to view animalism and Lockeanism as competitors, and so as offering alternative theories about the same phenomenon. But what is that phenomenon? On its standard interpretation Lockeanism gives an informative analysis of what it is for a person (one of us) to remain in existence over time, by identifying the psychological links that have to present over time.[5] It is, that is, a theory about our existence over time. But on the current reading of animalism it does not provide any such informative analysis of existence over time. It is a timeless identity thesis affirming that each of us is the same thing as an animal. This merely implies that our persistence conditions are the same as those for the animal each of us is, but the simple claim that is animalism does not say what those conditions are. It is therefore a mistake to regard animalism and Lockeanism as *accounts of the same thing*. It should be recognised that people who agree on animalism can disagree about our conditions for remaining in existence.[6] It turns out, I think, that this is a point that has some relevance to the reflections in Section 13.4.[7]

13.3 Arguments

Concerning animalism I have suggested that we interpret it as an identity thesis. As well as clarifying the relation between animalism and Lockeanism that also, I believe, illuminates the nature of the sort of arguments that have been devised or proposed to either reject it or accept it. Thus, the fundamental condition for *a* to be identical to *b*, to be, that is, the self-same thing, is that they are indiscernible, they must share the same properties. This is one way to state Leibniz's Law (of identity). Now the central way to argue *against* animalism (within the standard framework) is to claim to have located a property that we have but the animal lacks (or vice versa). Standard cases are where we start with an animal and a person (one of us) at the same place and time and then things move forward in such a way that it is judged to involve the presence of the animal but not of the person. For example, there is an accident which does not kill the animal but the patient enters a vegetative state, about which some claim that there is an animal but the person has ceased to exist. Another scenario is that the brain of someone is extracted and transplanted so that it still operates, but the rest of the body is left behind. About this it is judged by many that the person goes with the brain, but the animal is left behind. I call such arguments "dissociation arguments" because they allege that what I shall call the person (meaning, simply, one of us) and the animal initially co-located come apart, or, as one can say, "dissociate". There are fundamentally two types of dissociation arguments. There are those where it is alleged that at the end there is the animal without the person, and those where it is alleged that at the end there is the person without the animal.

Now how might a theory of animals help in adjudicating such arguments? They undoubtedly depend on judgements about animal persistence. In one case it is that the animal is still there in a vegetative state (although the person is gone). But it seems to me that in such cases the judgements about the animal are more or less obvious, and what is controversial is the judgement about the person. And so for most of these types of arguments biological insights are not needed and will not be forthcoming.

Now that seems obviously correct to me except, perhaps, in some versions of dissociative arguments, for example, in less extreme transplant cases where what is transplanted or preserved is not the brain alone. Thus we allow that we can reduce animals in certain ways; e.g., an animal can lose all its limbs and remain in existence. So we might look to biologists and their theories of animals to guide us as to how small an animal can be shrunk to.

The question is just how far can we shrink an animal with it remaining in existence. This matters since in such arguments the intuitively plausible thing to judge is that the person is still there, given the presence of mentation, so the critical question in deciding whether there is a dissociation is whether the animal is still there, although shrunken. A judgement from biology on this would help. However, it is hardly a question that any biologist would consider and how, if they did, can they determine the answer? Reflecting on this, it seems reasonable to suggest that biologists and life scientists operate something that might be called an "actuality constraint". They investigate their relevant domains as they actually find them. They develop their theories given what they actually find about animals. But frequently in philosophy the difficult cases are not actual ones. I am not suggesting this means that such issues cannot be adjudicated, but they seem to be issues outside the domain as empirically investigated by the scientist. This means, I am suggesting, that philosophers cannot really look to science to give answers.

I want, though, to highlight a second sort of alleged anti-animalist case – developed by Tim Campbell and Jeff McMahan – in their paper "Animalism and the Varieties of Conjoined Twinning" (2016). To give a sense of their type of argument, and its links to claims about animals and organisms, I want to present, in a summary form, one of their arguments. They consider the case of the Hensel twins and claim that there are two persons or subjects, but that there is only one organism or animal.[8] It follows that at least one subject is not an animal, and since they seem the same that neither is. But further, since we seem to be like the twins themselves, we are not animals either. There are, it seems to me, two questions that this argument raises. The first is: what should we actually say about the Hensel case, and cases like it? The second question is: is it correct to spread the conclusion to which we might incline about the Hensel case to ourselves? Thus, in relation to the second question, even if the Hensel twins are not themselves animals, they are not things which are totally unlike animals – maybe they are animalish parts of a single animal. If that is accurate, then, since we are not animalish parts of anything, we cannot be that kind of thing, and maybe we can count as the animal itself. However, is the description they offer of the Hensel case plausible? It is surely odd to baldly say that there is a single animal there with two heads. It is better to say that there is something that results from a process of fission of the kind that if it reaches its conclusion it results in identical twins but which terminates early resulting in incompletely separated twins. We have, that is, something that amounts to incompletely separated twins. Most importantly, this does seem to be a case where we should ask what biologists (and medical experts too, who have to deal with this sort of case) would say, because they speak in the light of an understanding of the processes whereby they came into existence. In this case we are not falling foul of the actuality constraint, since such cases are actual.

I have been exploring anti-animalist arguments and a possible role in relation to them for a biologically informed understanding of animals. But I want to ask now what contribution if any biology can make to the assessment of standard pro-animalist arguments.

The main argument that has had a significant influence is the so-called too-many-minds problem.[9] The argument aims to reduce the opposing claim to animalism to absurdity. Thus, let us assume that we are not animals. This means that where I am there is also an animal, which is not me. We can say that I am mentally endowed – which we can let stand as shorthand for the myriad of psychological attributions that are true of me. What psychological attributions can we make of the animal (here, but not me)? Surely, the obvious answer is that the animal here where I am is itself mentally endowed. It can see, act, work things out, etc. In fact, it would seem to possess the same range of psychological states that I do. It and I seem to be psychologically indistinguishable. If so, it seems to follow that there are two psychologically endowed and psychologically indistinguishable things here, where I am. The final claim in the argument is that that is an absurd claim, one which no one would seriously accept.

Now, I want to make two points about this argument. The first is that there really is nothing about its premises that needs certification or criticism by biology. Ethologists pursue in an intensive way the question what sort of mental states many of the animals around us possess, but they take it for granted that they do possess mental states, and there is no real question whether they do or not. Equally with human animals they are not sceptical about psychological ascriptions to them. So the argument does not engage with novel empirical investigations by biologists. The second point is that this argument is merely one argument in favour of animalism. Obviously there could be, indeed, there are, other arguments. It is a serious mistake to think that if the too-many-minds argument can be met, then animalism has been laid to rest or that its total support has been obliterated.

My conclusion at this point is that in philosophical debates about out nature (and whether we are animals) the arguments normally appeal to claims about animals which are not seriously disputable. In most arguments it is the judgements about ourselves which are controversial. In some other cases the claims about animals are disputable, but they fall outside the actuality constraint, and so again biology provides no real help. There are, finally, some arguments which rely on disputable claims about animals and on which, because they are actual cases, science can provide a guidance, and so we do need to heed what the life sciences say about them.

13.4 Issues about animals

What remains to consider though is what we might call the general nature of animals, and with that how to think about *our* general nature assuming that animalism is correct. At this point I want to engage with the proposal that a number of writers in this volume are sympathetic to that scientific and philosophical considerations should lead us to think of animals and ourselves as processes rather than as what tends to be called, somewhat portentously, "substances". I want to consider this idea and some of the issues that surround its assessment, but, sadly, in a rather preliminary and simple way, in a way that will strike proponents of these views as seriously underestimating their appeal. My hope is that the following elementary reflections might at least promote clarity.

Now, the simplest way to think of this issue is as follows. What we call "animals" are a very general type of thing present in nature. We would explain to a child what animals are by pointing to lots of them and explaining that such things are animals, and pointing to other things (e.g., trees, mountains, pools of water, cars) and explain that they are not animals. This process of instruction enables learners to latch onto what in nature the term applies to. This is merely a reiteration of the standard view that we have here a natural kind word. Now, although having gone through this process we understand what animals are, and so certainly know what claims about animals are about, we do not, thereby, get to know what the theoretical and fundamental nature of such entities is. Again, this parallels what can be said about our relation to water, gold, etc. *Science* tells us what such things fundamentally are, what their real nature is. Viewing the notion of an animal this way suggests then that empirical investigation by the relevant science can tell us what animals really are, what nature they really have. Now, it is the view that we are asked to consider that empirical investigation of animals together with philosophical reflection on those results yields the conclusion that they are properly thought of as processes. Putting the proposal this way, what we might call version 1, involves the idea that there certainly are animals but they can be characterised as processes. It is in effect an identity thesis – animals are the same thing as certain processes.

There is, however, a different way to read the proposal, which may be how some proponents of it read it. On version 2, as I shall call it, it is not that animals are processes but that, rather, there really are no animals but there are what we should recognise, given a correct appreciation of their nature, processes. The crucial difference between version 1 and version 2 is that a supporter of version 2 believes that our very notion of an animal has built into it in some way a commitment to the hypothesis that they are essentially things of a *non-process* kind. Using the terms that have been developed when thinking about different versions of physicalism, version 2 can be called "eliminativist", in that it is suggesting that the category of "animal" should be eliminated, or called a "disappearance theory", in that it is urging the disappearance of the animal category, at least from serious thought. A crucial difference between the two versions is that anyone supporting version 2 needs to persuade us that our concept of an animal has built into it something that is inconsistent with the idea that such things are processes. With that comes another difference. Someone arguing for version 2 can support it by arguing that the concept of an animal itself faces serious problems in application. This might lead us to think that there are

Nature of persons and nature of animals 241

deficiencies in it and so it requires replacement by the idea of a process. On the face of it producing examples where it is difficult to apply the concept of an animal does not encourage one to accept the claim that animals are processes. It is, rather, a strategy for generating scepticism about the concept.

It might be suggested that the role, as described above, of ostension in the generation of the concept of an animal means that eliminativism cannot be taken seriously, since what is being pointed at is undoubtedly there. But this inference is too speedy. It may be that the ostension engages with some imported conceptual structure that is committed to things that are false of what are in fact processes. So version 2 cannot itself be eliminated so rapidly.[10]

Having highlighted this contrast, I want to bring in another one. When one listens to arguments that supporters of the process view sometime propose, they allege that there is something conceptually problematic about the common-sense idea of what we might call material objects which remain in existence over time even though they change. Putting it very crudely, it is alleged that there is something close to a paradox in thinking that the continuing object is the *same* object and yet also thinking that it is not the same, that is, it has changed. It is alleged that this is more or less a contradiction. Another slightly different way of trying to elicit a paradox from ordinary thinking is to say that in the case of organisms there is no matter that is constantly present, since the matter in organisms is constantly changing, but if the organism is the same thing over time then there needs to be something that is the same at all times. So ordinary thinking requires there to be the same thing there for there to be genuine identity, but there is nothing the same permanently present.[11]

Now, it seems to me that these types of worries have found their way into the armoury of proponents of the process view, but it is not, of course, necessary for someone who accepts either or both of these criticisms of common-sense thinking to rebound into a process view. Any conception about objects and time that seems to avoid the paradox would be a candidate for adoption as a way of avoiding it. Another view that is offered as such a way is what is called "four-dimensionalism", the idea developed with considerable sophistication that we should be thinking of continuing existence as consisting of a sequence of time slices related, not of course by identity, but in some other causal or spatial ways. In the language of the debate, according to the normal conception material objects, including animals, endure over time, whereas a sequence of time slices constitutes something that perdures.

There is a massive literature within philosophy about these issues, but I want to rather bluntly suggest that the charge of incoherence or contradiction in normal thinking about enduring material objects should be rejected once we recognise two obvious components of ordinary thinking. The first is that in thinking that the organism in front of me is the same thing that I saw two days ago I am not committing myself to there being the same material constituents before me on both occasions. In the way we think it is one thing to think that in front of me is the same matter, and another to think that in front of me it is the same organism, and the latter claim does not entail the former claim. There is surely no problem at this point, and no amount of repeating the terms "the same thing" will generate the implication. The second obvious point is that there is no incoherence in the idea that a single thing can change over time. There is no contradiction in thinking that my dog is fatter today than it, the same dog, was yesterday. It was thin and is now fat. Holding these simple points in mind means that such allegations of incoherence in ordinary thinking are baseless. The upshot is that process theorists should not rely on them to support their conception.

Another direction of criticism of ordinary thinking about object persistence that I have encountered in discussion is that it is committed to the idea that objects have essences (or essential properties), which is an objectionable commitment. Responding to this in a very general way it can be said that it raises three questions. Is ordinary thinking about object persistence committed to essences (in some sense)? Is such a commitment objectionable? Do process theories of animals avoid such a commitment? Clearly, the first two questions are highly complex and hard, but I want to make a remark about the third one. Process theorists can be asked; are processes essentially processes? Might something that is a process – say the process of moving house – not be a process but be something else? It is, surely, rather hard to think that something that is a process might not have been a process. It seems to be essentially a process. We can also ask: might a process which is that of lighting a candle not have been that sort of process? Someone might reply: well, the process that began might not have resulted in a candle being lighted, so it need not have been that process. This reply invites the comment; in that case it would not have been that very same process, a process of lighting the candle, but another process, say, of dropping a match, a quite different process There is surely some pressure here on process theorists to make them at least think it not obvious that their own fundamental concept for thinking about persistence, that of a process, avoids some essentialist commitments. In which case a commitment to essentialism is not something that differentiates ordinary thinking about persistence from process theories.

There is, though, a good question that might be asked at this point. It might be asked: if we deflect the criticism that claims of identity over time require a permanent abiding core, which however does not exist, by saying that they do not require such a thing, what is the difference between ordinary thinking about continuants and the process proposal? Does not removing the requirement of a permanent core obliterate that distinction?

Now, this is not a question that only someone tempted to reject the process view in the cause of defending common sense needs to answer, because proponents of the process view invariably see it as an alternative to and improvement on another extant way of thinking, and so they need to have some defensible conception of what that way of thinking is, of what the concept of a persisting organism is according to the view to be rejected. As one might put it, when process theorists of either version say "Animals, or what we call 'animals', are processes" their meaning can be more fully explicated by adding "… and are not so and so's as has been thought until now". They must, therefore, have a conception of what thinking of them as so and so's amounts to. And it is surely beyond question that ordinary thinking about animals does not regard them as processes, since ordinary thinkers would deny they were processes if asked whether they were. However, articulating the pre-existing ordinary view is not an easy task.

I want to engage with this issue by looking at a brief clarification of the distinction proposed recently by Professor Wiggins. Wiggins says:

> Continuants [Wiggins's term for the contrasting category to processes] exist in time, have material parts and pass through phases. But such phases are not the material parts of the continuant. The phases are parts of the continuant's span of existence. Contrast processes. The phases of a particular historically dateable process *are* its parts.
> (Wiggins 2016, p. 270; emphasis in the original
> (= This volume, Chapter 9, p. 167))

Wiggins further points out that "[a] process can be rapid or regular or staccato, or steady. It can even be cyclical and lifelong". In contrast "an organism can be the proud possessor of eight fingers and two thumbs" (ibid.).

The last two contrasts concern what can be said about processes and what cannot, and what can be said about so-called continuants and what cannot. However, it is not clear how much weight to attach to these last contrasts. Taking the ascription to an animal of four fingers and a thumb, it is clear that some such an attribution must be makeable in some form or other even on the conception of animals as processes, since such a description obviously registers a feature that some animals possess but other do not. This means that the process theorist must provide an interpretation of such an ascription that is, however, suitable for a process. The challenge is to do that. In connexion to this example a first step one could note that a process can occur in a region and that may ground a shape attribution to the process, and then being four-finger shaped may be an attributable feature that a process possesses. We do attribute the features that Wiggins singles out, e.g., "staccato", to some processes, but if one is trying to think out the idea of organisms as processes, in some sense, it is not obvious that the strangeness of attributing the ones on this list to an animal flows from the idea of animals as processes, or, perhaps, rather, from special features of the selected predicates. There is another point here; when we think about processes, we tend to develop a vocabulary that picks out a basic sort of processes, for example, the weather condition that we call a "hurricane". If someone is tempted to propose that we think of animals as processes we have to suppose that the animal consists of many complex processes all sustaining its presence, and so the process that is an animal will be diverse and diffuse, and so some predicates that apply to some processes may not be sensibly attributed to that complex process that the animal is. Of course these remarks are speculative, but it is not obvious that process theorists cannot make progress with it. However, the task of what we might call re-interpreting ordinary predicates must have its limits; otherwise, the distinction between ordinary non-process thinking and process theories vanishes. In seeing animals as processes there must be some things that we previously said of them which can no longer be said, and things not previously said that are now said, otherwise there is no contrast.

In the first remarks Wiggins is proposing some deeper content to the contrast. He points out that material things exist in time, but process theorists might equally say that processes exist in time. He next adds that they have material parts, whereas a process does not. There is no doubt that we want to say something like this, at least the idea of a process seems to be the idea of a happening which itself does not have material parts. However, in one sense not all material continuants must have material *parts*, since there may be material simples, which can *be* parts but are not things *with parts*. And to label something "material" seems to presuppose the very notion that we are trying to clarify. But we can continue in the direction sketched by Wiggins. Thus, many of the material things around us are things that we can make by putting other objects together, as in making a desk. Or we can make an object by detaching it from other parts, as in carving some wood to make a spoon. But we do not talk of making a process, rather we start it. Again an object can fall apart, when its parts separate, whereas a process cannot fall apart. These contrasts reflect a difference between the way in which processes are spatial and objects are spatial. Processes occur at a place, whereas objects occupy and are in that space. Objects have other spatial properties, such as a shape. They might be round as a ball is. No process is round and no one would ask what shape the process of moving house has. Further, we can make or destroy an object, but we start or stop a process, but we do not talk of starting or stopping a desk or a spoon. These differences reflect the different mode of presence that objects and processes have. Noting these multiple contrasts it seems correct to say that objects are persisting things of one category, whereas processes are persisting things of another category. None of this is meant to show that there is anything wrong in suggesting that in some sense animals are properly thought of as processes, but it should not be thought that this proposal is supported by the general allegation that there is something incoherent or problematic about object persistence.

We come to a final question in this section: if we set aside what I am suggesting are confused criticisms of our normal way of thinking about organisms and animals as material objects, what sorts of reasons are there to assent to the process proposal (in either version 1 or version 2)?

I want to single out two reasons, formulated in a very general way. The first reason starts from the very plausible claim that such things as animals would not exist without the occurrence of extraordinarily complex processes. In the presentation of *arguments* appealing to this kind of fact the articulation of the necessary processes can be both extremely rich and highly instructive. But leaving the details aside, the question is: does it follow that we should either identify animals with the processes or sweep aside the concept of organism and describe biological reality in terms of processes? There are, it seems to me, considering the inference here two worries, one major and one far less major. The major worry is that the inference is fallacious. In general, one cannot identify an X with a process P just in case P is necessary for the existence of X. For one thing there may be things of other categories that are also necessary for the existence of X. Thus the existence of X may require some bodies or matter. And there is no reason to identify X with the necessary processes rather than as something made of the required matter. Now, it is hard to hang on to this insight given the details and complexity and interest of the processes that can be cited as necessary by the biologically informed. Of course, if there is no matter involved at all then something consisting of that will be no competitor to the identification with a process. But to defend the process view on that ground would require the very extreme claim that there is no matter at all required for animal persistence as a further premise in the argument, and normally no such claim is advanced. As far as I can see, though, if the argument under consideration is to be sound it must include some such premise.

I am suggesting then that there is no valid inference for the process theorist to rely on. There is, though, a second question about the argument. It is no doubt correct to say that organisms require processes for them to remain in existence as living organisms. After all, being alive is a highly complex and rich process. But this evident truth does not imply that organisms require processes to exist in all the states that they, organisms, can exist in. One possibility to consider is what is popularly called "suspended animation". Cannot an organism be frozen solid so that it still exists and maybe can be unfrozen and returned to a normal living condition? Another case to consider is whether the organism exists when it dies. Now, on the face of it, once it is dead there are no obvious processes that need to occur in it. Initially, after death that is not violent, it exists as a more or less complete corpse. What processes are needed for that? The more pressing question is whether the organism counts as being there although dead. Now, many philosophers, at least, do hold that death is the ceasing to exist of the animal. But it is by no means obvious that that is correct. Here I shall make simply two points. First, we do talk about dead animals in ways that seem to imply they are things that were alive, that is which implies that the same thing is still there, although dead. Think how people treat their dead pets. They refer to them by the name they used for their pet; they make past tense predications to them of what the living animal did; they look at them with the same feelings that perceiving their living pet generated. It seems quite untrue to human experience to suppose that they draw any distinction between the living and the dead animal, except acknowledging the crucial differences in capacity that dying makes. Second, we should consider the analogous case of the distinction between living and dead plants. When a plant dies, say, do we think that the dead thing is not the same plant as the one that used to be alive? We surely would ask a question such as: where did this plant come from? Where did it grow? These are questions that imply the self-same thing which is dead used to be alive and growing. I think that it is hard to deny that we do think about plants this way, but if so, it becomes harder to think we draw a sharp distinction between the animal that was alive and the animal that is dead. These are not knockdown considerations, but they should at least make us pause before simply agreeing that animals have to be alive and contain the processes constitutive of being alive. If they do not have to be then it becomes pressing to ask: what processes are the animal?[12]

In the arguments for the process view that one hears developed, one recurring theme is a stress on amazing cases that are found where, if we are operating with the concept of an animal, it is really hard to decide how many animals there are. All I wish to do is to comment on the general logic and relevance of such arguments. The first thing to note is that all such arguments can show is that the concept of an animal or organism is in trouble, and so perhaps should be abandoned. They do not show that the only available way to think is in terms of processes. Indeed, as such cases are normally described, they are couched in what might be called "object-involving descriptions". Thus, you might have something like: in this region there are 20 cell-like structures which seem to be organisms and yet the 20 cells also seem to constitute an organism. Such puzzle cases, even if insoluble using organism talk, do not show anything about the elimination of reference to all objects and replacement by process talk. On the contrary, the cases are described in quite different ways to that. This highlights an important question for process theorists: is the view that animals or organisms are processes and not objects, but that involved in the processes that animals are there are what we can call objects (which are not processes themselves) or is the view that even the so-called parts of animals (say cells) are processes too? The point is that if it is the former then the process view cannot be motivated by general problems with the general notion of object persistence. Second, in developing such difficulties it must not be assumed that an organism cannot be constituted by a collection of organisms. There seems to be nothing about organisms which prevents counting them in that way. Further, if we concede this possibility it implies that "organism" or "animal" is not a proper count term. If organisms can be parts of organisms, the question "How many organisms are there here?" is going to be difficult. Third, we need to balance the difficult cases against the easy cases of organism counting and recognition. Looking at a group of dogs it is completely easy to say how many dogs there are. Is it a requirement on the acceptability of employing a certain category to describe the world, where its application in the vast range of cases is quite straightforward, that there should not be some difficult cases? The implications of such difficulties are not obvious.

Nature of persons and nature of animals 247

I want to make one final point. We currently think in terms of objects, continuants if you like to so speak, in which complex processes occur. People thinking this way do not dismiss the idea of a process; indeed, they employ it centrally in their thought about certain objects, such as organisms. On the other side stands the proposal, developed in different ways, to replace the idea of organisms as objects and think of them, or of what is there, as processes. But a massive puzzle about this proposal is that in normal biological thought objects and processes are interlinked in the most fundamental way. Here is a quotation from *New Scientist* describing recent discoveries about cells.

> Single-cell profiling has already been used to investigate the earliest stages of pregnancy, as the fetal placenta implants into the mother's uterus. It revealed the location of fetal and maternal cells across the interface and the protein signals they use to communicate.[13]
> (*New Scientist*, 24 November 2018, p. 31)

What stands out here is that there is a concentration on objects of different kinds, especially cells, but also the mother, her uterus, and the fetus and placenta, and that the processes that are being investigated are ones specified in terms of interactions between these objects. This strongly suggests to me that we should recognise that objects are ineliminable from biology, and that our conception of processes is what we might call object involving. It is hard to see how we can replace talk of objects in biology by talk of processes when processes are picked out in terms of objects. The passage suggests another thought and a question that I shall voice briefly at the very end. There is no evidence to suggest that scientists studying the living are renouncing objects, including full organisms. It is, rather, philosophers reacting to developments who are proposing that. Is it not better to wait and see how the study of organisms evolves over time, rather than follow the advice of philosophers?

13.5 Conclusion

I have tried in this chapter to propose an interpretation of animalism that makes sense of it and of the arguments used in the debate about it. My hope is that philosophers of biology, and philosophically interested biologists, whose main focus is not on ordinary metaphysics, might get some clarification from that. In working through that I came to the tentative conclusion that the role of claims about animals in the standard philosophical debate about our persistence conditions does not really need validation by biologists, since they are obvious, or if they are not obvious they are usually not the kind of claim that biologists can shed any light on, being of a quite different character to the issues that they deal with. However, if we maintain that we are animals, then it is obvious that the correct theory of our general nature (as animals) is something about which we have to acknowledge science has the lead. In relation to the proposal that science should say that we are processes I argued that people who support this often make mistakes in their criticisms and characterisations of the alternative extant approach, and make some dubious inferences in supporting their own proposal. I also suggested that they have not taken on board the manifest dependence for the identification of processes on objects. But my discussion has been carried out at a very general level, and it may well be said that it underestimates the attractions of the proposal, which is a charge that I cannot with any confidence reject here and now. It may also be said that I have misunderstood the process proposal, and if so, I apologise.

Acknowledgements

I wish to thank John Dupré and Anne Sophie Meincke for the invitation to participate in the original conference, which was very stimulating, and for their patience, encouragement and the enormously helpful critical feedback they gave me during the construction of this chapter. I would also like to thank Stephan Blatti and Adam Ferner for stimulating discussion of some of these issues.

Notes

1 A possible explanation for this difference is that the kind of technical terms, of the sort I am talking about, that philosophers introduce are names for theories or claims, and with them it is hard to attach the name to anything more or less definite in the world, since theories are not, in the normal sense, located in the world and easily made out. Whereas substances or processes that are discovered and named in the sciences are easier to tie down.
2 This chapter was originally delivered in a lecture room which is what "here" stands for.
3 No one should expect that simple reflection could reveal to us what we are, or how the universe fundamentally is. It really would be surprising if philosophical metaphysics could reveal any such things.
4 I discuss this approach briefly and critically in Snowdon (2014, p. 21).
5 Locke offered a psychological analysis of what it is to be a person, in terms of possession of reason, reflection and self-consciousness, and defended as a consequence of that elucidation an analysis of what it is for a person to continue to exist (or to persist) in terms of psychological links over time, notably memory. Neo-Lockeans, such as Parfit and Shoemaker, liberalised the psychological analysis of persistence to avoid certain fairly obvious objections to the memory analysis.
6 To avoid confusion I wish to stress that what I am calling "animalism" should be regarded as incompatible with Lockeanism even when the latter is treated as an account of our persistence conditions, given the more or less undeniable truth that the proposed Lockean persistence conditions, in terms of psychological links over time, do not apply to animals who can remain in existence while

losing their mental powers. Further, other philosophers who espouse what they call "animalism", including Michael Ayers and Eric Olson, do as part of their total theory offer theories of animal persistence, and sometimes they are formulated in biological terms. Anyone proposing such theories can agree that what I and they call "animalism" is a distinct claim from their proposed theories of animal persistence.

7 Many of the ideas in this section were hammered out with the help of Stephan Blatti and receive a fuller exposition in the introduction to Blatti and Snowdon (2016).

8 The Hensel twins are a very unusual case, but that fact does not diminish the relevance of the argument if the premises in it are correct. Anyone who does not know of the Hensel case can find considerable information on the internet.

9 Versions of this argument can be found in Olson (1997, chapter 5, especially section V), and Snowdon (2014, chapter 4). The interest in animalism in philosophy over the last 30 years, compared to the almost total absence of the concept of an animal in discussions of personal identity before that, is due to the quite independent discovery by a number of people in the 1980s of this type of argument.

10 I wish to add in a note that there is what might be called version 3 of process theory, which is the claim that everything is a process. This idea is sometimes called "process philosophy". It would not be unwarranted to ask people who propose either version 1 or version 2 whether they actually also accept version 3. Crucially, if they do then they cannot regard empirically discerned facts about organisms and life as being the ultimate support for their claims. For the rest of the discussion here I shall ignore totally general process theories.

11 In Wiggins (2016, p. 272 (= This volume, Chapter 9, p. 169)) there is a quotation from a paper by Alexandre Guay and Thomas Pradeu (2016) where the second problem seems to make an appearance. They say: "The genidentity view is thus utterly anti-substantialist in so far as it suggests that the identity of X through time does not in any way presuppose the existence of a permanent 'core' or 'substrate' of X". I am taking this as evidence that some object to treating animals as fundamentally objects given the lack of a permanent core (*Editorial note*: The passage quoted by Wiggins appears in modified form on p. 318 of the published version Guay and Pradeu's chapter.).

12 I discuss the issue of the relation between life and animal existence at greater length in Snowdon (2014, chapter 5), though without relating it there to a consideration of the cogency of the process proposal. Professor Dupré has made to me the perfectly correct point that the considerations here do not show that animals are not processes but merely that the processes they are need not be processes involved in being alive. That is true, and if it is to have a more general bite then that we would need to investigate properly whether there is any case for thinking that there must be processes in all the conditions in which it is correct to think the animal exists. That is not something investigated here.

13 The overall title of the article is "37 Trillion Pieces of You". Clearly a piece is not a process, but rather an object.

References

Blatti, S. and Snowdon, P. F. (Eds.) (2016) *Animalism,* Oxford: Oxford University Press.

Campbell, T. and McMahan, J. (2016) 'Animalism and the Varieties of Conjoined Twinning', in: Blatti, S. and Snowdon, P. F. (Eds.), *Animalism,* Oxford: Oxford University Press, pp. 229–252.

Guay, A. and Pradeu, T. (2016) 'To Be Continued. The Genidentity of Physical and Biological Processes', in: Guay, A. and Pradeu, T. (Eds.), *Individuals Across the Sciences*, Oxford: Oxford University Press, pp. 317–347.

New Scientist No 3205, '37 Trillion Pieces of You', 24 November 2018, pp. 28–31, https://www.newscientist.com/article/mg24032050-100-37-trillion-pieces-of-you-the-plan-to-map-the-entire-human-body/.

Olson, E. T. (1997) *The Human Animal. Personal Identity Without Psychology*, Oxford: Oxford University Press.

Snowdon, P. F. (2014) *Animals, Persons, Ourselves*, Oxford: Oxford University Press.

Wiggins, D. (2016) 'Activity, Process, Continuant, Substance, Organism', *Philosophy* 91, pp. 269–280 (reprinted as Chapter 9 in this volume).

14 Processual animalism

Towards a scientifically informed theory of personal identity

Anne Sophie Meincke

14.1 Introduction

The problem of personal identity as discussed in contemporary analytic metaphysics consists in determining what it takes for a person to persist through time. What are the constitutive criteria of a person's diachronic identity? Traditionally, the focus has been on psychological criteria – memory, personality traits, consciousness – in accordance with a purely psychological notion of a person which could be exemplified by human people, angels, God, robots and aliens alike. This psychological or mentalist approach to personal identity and personhood is paradigmatically expressed in John Locke's influential definition of a person as "a thinking intelligent being, that has reason and reflection, and can consider itself as itself, the same thinking thing, in different times and places" (Locke 1987, p. 335). Locke explicitly distinguishes "personal identity" from the "identity of man", arguing that we can conceive of scenarios where these two come apart. His story about the prince whose soul and, thereby, consciousness switch into a cobbler's body[1] has been followed in the philosophical debate by an abundance of thought experiments of varying degrees of extravagance (e.g., brain transplants, tele-transport, brain-state transfer), meant to show that physical aspects of a person's existence matter, if at all, to personal identity only as the material basis for the realisation of the relevant psychological properties. Up until recently, personal identity, in one way or the other, has widely been believed to be inextricably tied to sentience, rationality and psychology (see also Meincke 2019c).[2]

"Up until recently" since things are changing now, given the rise of so-called animalism (van Inwagen 1990, Olson 1997, Wiggins 2001, Snowdon 2014, Blatti and Snowdon 2016). Animalists claim that we are animals, namely human animals and, that is, a species of organism. Accordingly,[3] the constitutive conditions of our identity through time are thought to be primarily or purely biological rather than (purely) psychological.[4] Animalists challenge the mentalist standard interpretation of the notorious puzzle cases, insisting that we do not "go where the brain goes" or, more precisely, "where the cerebrum goes",[5] that we do not continue to exist through tele-transport, brain-state transfer, etc. This criticism leaves open the possibility of prudentially or morally relevant relations between, say, the person whose cerebrum is surgically removed and the person who receives this cerebrum through transplantation (Olson 1997, chapter 3). Animalism calls for a change of focus within the debate on personal identity: insofar as we are concerned with the ontological problem of "our" identity through time, we should replace the traditional question "What are the persistence conditions of persons?" with the question "What are the persistence conditions of human animals?" (Olson 1997, p. 23ff.). "Personal identity", in some sense, is a misnomer for what we really ought to investigate (Olson 1997, p. 26f.): biological identity.

Animalism is an important novel approach to the problem of personal identity in contemporary analytic metaphysics. Its emphasis on the biological dimension of our persistence through time certainly takes us a step closer to an understanding of what is *human* about human persons and their identity through time – something that, within the traditional mentalist framework, has not even come into sight as a worthwhile project. The animalist approach also raises hopes to overcome the stalemate between the so-called Complex View and the so-called Simple View of personal identity.[6] According to the Complex View, personal identity is a complex phenomenon that can be analysed in terms of more basic empirical relations of continuity, typically – in accordance with the mentalist approach – relations of psychological continuity. However, proponents of the Complex View struggle to provide an analysis that would persuasively handle hypothetical cases of branching and reduplication of mental lives while holding on to the thesis that psychological continuity constitutes personal identity in the sense of numerical identity. In response, Derek Parfit (1987) has famously argued that when it comes to accounting for our survival, we should be content with a version of psychological continuity itself, i.e., with a relation weaker than numerical identity. Proponents of the Simple View, on the other hand, insist that personal identity is strict, numerical identity. They avoid the difficulties of the Complex View by treating personal identity as a simple phenomenon not admitting of an analysis in terms of empirical continuity relations. Personal identity is believed to be a "further fact" over and above any empirical facts about the world and, that is, effectively: a mystery. The debate thus is stuck in a dilemma: personal identity is either explained away or not being explained at all (Meincke 2015a, 2018c, 2019c). In contrast, animalism's appeal to biological identity appears to tell us informatively what our identity through time consists in, without thereby sacrificing its robust, numerical character; and it does so, as it seems, in accordance with science, i.e., biology.

In what follows, I assess animalism by taking a closer look at its core tenet that personal identity, properly understood, is biological identity. As it turns out, animalism largely fails to live up to the hopes it gives rise to. More particularly, I argue for the following three critical claims – (i) the *Harmless Claim*: animalism has not yet sufficiently explicated its key notion of biological identity; (ii) the *Not-so-Harmless Claim*: a large part of what animalists do say about biological identity conflicts with what biologists and philosophers of biology say about biological identity; (iii) the *Radical Claim*: animalism cannot provide a convincing account of personal identity as long as the notion of biological identity employed is based on the metaphysical assumption that organisms are substances, or things composed of (smaller) things. This is to say that animalism can do better if it adopts a different ontological framework. I show how, due to a shared commitment to thing ontology, attempts made by animalists Eric Olson and Peter van Inwagen to explicate biological identity in terms of biological continuity and sameness of life run into the same dilemma faced by the competing psychological theories of personal identity. I then propose what I call processual animalism: the view that human persons, qua organisms, are a particular kind of biological process. I explain how processual animalism delivers a convincing, scientifically informed account of biological identity and, on that basis, of personal identity that overcomes the traditional dilemma rather than repeating it.

14.2 Animalism on biological identity: two criteria

14.2.1 Biological continuity

According to Olson (1997, p. 20), animalism, also called "the Biological Approach", offers "a distinct physicalistic criterion" of our persistence, "namely continuity of life-sustaining, vegetative functions". Olson formulates this criterion as follows:

> If x is an animal at t and y exists at t^*, $x = y$ if and only if the vital functions that y has at t^* are causally continuous in the appropriate way with those that x has at t.
>
> (1997, p. 135)

Or less formally:

> Like other animals, we persist as long as our life-sustaining functions remain intact. One survives, at any point in one's career, just in case one's circulation, respiration, metabolism, and the like continue to function, or as long as those activities have not irreversibly come to a halt, or as long as one's capacity to direct and regulate those functions is not destroyed.
>
> (Olson 1997, p. 89)

The Biological Continuity Criterion entails that no psychological continuity is needed for our persistence. Animals like us contingently acquire mental states, but they don't have to in order to be what they are: a type of organism. Likewise, animals can lose mental capacities without ceasing to exist. This explains both how it is possible that each of us once was a fetus and might continue to exist in a persistent vegetative state. Assuming, as Olson (1997, p. 73f.) does, that neither a human fetus nor a so-called human vegetable has any mental states, this is something the competing psychological versions of a Complex View of personal identity struggle to explain.

The Biological Continuity Criterion, though prima facie plausible, faces a problem: the branching problem. We know of sexually reproducing animals that each such animal develops from a fertilised egg, a zygote, passing through different developmental stages: morula, blastula, gastrula and so on. Humans are sexually reproducing animals like others; but unlike many others, humans occasionally produce more than one new human animal from a single fertilised egg. In about three of a thousand pregnancies, the zygote develops into two (or even three) morulas which then develop into monozygotic twins (or triplets). According to the Biological Continuity Criterion, biological continuity is necessary and sufficient for biological identity in the sense of numerical identity. However, this leads to contradiction in the branching case: since both twins are biologically continuous with the original zygote, we would have to conclude that both are numerically identical with it, from which, per transitivity of the relation of numerical identity, we would have to infer further that both twins are numerically identical with one another.

Have we ever been a zygote? The prevailing answer given by animalists in the light of above contradiction is No. No human animal once was a zygote. Instead, human organisms are claimed to start existing generally not before twinning ceases to be possible (Wiggins 2001, p. 239), i.e., at roughly 12 (Wiggins 2016, p. 276) or 14 to 16 days after conception (van Inwagen 1990, p. 152ff., Olson 1997, p. 90ff.). This claim – let's call it the "Fourteen Days Claim" – is problematic for several reasons. First, it looks like an *ad hoc* stipulation made solely for the purpose of avoiding the breakdown of the Biological Continuity Criterion in a case of branching. Note that the situation is parallel to the much-discussed branching problem for mentalist Complex Views of personal identity employing a Psychological Continuity Criterion: in the famous "split brain" case, or any of its variants, applying the Psychological Continuity Criterion results, per transitivity of identity, in identifying the two post-operative persons both with the original person and with one another. Proponents of the Psychological Continuity Criterion sometimes resort to a non-branching clause according to which psychological continuity is constitutive of personal identity only if it does not take a branching form (Shoemaker 1970, p. 278f.). In the same way, animalists may want to claim that a human animal y at t^* that is biologically continuous with a zygote x at t is identical with the zygote x at t^* if and only if there is no other human animal z at t^* that is biologically continuous with the zygote x at t^*. However, this move seems no less *ad hoc* than the claim that human animals in general start existing only 14 to 16 days after conception. As has been discussed extensively with respect to the parallel case of psychological accounts of personal identity, it also comes at the cost of rendering biological identity extrinsic: the answer to the question of when I began to exist would become dependent on extrinsic facts, such as the existence of a monozygotic twin.

The second problem is that it is hard to make sense of the Fourteen Days Claim's implication that even when no twinning occurs – which holds for the majority of human pregnancies and even more so for other sexually reproducing animal species – literally a new entity comes into existence at a certain point of development: what we thought were stages within one single continuous developmental process has to be reinterpreted as a replacement of entities. How can this be squared with the Biological Continuity Criterion? There are two possible answers: either the case at issue is no case of biological continuity and, hence, the Biological Continuity Criterion is not applicable; or the case at issue is a case of biological continuity but the Biological Continuity Criterion actually formulates only a necessary, not a sufficient condition of biological identity.

Olson (1997, p. 90ff.) seems to take the first route, arguing that the cleavages undergone by the zygote must be understood as producing a multiplicity of organisms rather than as developmental changes undergone by one and the same organism that is first unicellular and then multicellular. Only with the formation of the primitive streak a multicellular organism comes into existence; before that there is "merely a mass of living cells stuck together" (Olson 1997, p. 91). To support this argument, Olson (1997, p. 93) compares the cleaving zygote with an amoeba undergoing binary fission: just as the amoeba stops existing as soon as it divides, being replaced with two daughter amoebae, the zygote stops existing as it cleaves, being replaced with two daughter cells. This is not meant to deny that there is continuity here – the process of zygote cleavage is as continuous as is the process of amoeba fission; however, only in the sense of spatiotemporal continuity, not in the sense of biological continuity. Spatiotemporal continuity, Olson argues, cannot be sufficient for biological identity because if it were, given the spatiotemporal continuity between the fertilised egg with the unfertilised egg and the sperm, we would, by transitivity of identity, have to identify these with one another too so "that you and I and every other living organism existed from the beginning of life on earth" (Olson 1997, p. 93).

Van Inwagen (1990, p. 149ff.), who inspired Olson's discussion, presents the case of the cleaving zygote and the, supposedly analogous, case of the splitting amoeba in a slightly different way which indicates that he leans towards the second route of combining the Fourteen Days Claim with a Biological Continuity Criterion of biological identity. Apart from treating the physical separation of the two daughter amoebae resulting from binary fission as a form of spatiotemporal *dis*continuity, van Inwagen believes that the original amoeba's life ends some time before the event of physical separation and, that is, "without a readily apparent break in the continuity of the processes of life" (van Inwagen 1990, p. 151). Assuming that "the continuity of the processes of life" means biological continuity and that the "end" of an amoeba's "life" amounts to the end of that amoeba's existence, this is to say that biological continuity is necessary but not sufficient for biological identity. Accordingly, the fact that embryo development appears to be one single biologically continuous process does not warrant the conclusion that there is one single organism present throughout this process. Instead, "if an embryo is still capable of twinning, then it is a mere virtual object" (van Inwagen 1990, p. 154) and that means it doesn't exist. A gastrula, van Inwagen (1990, p. 157) argues, since it fails to restore its original structure if injured, has no overall metabolism and does not grow in a directed manner, "is not an organism" but "like a mass of swarming adders or a bliger[7] or a chair; it is not really there; it does not exist; the cells arranged in gastrula fashion compose nothing" (ibid.).

Despite minor differences, both Olson and van Inwagen ask us to conceive of the coming into existence of a single human organism as a somewhat[8] sudden emergence from a multiplicity of organisms which are human organisms too but not in the (proper) sense of being "human beings, *Homo sapiens*" (Olson 1997, p. 92). This is not convincing. Surely, coming-to-be mothers will be surprised to learn that, in the early stages of their pregnancy, they are not carrying a developing embryo in their womb but rather a fast increasing number of tiny would-be human organisms which at 14 to 16 days after conception abruptly give way to a single new human organism.[9] The empirical claims made in support of this picture are, to say the least, disputable. Is the growth of an embryo at the gastrula-stage really just "the sum of the uncoordinated growth and fission of its component cells" (van Inwagen 1997, p. 157)? We know that a cell's position in the morula and blastula partly determines its later developmental fate and that cell differentiation is sensitive also to other epigenetic factors (Wolpert 1969, 2011, Jaeger and Reinitz 2006, Jaeger, Irons and Monk 2008; see also the discussion in Meincke 2015b, 2018b). Contrary to what Olson and van Inwagen seem to assume, the plasticity of early-stage embryonic cells exactly does not speak against their coherence and interaction with one another. Instead, especially in the regulative type of cellular development that we find in mammalian embryos, plasticity appears to be highly functional with respect to the process of cell differentiation in that cells are able to respond in a flexible way to extracellular cues and to stand in for one another in the case of disturbances. Early embryonic development is driven by gene regulatory networks such that each cell's fate is conditional upon the neighbouring cells' fates (Arias, Nichols and Schröter 2013).

If, as van Inwagen (1990, p. 153f.) stresses, the question of when a multicellular organism begins to exist "has no answer that can be discovered apart from a very detailed examination of the facts of embryonic development", then we are well advised to put the Fourteen Days Claim to bed. Multicellular organisms don't appear out of the blue. The facts of embryonic development suggest continuity all the way through, including a continuous process of gradual increase of unity. A morula may have less unity than a two-week-old embryo; but it does have some unity. Generally, it would be rather surprising if members of the species homo sapiens started to exist 14 to 16 days later than members of other sexually reproducing species, just because in humans occasionally more than one embryo (and, later, child) develops from a zygote. One would reasonably expect the same gradual development here and there.

Animalism's Biological Continuity Criterion of biological identity thus turns out to be in no better position to account for "our" identity than its competitors' Psychological Continuity Criterion of personal identity: it is prone to the branching problem just the same. Note that it also faces the reverse problem: tetragametic chimerism, the fusion of two nonidentical twins, which is common in some primates, especially marmosets, and which occurs in humans too (apparently more often than assumed in the past). Additionally, it is far from clear that asexual reproduction through binary fission such as in amoeba can be interpreted in the ways suggested by Olson and van Inwagen; the rumour persists that amoebae are theoretically immortal. Or think of budding, where a part of an organism's body develops into a new body, which occurs not only in plants but also in many animals. All these scenarios present cases where biological continuity arguably obtains but where the inference to numerical identity suggested by the Biological Continuity Criterion incurs the contradiction that numerically distinct entities would have to be identified with one another. If there are no good reasons to deny biological continuity for certain phases of embryonic development, it seems one would have to appeal to something else, some factor above and beyond biological continuity, to mark out the beginning of a new organism within what looks like a perfectly continuous process.

14.2.2 Life

Olson (1997, p. 136) acknowledges that determining "which causal connections are 'appropriate'" can be difficult. However, apart from regarding this as a challenge faced by "any theory about the persistence of concrete objects", he insists that the main point is sufficiently clear: animals persist as long as they are alive; and they cease to exist when they die.[10] Olson (1997, p. 137) then presents an alternative formulation of the animalist identity criterion, namely "a criterion of identity for organisms in terms of lives":

> An organism persists just in case the metabolic process that is its individual biological life continues to impose its characteristic organization on new particles.

Or put more formally (Olson 1997, p. 138):

> For any organisms x and y, $x = y$ if and only if x's life is y's life.

Olson relies here on van Inwagen's idea of a life as a unifying event with compositional power. Assuming "that matter is ultimately particulate" (van Inwagen 1990, p. 5), van Inwagen defends the view that the only instance in which elementary particles compose some material object is the composition of living material objects:

> ($\exists y$ the xs compose y) if and only if the activity of the xs constitutes a life.
> (van Inwagen 1990, p. 90)

According to this so-called *Special Composition Thesis*,[11] there are no composite objects other than organisms: no tables, cars, volcanoes, rocks, planets, etc. All material objects that exist in our world are either mereological simples or living beings. To explain the unique ontological status of biological life, van Inwagen (1990, p. 84ff.) introduces the analogy of a club that constantly recruits members through press-ganging and expels them after exploiting their resources. In a modified version of this analogy (van Inwagen 1990, p. 85f.), the club's members are unconscious automata, some of which are sent out to capture wild automata, which, when brought in, are each taken apart so that their components can be used "to construct an automaton that is physically suited for membership" (van Inwagen 1990, p. 86). In the same way, material particles are drawn into, disassembled and expelled by the "homeodynamic event" called "life" (ibid.).[12] Organisms thus are material objects for which it is possible "to be composed of different elementary particles at different times" (van Inwagen 1990, p. 6); they persist like a club "persists through all its changes of membership" (van Inwagen 1990, p. 85). Van Inwagen's criterion of biological identity accordingly reads as follows:

> If an organism exists at a certain moment, then it exists whenever and wherever – and only when and only where – the event that is its life at that moment is occurring; more exactly, if the activity of the xs at t_1 constitutes a life, and the activity of the ys at t_2 constitutes a life, then the organism that the xs compose at t_1 is the organism that the ys compose at t_2 if and only if the life constituted by the activity of the xs at t_1 is the life constituted by the activity of the ys at t_2.
> (van Inwagen 1990, p. 145)

There is an obvious problem with van Inwagen's and Olson's appeal to life as a criterion of biological identity: it seems to presuppose what it is meant to explain – the very identity of the organism. To say that organism *a* identified at t_1 and organism *b* identified at t_2 are the same individual organism because *a*'s life is also *b*'s life is to say that *a* and *b* are identical because their lives are identical. The double appearance of identity in explanandum and explanans renders the criterion circular – unless we can give an independent criterion for the identity of lives.

Van Inwagen acknowledges this problem: the criterion – van Inwagen calls it *Life* – "tells us when organisms persist; but in order to apply it, we should have to know when lives persist" (van Inwagen 1990, p. 145). He goes on to explain that biological continuity of the sort Locke (1987, p. 330f.) required for organisms is indeed necessary for the persistence of lives: a life can be suspended (e.g., through freezing the organism) but not be disrupted without its (and, hence, the organism's) identity thereby being destroyed (van Inwagen 1990, p. 145–149). However, this does not imply that biological continuity is also sufficient for the identity of lives. Instead, there seem to be cases in which the former obtains but the latter does not, such as the case of the monozygotic twins which have distinct lives despite continuously tracing back to a common origin: the zygote (van Inwagen 1990, p. 149).[13]

At this point, surely, one would like to know what *is* sufficient for the identity of lives if not biological continuity. Van Inwagen's reply is that there might not be an answer to this question. However, he emphasises that this is unavoidable. Assuming that "one can lay down the conditions governing the persistence of the objects in a certain category only if one is allowed to presuppose the persistence of the objects in some other category", "[o]ne may, of course, go on to lay down the conditions governing the persistence of the second category", and so on. But one cannot do this ad infinitum;

> [e]ventually, one must be content with describing the persistence of some sort of object by reference to the persistence of some other sort of object in the absence of any general and explicit statement of the conditions of persistence for objects of that other sort.
> (van Inwagen 1990, p. 145; see also 157f.)

That is, in the case of the persistence of organisms, we have to rely on the persistence of lives in the absence of any informative explanation of what this persistence depends on. Note the astonishing parallel with the mentalist Simple Views of personal identity which, in order to evade difficulties like the branching problem, resort to postulating some deep, further fact over and above the empirical facts about a given (e.g., branching) scenario, such as the continuing existence of an immaterial, indivisible soul. However, to learn that "the person goes where the soul goes" (Swinburne 1984, p. 27ff.) clearly could count as an informative explanation of personal identity only if we knew the identity conditions of souls; but we do not appear to know them, and it is not clear if we could ever know them even if we knew for sure (which we don't) that there are such things as souls.[14]

According to van Inwagen, there is more to be said in defence of the Life Criterion of biological identity. Apart from the continuity involved, which in many cases indeed makes it possible to safely conjecture the presence of a single, self-identical life (van Inwagen 1990, p. 156), there is, van Inwagen claims, the fact that "a life is a reasonably well-individuated event" (van Inwagen 1990, p. 87). Indeed, biological continuity tends to be produced by life itself by virtue of its self-directing nature (ibid.). Most importantly, unlike a wave that can be in superposition with another wave, a life is "a jealous event" (van Inwagen 1990, p. 89) in the sense that it does not allow any of the simples that constitute it at a certain time to constitute, at the same time, another, numerically distinct life unless it and the other life are related to one another through subordination. This is to say that there is always a one-to-one relation between organisms and lives: "for any x, if the activity of those xs constitutes a life, then those xs compose exactly one organism" (van Inwagen, 1990, p. 91). This so-called *Uniqueness Thesis* follows both from the *Special Composition Thesis*, according to which the constitution of lives through the activities of elementary particles amounts to the composition of an object – an organism, and from the jealous and subordinating character of life – van Inwagen (1990, p. 89f.) compares life to a brigand or a bank robber who would not tolerate another bank robber simultaneously using the same bank. Van Inwagen (1990, p. 89) admits that there are cases that seem to involve overlap of two lives without one being subordinate to the other, for instance, conjoined twins; but he believes this impression can be debunked as illusory. We are, therefore, assured that the impossibility to identify sufficient persistence conditions for lives is not only unavoidable but also innocuous. The latent circularity of a criterion of biological identity in terms of the identity of lives does not impinge on the criterion's explanatory utility thanks to the individuality of life:

> It is this feature of lives, their *seeming* to be *well individuated*, that made it possible for Locke to explain the identity of a man in terms of the identity of a life and thereby to offer something that we can at least take seriously as a possible explanation of human identity. If lives did not at least *appear* to be *well-individuated events*, Locke's explanation would not even be worth considering; we should all regard it as an explanation of the obscure through the no less obscure.
>
> (van Inwagen 1990, p. 87; emphases added)

If this argument is valid, it implies that the explanation of biological identity in terms of the identity of lives in fact would have to be deemed useless, should it turn out that biological lives are less well individuated than hitherto assumed. As it happens, the on-going debate about biological individuality in the philosophy of biology may well convince us that exactly this is the case. While philosophers of biology have not yet reached an agreement about how to conceptualise biological individuality most appropriately (Bouchard and Huneman 2013, Clarke 2010, 2013, Lidgard and Nyhart 2017), one thing that seems widely accepted is that biological life is far from being well-individuated – this, in fact, being one of the reasons why it is so difficult to determine what exactly is required for something to count as a "biological individual". A growing number of biologists and philosophers of biology (e.g., Godfrey-Smith 2009, Huneman 2014, Huneman this volume,

Queller and Strassmann 2016) argue that there is a plurality of requirements which are met at various degrees of strength by different living systems, which is to say that biological individuality and identity[15] as we find it in the living world is not all-or-nothing but rather comes in degrees. Organisms are *more or less* "individuals". Accordingly, it is also far from clear that there is always an exact one-to-one relation between lives and organisms – if lives and organisms can be meaningfully distinguished from one another at all.[16]

There are at least three respects in which recent findings in biology and their reflection in the philosophy of biology challenge van Inwagen's and Olson's picture of biological lives.

First, biological individuals are located within "space of individuality" defined by both conflict and cooperation (and possibly other factors, see Huneman this volume). This implies that van Inwagen's and Olson's characterisation of life in terms of suppression is one-sided, to say the least. In reality, biological life is much less about suppression and competition, and much more about cooperation (Dupré and O'Malley 2009). Multicellular organisms are likely to have emerged either from the cooperation of different species of single-cell organisms, according to the so-called Endosymbiotic Theory most prominently defended by Lynn Margulis, formerly Lynn Sagan (first 1967), or from the cooperation of many organisms of the same species, according to the Colonial Theory (which was proposed by Ernst Haeckel in 1874). Whichever story may turn out to be true, there was and is no life that mysteriously forces multiple cells into one single organism, suppressing their single lives, as insinuated by van Inwagen's analogy of the press-ganging club.[17] This is impressively illustrated by the numerous extant intermediate stages between groups of cooperating organisms and true multicellular organisms, such as the so-called modular or colonial organisms. Corals, for instance, typically live in compact colonies of many genetically identical individual polyps. Each polyp is a sac-like animal. What is "the" organism here? The single polyp or the colony as a whole? And how many lives are there? The relevant life processes run through all the polyps as they are physically connected, and new polyps are generated by existing polyps budding or dividing. At the same time, single polyps can die or even wander off and found a new colony (through fragmentation or bailout). It seems that we would be as justified in saying the coral reef is one organism hosting many lives as we would be in saying that there are many organisms (the polyps) participating in one life (the life of the coral reef). Other examples of this sort are abundant (on the Portuguese man-of-war, see Oderberg this volume).

Second, biology recently has come to discover the ubiquity of symbiosis. That is to say, symbiosis has not only been a major driving force in the early evolution of life on Earth (see Moreno this volume), but also is the rule, rather than the exception, for life as present today. Nearly 60% of the cells in the human body are microbial. There wouldn't be any corals without their various algae endosymbionts. Symbiotic relationships of various kinds with various species (but especially with microbes, see McFall-Ngai 2015, Chiu and Eberl 2016) shape the anatomy of organisms of all taxa, guide their development through the entire life cycle, are crucial to their physiological organisation including metabolism and gene expression, co-create the immune system and thus contribute to an organism's fitness (Gilbert, Sapp and Tauber 2012, Pradeu 2012, Gilbert and Tauber 2016). Organisms are, as Gilbert, Sapp and Tauber (2012, p. 331) put it, "anatomically, physiologically, developmentally, genetically, and immunologically multigenomic and multispecies complexes" and, i.e., they are so-called "holobionts": "integrated organism[s] comprised of both host elements and persistent populations of symbionts" (ibid., 327f.; first Margulis 1991). According to the hologenome theory of evolution (first Rosenberg et al. 2007, Zilber-Rosenberg and Rosenberg 2008; see also Rosenberg and Zilber-Rosenberg 2018), it is these larger unities – the holobionts – that are subject to natural selection and, in that sense, "evolutionary individuals" (for a critical discussion see Skillings 2016). Again, difficult questions arise as to the relationship between lives and organisms. Is the holobiont as an "integrated organism" made up of the lives of the symbionts? Or should we think of the multispecies organisms that are intertwined in symbiotic relationships, especially in obligate forms of mutualism, as sharing one life with one another?

Finally, research on the behaviour of eusocial insects, such as termites, ants and bees, has led to a revival of Wheeler's (1911) concept of the superorganism: the division of labour practised by these insects is so specialised that individual members of the colony cannot survive on their own for long. An ant colony is comparable to a multicellular organism in terms of unity and interdependency, with the analogues of the cells being holobionts (in the case of ants, each consisting of the ant plus certain fungi and bacteria; see also Dupré this volume). It has been argued that such superorganisms undergo natural selection at the group level (Wilson and Sober 1989, Hölldobler and Wilson 2009) and count as biological individuals (Haber 2013; though Haber remains sceptical of the concept of superorganism). It thus appears we can ask the same questions about organisms, lives and their relationships as we did before, only again one more level up; and just in the same way it remains unclear if the case at hand is one of many organisms sharing one single life or many lives composing one organism. The only thing that is clear is that *some* higher-level unity is present; however, this higher-level unity does not entirely suppress the lower-level unities.

Colonial organisms, holobionts and superorganisms exemplify the variety of possible locations within the multi-dimensional space of individuality. Note that these locations are variable not only on evolutionary time-scales. Individual members of certain species may move through this space during their life-time, becoming less or more individual in different ways, depending on their body structure, immunology, life-style, etc. A striking example, which, despite its familiarity, is often overlooked, is mammalian pregnancy. Is a six-week pregnant woman one or two individuals? Does the fetus, not being viable outside the womb, participate in the mother's life? Does the fetus even count as an organism in its own right given its intimate connections with the maternal organism (Kingma 2018, 2019, forthcoming, this volume, Dupré this volume, Meincke under review)?

Biology debunks the idea that biological life is a well-individuated event and that there is always an exact one-to-one relation between lives and organisms. It is not true that, as Olson (1997, p. 137) claims, "[l]ives are easy to count", i.e., that "there is a clear difference between a situation that contains one life and a situation that contains two" – not even "in most cases". Accordingly, it will not help to appeal to the identity of lives in order to account for the identity of organisms through time, without providing any informative explanation as to what constitutes the identity and individuality of lives. Contrary to van Inwagen's and Olson's hopes, the Life Criterion of biological identity remains trapped in circularity[18] and, if anything, explains the obscure through the even more obscure.

14.3 Biological identity reconceived: processual animalism

Animalism has not yet provided a satisfactory account of our identity through time to the extent that it has not yet given us a satisfactory account of biological identity. This is partly due to the fact that animalists have not said so much about biological identity at all. Their main concern has been on rejecting the intuitions supporting the Psychological Theory of personal identity, in particular the "We go where our brain goes" intuitions (van Inwagen 1990, chapter 15, Olson 1997, chapter 3), and on defending the foundational animalist claim that we are human animals (see especially Snowdon 2014; also Olson 1997, chapter 5, 2007, chapter 2). This is understandable given the historical predominance of the mentalist view of the human person. However, in order to assess animalism's implications with respect to our identity through time, we should get to know more about the key concept of biological identity. What does biological identity consist in?

This criticism appears to be harmless – therefore I call it the *Harmless Claim* – as one could reasonably suspect that the weakness it identifies can easily be remedied by just adding more details after the basic picture has been sketched. But here is my second, the *Not-so-Harmless Claim*: much of what animalists do say about biological identity conflicts with what biologists and philosophers of biology say about biological identity. Claims, such as that the morula is just a mass of disconnected cells and, hence, does not exist, and that biological life, in virtue of its subordinating power, is well-individuated, have little or no foothold in empirical facts about organisms, their organisation and their development.

Could animalists address this second concern by just leaving the details to the biologists? Van Inwagen (1990, p. 84) concedes that "[i]n the last analysis, it is the business of biology to answer [the] question" of what life is. Maybe it was wrong from the start to interfere with the biologists' (and possibly also the philosophers of biology's) specialist expertise? Maybe not too much empirical detail is even needed for the purpose of a metaphysical account of biological identity? Unfortunately, this strategy won't succeed. As we have seen, van Inwagen's and Olson's metaphysical claims that human organisms do not start existing until 14 to 16 days after conception and that biological identity consists in the identity of lives remain unconvincing exactly because they rely on the above mistaken empirical assumptions about morulas and lives. This indicates that the information contributed by biologists and philosophers of biology does not simply add details to the basic metaphysical picture drawn by animalists but in fact fundamentally calls into question the adequacy of this picture. Indeed, I want to defend the *Radical Claim* that, ultimately, the commitment to an inappropriate ontological framework is responsible for animalism's failure to deliver a convincing account of our identity through time.

Evidence for something being fundamentally wrong with animalism comes from the return of the very dilemma an animalist approach to our identity through time had appeared to avoid. As we have seen, the animalist appeal to biological continuity as a necessary and sufficient condition of biological identity does not fare any better than the mentalist appeal to psychological continuity as a necessary and sufficient condition of personal identity; fission and fusion – which in biology are not mere thought experiments but phenomena as real and frequent as anything can be – cause the very same troubles, to which animalism responds with structurally the same move made by the (mentalist) Simple Views of personal identity: the appeal to some fact over and above continuity, a "further fact" which for principal reasons does not admit of informative analysis. Avoiding the looming threat of identity getting drowned in continuity comes at the price of explanatory mysticism here and there.

Elsewhere (Meincke 2018c) I have argued that the dilemma of personal identity can directly be traced back to the antagonism between two ontological views of the person – the substance ontological view of persons as mental substances or "souls" and the Humean view of persons as bundles of mental events; two views which, despite their differences, coincide in the conviction that a person is some sort of *thing* – either a bigger *thing* or a bundle of small *things*. A parallel story applies to the dilemma of biological identity for animalism. Animalism is characterised by a peculiar combination of two historically opposed views: an Aristotle-style substance concept on the one hand and atomism on the other. To wit, on the one hand, van Inwagen's and Olson's confidence in the individuating power of biological life, which supposedly guarantees a one-to-one relation between organisms and lives, corresponds with Aristotle's idea of an organism as a living substance, i.e., as a concrete particular individuated by a soul. (Van Inwagen (1990, p. 83) explicitly compares his concept of life with Aristotle's concept of ψυχή.)[19] Animalism's explanatory mysticism – the first horn of the dilemma – thus doesn't come as a surprise: substances, and – more generally – things, are numerically self-identical by definition. On the other hand, according to van Inwagen and Olson, if there were no individuating lives (no souls, so to speak), we would be left with nothing but elementary particles of matter, i.e., with atoms. Thus, eliminativism, the other horn of the dilemma, is likewise predictable: without the numerical identity of living substances – organisms – there would not be any identity apart from that of the elementary particles. As indicated, historically atomism and substance ontology are competing views of reality: Aristotle developed his concept of a substance exactly in order to overcome atomism as defended by the pre-Socratics, such as Democritus. However, in one important and fundamental respect the two competing ontologies resemble each other – things are taken to be the building blocks of reality: either bigger things, composed of smaller things (substances), or just smaller things (atoms).

I call "thing ontology" any ontology according to which the world most fundamentally[20] is inhabited by "things", i.e., by entities for the identity of which change is not essential. Within such an ontology, any account of our identity – whether mentalist or animalist – is bound to oscillate between mysticism and eliminativism. Animalism, in other words, cannot provide a convincing account of our identity through time as long as the notion of biological identity employed is based on the metaphysical assumption that organisms are substances or things composed of (smaller) things.[21] However, this is not to say that animalism, in principle, cannot be a persuasive theory of personal identity. Indeed, I want to show that it can, namely if it switches to a process ontological framework, according to which reality most fundamentally, and biological reality specifically, is constituted by processes, i.e., entities for the identity of which change is essential. Recent progress in biology, especially in systems biology, developmental biology, evolutionary biology, ecology and their fusions ("eco-evo-devo"), suggests that organisms are processes: they exist only insofar as they constantly change. A growing number of philosophers of biology, biology-inspired metaphysicians and biologists (Bickhard 2011, Dupré 2012, 2014a, 2014b, 2017, this volume, Jaeger and Monk 2015, Meincke 2018c, 2019a, 2019b, 2019c, Nicholson and Dupré 2018, Simons 2018) therefore are promoting process ontology as the most appropriate ontological framework for research on organisms in science and philosophy.

For reasons of space, let us consider here only metabolism, the exchange of matter and energy with the environment. An organism exists and persists only if and so long as this exchange takes place – the end of metabolism means death: the constant renewal of its material constitution breaks down, the body disintegrates. Van Inwagen, as we have seen, tries to capture this special feature about organisms – that they, as he puts it, can be "composed of different elementary particles at different times" (van Inwagen 1990, p. 6) – with the analogy of the press-ganging or automata club that "persists through all its changes of membership" (van Inwagen 1990, p. 85; see Section 14.2.2). But he does not give us an answer to the question of what it is that makes the club the same club over time. All we are told is, "it is the same club that keeps press-ganging members", which translates to: it is the same life that keeps drawing in particles and suppressing other lives. The diachronic identity of life is presupposed rather than explained, this in accordance with van Inwagen's conviction that any explanation of an object's persistence at some point has to presuppose, at the expense of any further explanation, the persistence of some other object (van Inwagen 1990, p. 145; see Section 14.2.2).[22] From a process perspective, the very idea that the persistence of an object should, or could, be explained in terms of the persistence of some other object is dubious. What should this other object be in the case of the organism? It can't be any of the organism's material components as these are constantly changing (van Inwagen agrees with us on this point); but it can't be "the organism's life" either (we insist) because "the organism's life" is no different from the organism itself. An organism *is* a life; it is a *living process*. Therefore, what van Inwagen should have said about the press-ganging club's persistence is that it *consists* in the process of changing members just as the organism's persistence *consists* (among other things) in metabolism. He didn't say this – possibly for at least one legitimate reason: when it comes to persistence, the analogy between clubs and organisms breaks down because clubs can perfectly well persist without changing their members while organisms cannot persist without changing their material constitution. In other words, clubs are things,[23] organisms are processes.

Processual animalism understands biological identity not in terms of the brute fact of some other thing's identity but in terms of the stabilisation of biological processes. More precisely, biological individuals are understood to be those orchestrated complexes of processes that stabilise themselves at various levels of complexity through continuous interaction with surrounding processes, namely in such a way that processes occurring at lower levels influence and structure processes occurring at higher levels (upward causation) and *vice versa*, processes occurring at higher levels influence and structure processes occurring at lower levels (downward causation). The concept of interactive self-stabilisation assumed by processual animalism thus entails a commitment to what van Inwagen (1990, p. 90) calls holism: "the thesis that the properties of organisms are not wholly determined by, do not wholly supervene upon, the properties of their parts". What parts a biological individual has at a given time depends on the overall process of interactive self-stabilisation, which – qua being interactive – is co-shaped by the environment. Living systems, maintaining a fragile far-from-equilibrium stability, are thermodynamically open systems: they depend for their survival on environmental resources of matter and energy. Therefore – contrary to van Inwagen's (ibid.) surmise – "to reckon the properties of [these] wholes from the complete truth about the intrinsic properties of and the relations that hold among the parts that compose the wholes" is impossible in principle. This insight is already inherent in the autopoiesis theory of life (Maturana and Varela 1980)[24] and even more prominent in the recent theory of biological autonomy (Moreno and Mossio 2015), which both endorse a process view of life (Meincke 2019a).

Organisms persist through time as long as stabilisation and functional integration can be maintained. The conditions of process orchestration and stabilisation can be scientifically investigated. Favourable conditions can be distinguished from conditions that put living organisation under threat. Far from defying an informative analysis, the existence and persistence conditions of lives is a heavily researched subject of a whole bunch of empirical sciences, including the medical sciences. Surely, biological identity, reconceived in terms of process stabilisation, is weaker than the identity of a substance in that it is not a logical or metaphysical presupposition but rather what results from constant efforts. Biological identity is gradual, hard-won and vulnerable to disturbance. Yet, it is a robust phenomenon: it is stronger than mere continuity.[25] In the biological domain, there is continuity all over the place. However, this does not mean that biological reality is an amorphous lump. What enables us to identify certain complexes of continuous processes as organisms is the organisation of these processes, reaching a certain level of stability and functional integration of a certain kind.[26] This organisation can be described by biology as part of a scientific investigation into the mechanisms of organic stabilisation and destabilisation.

A processual understanding of biological identity along these lines gives us a handle on biological branching cases, such as twinning in mammals. This includes acknowledging that twinning doesn't occur in a void but as part of a process of gestation, also known as pregnancy.[27] The process account of mammalian pregnancy which I suggest in Meincke (under review) conceptualises a pregnant organism as an asymmetrically bifurcating hypercomplex process, i.e., as a higher-level process that incorporates and actively maintains an asymmetrical dynamical internal relation between organised complexes of lower-level processes, namely through processes of both mutual stabilisation and successive disentanglement. Twin pregnancies, then, are to be understood as cases in which the ontologically dependent lower-level part of the bifurcating hypercomplex process bifurcates itself so that two fetal lower-level processes jointly participate in the overall higher-level process of bifurcation. As already indicated in Section 14.2.2, mammalian pregnancy impressively illustrates the graded character of biological individuality. The point is that this holds true for singleton and twin pregnancies alike. There is thus nothing special about twinning when it comes to the lack of neat boundaries between biological individuals both synchronically and diachronically.[28] Just as little as we can tell if a mammal pregnant with a single offspring is one or two individuals can we tell if a mammal pregnant with monozygotic twins is one or three individuals. The most plausible answer in both cases is: neither the one nor the other but something in between.[29] This entails that, though the origins of any given mammal in either case can be traced back to a particular hypercomplex process, there is no answer to the question of when *exactly* a mammalian organism comes into existence. But only from a thing ontological viewpoint could this appear to be a problem. *Things* either exist or they don't; whereas for *processes* it is fine – and normal – to emerge gradually and to fade away gradually. No need to postulate that you start to exist in a certain determinate moment around 14 days after conception; no need to project the logics of numerical identity onto biological reality.[30]

As I have argued in more detail elsewhere (Meincke 2019c), a process account of biological identity can serve as an ontological basis for a theory of our identity through time that takes into account the specifically human and, that is, the personal aspects of this identity. Processual animalism thus opposes the common separation between biological identity and personal identity supported by both animalists and mentalists.[31] Our identity through time is both biological *and* personal, with the personal dimension being based on the biological dimension. This entails the claim that what is special about us – what makes us persons – includes, but is not limited to, mental states and capacities. These are found in animals too. Indeed, there are good reasons to agree with enactivism (Maturana and Varela 1980, Varela, Thompson and Rosch 1991, Varela 1997, Thompson 2007, Bich and Moreno 2016) that cognition is an essential dimension of the functioning of most, or even all,[32] living systems. Any biological identity condition will have a grain of "psychology" in it,[33] be it only for its agentive character.[34] When it comes to the identity of human persons specifically, we should, as I have suggested, look not only at the social, historical and narrative elements of human life but take particularly seriously the fact that persons have – or rather *are* – personalities. Personal identity most distinctively consists in a process of developing and maintaining a personality through interaction with other persons (see also Meincke 2016). Personal identity, just like biological identity, is the result of interactive self-stabilisation.[35]

14.4 Conclusions

In this chapter, I have argued that animalism remains unsatisfactory as it stands because it has not yet provided a notion of biological identity that is not only sufficiently elaborate but also in line with biology. Both criteria of biological identity discussed suffer from this deficiency in characteristic ways. First, mere biological continuity, as required for biological identity according to the Biological Continuity Criterion, is unable to handle cases of branching, which – unlike the parallel branching scenarios faced by the Psychological Continuity Criterion of personal identity – are real and ubiquitous in biology. The Fourteen Days Claim, animalists' response to the branching problem, either entails an implausible denial of biological continuity even when no branching occurs, as in the case of the non-twinning developing zygote, or requires degrading biological continuity to a merely necessary, but not sufficient, condition of biological identity. Second, accounting for biological identity in terms of the sameness of life, as attempted by the Life Criterion, suffers from a circularity that cannot be declared to be innocuous by reference to the supposedly well-individuated character of biological life. As is to be learnt from the on-going debate on biological individuality and identity in the philosophy of biology, biological life simply is not well-individuated. Biological identity has rather fuzzy boundaries and comes in degrees.

I have further argued that the return of the dilemma of personal identity in the form of an analogous animalist dilemma of biological identity reveals animalism's commitment to thing ontology and thereby the dilemma's inevitability. If our identity through time – and biological identity in general – is to be accounted for as an identity of things in terms of the identity of other things, it seems that the only way of avoiding giving up on a robust notion of biological identity in the face of the all-pervasiveness of biological continuity[36] is to resort to an unexplained primitive identity, such as to the identity and supposedly unifying, identity-creating power of biological life. However, I have emphasised that recent developments in biological research provide compelling reasons to dismiss thing ontology in favour of process ontology. Processual animalism, accordingly, holds that organisms are processes rather than things, and reconceives biological identity in terms of interactive process self-stabilisation and gradual functional integration. Biological identity, as understood by processual animalism, is weaker than substance identity but still robust enough, and amenable to scientific investigation. Such a processual and scientifically informed notion of biological identity then not only resolves the difficulties standard animalism has with branching cases but also facilitates a more comprehensive account of our identity through time including its specifically personal aspects – an account that neither eliminates nor mystifies personal identity.

Acknowledgements

This chapter is based on a talk first given in June 2016 at the Institute of Philosophy, London, UK, at the conference "Biological Identity" which I co-organised with John Dupré. I am grateful to the Institute of Philosophy for funding the conference with their annual conference grant as well as to the British Society of Philosophy of Science and the Mind Association for additional funding. Different versions of the talk were also delivered in July 2017 at the Biennial Conference of the International Society for the History, Philosophy and Social Studies of Biology (ISHPSSB) at the University of São Paulo, Brazil, and in September 2017 at the conference "The Vagueness of the Embryo: Phenomena of Indeterminacy in Bioscience", funded by the German Federal Ministry of Education and Research (BMBF), at the Technical University of Dresden, Germany. I wish to thank the organisers for inviting me and the audiences for stimulating discussions. The ideas presented in this chapter were first developed while I was working on the project "A Process Ontology for Contemporary Biology" (PROBIO) funded by the European Research Council under the European Union's

Seventh Framework Programme (FP7/2007-2013, grant agreement number 324186, PI: John Dupré). The chapter was completed while I was working on the project "Better Understanding the Metaphysics of Pregnancy" (BUMP) under the European Union's Horizon 2020 research and innovation programme (grant agreement number 679586, PI: Elselijn Kingma). I am grateful to John Dupré, Eric Olson and Menno Lievers as well as to Elselijn Kingma and the BUMP group, especially Teresa Baron and Alexander Geddes, for helpful comments on earlier versions of this chapter.

Notes

1 The point of this story is the transferral of the prince's consciousness rather than of his soul, given that, according to Locke, sameness of the soul is neither necessary nor sufficient for sameness of consciousness and, thus, for personal identity.
2 Unlike his neo-Lockean contemporaries, Bernard Williams argued that bodily identity, or spatio-temporal continuity, is a necessary condition of personal identity (see Williams 1956/57). However, Williams also thought spatio-temporal temporal continuity to be sufficient for personal identity only together with psychological continuity. See Meincke (2015a), chapter 2.3.1(a) for a detailed discussion of Williams's so-called bodily criterion of personal identity, as well as Meincke (2010) for a discussion of the relationship between the bodily criterion and animalism.
3 Olson (2015) argues that the view called "animalism" is restricted to the claim that we are animals. This is to insist that animalism as such *does not imply* that the conditions of our identity through time are biological. According to Olson, the latter can be inferred only from the conjunction of animalism – or "weak animalism" – with independent claims about the nature of animals, such as the claim that animals have purely biological identity conditions. Olson (2015) calls the conjunction of (weak) animalism with independent claims about animals "strong animalism" or, in the particular case of identity through time, the "biological approach" (Olson 1997, pp. 16ff.). In what follows, I will ignore this distinction and use the term "animalism" in the sense of what Olson calls "strong animalism" or the "biological approach", which (as Olson (2015, p. 99ff.) himself seems to acknowledge) is the interesting and distinctive position competing with the psychological approach to an explanation of our identity through time. Additionally, for reasons that I do not have space here to expound, I doubt that a coherent and convincing definition of animalism as such can be given that does not *entail* the strong thesis that we are animals *fundamentally* or *essentially*; something Olson rejects (Olson 2015, especially pp. 94–97).
4 Wiggins (1980; see also 2001) takes our identity to be dependent on both biological and psychological conditions, while Olson (1997) insists on purely biological identity conditions. Olson (1997, p. 20f.) therefore doubts that Wiggins is rightly categorised as an "animalist".
5 Olson (1997, pp. 42–46) argues that whole-brain transplants, by moving the brain stem from one place to another, thereby also move the animal, whose life is regulated by the brain stem, from one place to the other; whereas transplants of the cerebrum only, by leaving the brain stem behind, thereby also leave the animal behind. See, however, note 18.
6 This terminology goes back to Derek Parfit, e.g., (1982).

7   As van Inwagen (1990, chapter 10, pp. 104ff.) explains, "bliger" is a made-up word for "black tiger", an animal spotted from the distance by the inhabitants of the imaginary land Pluralia which, however, later is revealed by zoologists to be a temporary assemblage of six animals of three different species. The "story of the bligers" is meant to show that just as little as those six animals compose a single thing called "bliger" so do elementary particles not compose a single thing such as a chair.
8   Van Inwagen (1990, p. 154) acknowledges that there may be "no perfectly sharp answer to the question when a life begins" (which leads him to the claim that, as a result of life possibly being vague, composition and existence can be vague too; see van Inwagen 1990, chapters 18 and 19); but he is confident that there is at least "a moderately sharp answer".
9   According to van Inwagen (1990, p. 152f.), expectant mothers would even have to accept that they initially do conceive an individual organism through copulation – the zygote that results from a fusion of egg and sperm – but that this individual organism is destroyed once the first mitotic division has occurred so that the womb instead becomes populated by a rapidly growing collection of cells (or rather simples) which fail to compose an individual organism. In other words, if van Inwagen is right, every pregnancy involves the death of the original child, thus resulting (if all goes well) in the birth of a child numerically different from the one conceived. I doubt that van Inwagen's (1990, p. 153) assurance that the "lost child", i.e., the zygote, despite being "an organism, and hence a real object", is "just a cell" will put pregnant women's minds at ease.
10 Most animalists subscribe to this, with the exception of Snowdon (2014) and (this volume).
11 The Special Composition Thesis is meant to be an answer to the Special Composition Question: "When is it true that $\exists y$ the xs compose y? [...] Less formally, in what circumstances do things add up to or compose something? When does unity arise out of plurality?" (van Inwagen 1990, p. 30f.). The answer it provides – sometimes dubbed "organicism" – is a version of compositional restrictivism which is opposed to both compositional universalism (composition always happens) and compositional nihilism (composition never happens).
12 Without mentioning it explicitly, van Inwagen refers here to metabolism as the basic process that constitutes and maintains an organism. As indicated by the modified version of the analogy, metabolism involves catabolic and anabolic processes: incoming compounds are broken down and new compounds are built. It is unclear how the concept of metabolism can be squared with van Inwagen's Special Composition Thesis. Surely, the chocolate bar I'm eating while writing this note is not alive. But if so and, that is, if the chocolate bar actually doesn't exist, how can I digest it? How can my metabolism break down the sugars contained in it? The existence of metabolism, especially of the catabolic reactions it involves, seems to presuppose the existence of non-living material composite objects (with respect to the anabolic reactions we may say that the new compounds these build are alive by virtue of being part of a living organism).
13 Other cases discussed by van Inwagen are cell division, as occurring in amoebae (see 1990, p. 150f., and the discussion above in Section 14.2.1), and metamorphosis (1990, p. 155).
14 Interestingly enough, Swinburne (1984) makes the attempt to explain the identity of souls in terms of the continuity of immaterial "soul stuff"; see my critical discussion in Meincke (2015a, p. 92ff.).
15 The concepts of "individuality" and "identity" as used in the debate in philosophy of biology overlap to some extent. However, "individuality" emphasises the synchronic aspect of identity ("How many xs are at place l at time t?") in contrast

with the diachronic aspect ("Are x, identified at t, and y, identified at t^*, numerically the same individual?"), which is comprised by "identity".

16 As I'm going to explain, I don't think they can.
17 Compare also Olson's (1997, p. 137) military analogy: "Like an army or a totalitarian state, a life imposes 'total obedience' upon the materials whose activities constitute it".
18 The same holds for a third animalist identity criterion, proposed by Olson (1997): the Brain Stem Criterion, according to which biological identity depends on the continuous functioning of the organism's brain stem. This criterion reflects Olson's desire to say something more informative about what constitutes the identity of lives (Olson 2016); however, Olson has meanwhile abandoned the Brain Stem Criterion due to its empirically inappropriate presumption that the brain stem is the central organ that maintains life (ibid.). See Meincke (2015a, p. 137ff.) for a critical discussion of the Brain Stem Criterion.
19 He even declares that the picture of the living organism he presents, "stripped of its atomism, [...] would be Aristotle's picture" (van Inwagen 1990, p. 92). But this is not true given that he rejects holism (see van Inwagen 1990, p. 90, and the discussion below), something which is essential to Aristotle's hylomorphist concept of a substance: according to Aristotle, a living substance (an organism), like every substance, consists of matter and form, with the form (the soul) trans-forming matter such that a living substance, even if there were such things as atoms, is never composed of atoms but rather of functional parts.
20 "Most fundamentally" means that everything else that exists is ontologically dependent on things. This typically includes the thesis that process and change are parasitic upon things – if they actually exist at all, see Meincke (2019b).
21 Bear in mind the technical sense of "thing" introduced, denoting an entity for the identity of which change is not essential. To deny that organisms are things composed of smaller things thus is neither to deny that organisms are material entities nor to rule out the empirical possibility of matter being constituted by ultimate elementary particles ("atoms"). The point rather is that organisms only exist by virtue of change; neither at a time nor through time do they exist independently of whether or not they change, as I'm going to explain in a moment. This entails a holistic concept of "composition", as I'm going to explain too, but see already note 19.
22 Oddly enough, life, though ontologically categorised by van Inwagen as an event, is made to play the role of an object here. Treating events like things is typical of thing ontological reasoning.
23 This is not meant to exclude the possibility that, in a more fine-grained ontological analysis, clubs would come out as processes too (indeed, this is what a process ontologist would expect). However, clubs would still need to be considered a different kind of process then, with a different structure and different unifying mechanisms.
24 The neologism "autopoiesis" translates to "self-construction" and is meant to capture both the productive and self-recursive character of the organisation of living systems. See Meincke (2019a) for a detailed discussion.
25 Organisms, to put it in metaphysical terms, are processes that qualify as "continuants", i.e., entities that maintain their identity while – and through – changing. The claim that some processes, such as organisms, are continuants challenges the long-standing consensus among metaphysicians that only substances are, and can be, continuants. It is connected to the equally revolutionary claim that the mode of persistence of organisms, qua processual continuants, is correctly described by neither of the two main theoretical approaches to the problem of persistence, perdurance and endurance, to the extent that these approaches are

inherently committed to a thing ontological view of reality. For a critical discussion of current accounts of persistence, see Meincke (2019b).
26 For an attempt to account for the diachronic identity of organisms in terms of organisational continuity, see DiFrisco and Mossio (this volume).
27 Not by accident have metaphysicians of two millennia largely ignored this fact. Their astounding ignorance corresponds with the prevalence of what Kingma (2018, 2019, forthcoming, this volume) has aptly called the Containment View of mammalian pregnancy, according to which the fetus resides in the maternal organism like in a container. On this view, the womb literally appears as a void easy to abstract away from when looking at fetal development. Furthermore, the prevalence of the Containment View has itself been crucially facilitated by the traditional hegemony of thing ontology and is its paradigmatic expression: fetus and maternal organism are conceptualised as things – neatly bounded, ontologically independent concrete particulars with primitive identities – one being contained in the other.
28 It is worth remembering in this context that the development of all placental mammals involves the differentiation of the blastula into the embryoblast (inner cell mass), from which the embryo develops, on the one hand, and the trophoblast (outer cell mass), which gives rise to the placenta, on the other. In this sense even singleton (placental) mammalian pregnancies pose a "branching problem".
29 Note that this does not imply that a mammal pregnant with twins is two individuals. Hypercomplex processes are subject to a principle of non-additivity: a mammal pregnant with a singleton is something in between one and two, a mammal pregnant with twins is something in between one and three, and so on.
30 Another version of such a projection would be the claim that (monozygotic) twin pregnancies are to be analysed within a four-dimensionalist framework as cases of overlap or cohabitation. To be clear, the processual concept of biological identity proposed here is strongly opposed to four-dimensionalism. To claim that twins share their earliest temporal parts with one another is to claim that there are two individuals from the start. In contrast, my view is that twinning is a process of bifurcation comprised in a larger, hypercomplex process of bifurcation, i.e., that there is a transition from one to two processes as part of an overall transition from one hypercomplex process to three processes.
31 Animalists may want to object that, by accounting for *our* identity, they *are* delivering a theory of personal identity. Very well; but let's not forget that animalists like Olson argue that our identity, ontologically speaking, consists in "some brute-physical continuity" (Olson 2015, e.g., p. 85), such as biological continuity, while at the same time claiming that psychological continuity both grounds personal identity in a practico-ethical sense (Olson (1997, chapter 3.VII) calls the latter the "same person as" relation) *and* may go separate ways from biological continuity, e.g., in brain transplant scenarios.
32 That all living systems, including those without a nervous system, are cognitive systems was Maturana and Varela's radical claim. This is echoed in recent research in plant cognition (Calvo 2016; see the discussion in Meincke 2019d).
33 This entails rejecting the assumption common among animalists that fetuses and people in a persistent vegetative state do not have any psychological states (Olson 1997, p. 24 and 73ff.). Latest research on fetal psychology (Reissland and Hopkins 2010, Gliga and Alderdice 2015) and on PVS (McCullagh 2004, Shea and Bayne 2010, Liberati, Hünefeldt and Olivetti Belardinelli 2014) provides empirical support for this rejection.
34 According to Moreno and Etxeberria (2005), metabolism is a minimal form of agency. On the recently emerging debate about bio-agency in philosophy of biology, see Meincke (2018a).

35 The processes of biological and personal interactive self-stabilisation overlap during the early phase of what I call "postnatal pregnancy": the continuous ontological entanglement between mother and child, especially in the form of physiological and immunological entanglement through breastfeeding (see Meincke under review).

36 Note that continuity, in the thing ontological picture, remains ontologically superficial, namely secondary to discrete things such as atoms.

References

Arias, A. M., Nichols, J. and Schröter, C. (2013) 'A Molecular Basis for Developmental Plasticity in Early Mammalian Embryos', *Development* 140, pp. 3499–3510, doi:10.1242/dev.091959.

Bich, L. and Moreno, A. (2016) 'The Role of Regulation in the Origin and Synthetic Modelling of Minimal Cognition', *BioSystem* 148, pp. 12–21.

Bickhard, M. (2011) 'Systems and Process Metaphysics', in: Hooker, C. (Ed.), *Handbook of Philosophy of Science. Philosophy of Complex Systems*, Vol. 10, Amsterdam: Elsevier, pp. 91–104.

Blatti, S. and Snowdon P. F. (Eds.) (2016) *Animalism. New Essays on Persons, Animals, and Identity*, Oxford: Oxford University Press.

Bouchard, F. and Huneman, P. (Eds.) (2013) *From Groups to Individuals: Evolution and Emerging Individuality*, (*The Vienna Series in Theoretical Biology*), Cambridge, MA: MIT Press.

Calvo, P. (2016) 'The Philosophy of Plant Neurobiology: A Manifesto', *Synthese* 193(5), pp. 1323–1343.

Chiu, L. and Eberl, G. (2016) 'Microorganisms as Scaffolds of Biological Individuality: An Eco-immunity Account of the Holobiont', *Biology and Philosophy* 31, pp. 819–837.

Clarke, E. (2010) 'The Problem of Biological Individuality', *Biological Theory* 5, pp. 312–325.

Clarke, E. (2013) 'The Multiple Realizability of Biological Individuals', *The Journal of Philosophy* 110(8), pp. 413–435.

DiFrisco, J. and Mossio, M. (this volume) 'Diachronic Identity in Complex Life Cycles: An Organizational Perspective', in: Meincke, A. S. and Dupré, J. (Eds.), *Biological Identity. Perspectives from Metaphysics and the Philosophy of Biology*, (*History and Philosophy of Biology*) London: Routledge, pp. 177–199.

Dupré, J. (2012) *Processes of Life: Essays in the Philosophy of Biology*, Oxford: Oxford University Press.

Dupré, J. (2014a) 'Animalism and the Persistence of Human Organisms', *The Southern Journal of Philosophy* 52, Spindel Supplement, pp. 6–23.

Dupré, J. (2014b) 'The Role of Behaviour in the Recurrence of Biological Processes', *Biological Journal of the Linnean Society* 112, pp. 306–314, doi:10.1111/bij.12106.

Dupré, J. (2017) 'The Metaphysics of Evolution', *Interface Focus*, Published online, August 18, 2017, http://rsfs.royalsocietypublishing.org/content/7/5/20160148\.

Dupré, J. (this volume) 'Processes Within Processes: A Dynamic Account of Living Beings and its Implications for Understanding the Human Individual', in: Meincke, A. S. and Dupré, J. (Eds.), *Biological Identity. Perspectives from Metaphysics and the Philosophy of Biology*, (*History and Philosophy of Biology*), London: Routledge, pp. 149–166.

Dupré, J. and O'Malley, M.A. (2009) 'Varieties of Living Things: Life at the Intersection of Lineage and Metabolism', *Philosophy and Theory in Biology*, http://hdl.handle.net/2027/spo.6959004.0001.003.
Gilbert, S. F., Sapp, J. and Tauber, A. I. (2012) 'A Symbiotic View of Life: We have Never Been Individuals', *The Quarterly Review of Biology* 87(4), pp. 325–341.
Gilbert, S. F. and Tauber, A.I. (2016) 'Rethinking Individuality. The Dialectics of the Holobiont', *Biology and Philosophy* 31(6), pp. 839–853.
Gliga, T. and Alderdice, F. (2015) 'New Frontiers in Fetal and Infant Psychology', *Journal of Reproductive and Infant Psychology* 33(5), pp. 445–447.
Godfrey-Smith, P. (2009) *Darwinian Populations and Natural Selection*, Oxford: Oxford University Press.
Haber, M. H. (2013) 'Colonies are Individuals: Revisiting the Superorganism Revival', in: Bouchard, F. and Huneman, P. (Eds.), *From Groups to Individuals: Evolution and Emerging Individuality*, (*The Vienna Series in Theoretical Biology*), Cambridge, MA: MIT Press, pp. 195–218.
Hölldobler, B. and Wilson, E. O. (2009) *The Superorganism: The Beauty, Elegance, and Strangeness of Insect Societies*, London: W. W. Norton.
Huneman, P. (2014) 'Individuality as a Theoretical Scheme. II. About the Weak Individuality of Organisms and Ecosystems', *Biological Theory* 9(4), 374–381.
Huneman, P. (this volume) 'Biological Individuals as 'Weak individuals' and Their Identity: Exploring a Radical Hypothesis in the Metaphysics of Science', in: Meincke, A. S. and Dupré, J. (Eds.), *Biological Identity. Perspectives from Metaphysics and the Philosophy of Biology*, (*History and Philosophy of Biology*), London: Routledge, pp. 40–62.
Jaeger, J., Irons, D. and Monk, N. (2008) 'Regulative Feedback in Pattern Formation: Towards a General Relativistic Theory of Positional Information', *Development* 135, pp. 3175–3183.
Jaeger, J. and Monk, N. (2015) 'Everything Flows: A Process Perspective on Life', *EMBO Reports* 16(9), pp. 1064–1067.
Jaeger, J. and Reinitz, J. (2006) 'On the Dynamic Nature of Positional Information', *BioEssays* 28, pp. 1102–1111.
Kingma, E. (2018) 'Lady Parts: The Metaphysics of Pregnancy', *Royal Institute of Philosophy Supplement* 82, pp. 165–187.
Kingma, E. (2019) 'Were You a Part of Your Mother?', *Mind* 128(511), pp. 609–646, doi:10.1093/mind/fzy087.
Kingma, E. (forthcoming) 'Nine Months'. *Journal of Medicine and Philosophy*.
Kingma, E. (this volume) 'Pregnancy and Biological Identity', in: Meincke, A. S. and Dupré, J. (Eds.), *Biological Identity. Perspectives from Metaphysics and the Philosophy of Biology*, (*History and Philosophy of Biology*), London: Routledge, pp. 200–213.
Liberati, G., Hünefeldt, T. and Olivetti Belardinelli, M. (2014) 'Questioning the Dichotomy between Vegetative State and Minimally Conscious State: A Review of the Statistical Evidence', *Frontiers in Human Neuroscience* 8, Article 865, doi:10.3389/fnhum.2014.00865.
Lidgard, S. and Nyhart, L. K. (Eds.) (2017) *Biological Individuality: Integrating Scientific, Philosophical, and Historical Perspectives*, Chicago, IL and London: The University of Chicago Press.
Locke, J. (1987) *An Essay Concerning Human Understanding*, Nidditch, P. (Ed.), (*The Clarendon Edition of the Works of John Locke* 7), Oxford: Clarendon Press.

Margulis, L. (1991) 'Symbiogenesis and Symbionticism', in: Margulis, L. and Fester, R. (Eds.), *Symbiosis as a Source of Evolutionary Innovation: Speciation and Morphogenesis*, Cambridge, MA: The MIT Press, pp. 1–14.

Maturana, H. and Varela, F. (1980) *Autopoiesis and Cognition. The Realization of the Living*, Dordrecht: Reidel Publishing.

McCullagh, P. (2004) *Conscious in a Vegetative State? A Critique of the PVS Concept*, New York: Kluwer.

McFall-Ngai, M. J. (2015) 'Giving Microbes their Due – Animal Life in a Microbially Dominant World', *The Journal of Experimental Biology* 218, pp. 1968–1973.

Meincke, A. S. (2010) 'Körper oder Organismus? Eric T. Olsons Cartesianismusvorwurf gegen das Körperkriterium transtemporaler personaler Identität', *Philosophisches Jahrbuch* 117, pp. 88–120.

Meincke, A. S. (2015a) *Auf dem Kampfplatz der Metaphysik. Kritische Studien zur transtemporalen Identität von Personen*, Münster: Mentis.

Meincke, A. S. (2015b) 'Potentialität und Disposition in der Diskussion über den Status des menschlichen Embryos: Zur Ontologie des Potentialitätsarguments', *Philosophisches Jahrbuch* 122, pp. 271–303.

Meincke, A. S. (2016) 'Personale Identität ohne Persönlichkeit? Anmerkungen zu einem vernachlässigten Zusammenhang', *Philosophisches Jahrbuch* 123, pp. 114–145.

Meincke, A. S. (2018a) 'Bio-Agency and the Possibility of Artificial Agents', in: Christian, A., Hommen, D., Retzlaff, N. and Schurz, G. (Eds.), *Philosophy of Science. Between the Natural Sciences, the Social Sciences, and the Humanities*, (European Studies in Philosophy of Science 9), Cham: Springer, pp. 65–93.

Meincke, A. S. (2018b) 'Haben menschliche Embryonen eine Disposition zur Personalität?', in: Rothhaar, M., Hähnel, M. and Kipke, R. (Eds.), *Der manipulierbare Embryo. Potentialitäts- und Speziesargumente auf dem Prüfstand*, Münster: Mentis, pp. 147–171.

Meincke, A. S. (2018c) 'Persons as Biological Processes. A Bio-Processual Way Out of the Personal Identity Dilemma', in: Nicholson, D. J. and Dupré, J. (Eds.), *Everything Flows. Towards a Processual Philosophy of Biology*, Oxford: Oxford University Press, pp. 357–378.

Meincke, A. S. (2019a) 'Autopoiesis, Biological Autonomy and the Process View of Life', *European Journal for Philosophy of Science* 9:5, doi:10.1007/s13194-018-0228-2.

Meincke, A. S. (2019b) 'The Disappearance of Change. Towards a Process Account of Persistence', *International Journal of Philosophical Studies* 27(1), pp. 12–30, doi:10.1080/09672559.2018.1548634.

Meincke, A. S. (2019c) 'Human Persons – A Process View', in: Noller, J. (Ed.), *Was sind und wie existieren Personen?* Münster: Mentis, pp. 57–80.

Meincke, A. S. (2019d) 'Review of: Chauncey Maher: Plant Minds, London: Routledge, 2017', *British Journal for the Philosophy of Science: Review of Books*, www.thebsps.org/2019/02/chaunceymahers-plantminds-meincke/.

Meincke, A. S. (under review) 'One or Two? A Process View of Pregnancy'.

Moreno, A. (this volume) 'The Role of Individuality in the Origin of Life', in: Meincke, A. S. and Dupré, J. (Eds.), *Biological Identity. Perspectives from Metaphysics and the Philosophy of Biology*, (History and Philosophy of Biology), London: Routledge, pp. 86–106.

Moreno, A. and Etxeberria, A. (2005) 'Agency in Natural and Artificial Systems', *Artificial Life* 11, pp. 161–175.

Moreno, A. and Mossio, M. (2015) *Biological Autonomy. A Philosophical and Theoretical Enquiry*, (*History and Theory of the Life Sciences* 12), Dordrecht: Springer.
Nicholson, D. J. and Dupré, J. (Eds.) (2018) *Everything Flows. Towards a Processual Philosophy of Biology*, Oxford: Oxford University Press.
Oderberg, D. (this volume) 'Siphonophores: A Metaphysical Case Study', in: Meincke, A. S. and Dupré, J. (Eds.), *Biological Identity. Perspectives from Metaphysics and the Philosophy of Biology*, (*History and Philosophy of Biology*), London: Routledge, pp. 22–39.
Olson, E. T. (1997) *The Human Animal. Personal Identity Without Psychology*, Oxford: Oxford University Press.
Olson, E. T. (2007) *What Are We? A Study in Person Ontology*, Oxford: Oxford University Press.
Olson, E. T. (2015) 'What Does it Mean to Say That We Are Animals?', *Journal of Consciousness Studies* 22(11–12), pp. 84–107.
Olson, E. T. (2016) 'The Role of the Brainstem in Personal Identity', in: Blank, A. (Ed.), *Animals: New Essays*, München: Philosophia, pp. 291–302.
Parfit, D. (1982) 'Personal Identity and Rationality', *Synthese* 53(2), pp. 227–241.
Parfit, D. (1987) *Reasons and Persons*, 3rd ed., Oxford: Oxford University Press.
Pradeu, T. (2012) *The Limits of the Self: Immunology and Biological Identity*, New York: Oxford University Press.
Queller, D. C. and Strassmann, J. E. (2016) 'Problems of Multi-species Organisms: Endosymbionts to Holobionts', *Biology and Philosophy* 31(6), pp. 855–873.
Reissland, N. and Hopkins, B. (Eds.) (2010) *Special Issue: Towards a Fetal Psychology, Infant and Child Development* 19(1), https://onlinelibrary.wiley.com/toc/15227219/19/1
Rosenberg, E., Koren, O., Reshef, L., Efrony, R. and Zilber-Rosenberg, I. (2007) 'The Role of Microorganisms in Coral Health, Disease and Evolution', *Nature Reviews Microbiology* 5, pp. 355–362.
Rosenberg, E. and Zilber-Rosenberg, I. (2018) 'The Hologenome Concept of Evolution After 10 Years', *Microbiome* 6:78, doi:10.1186/s40168-018-0457-9.
Sagan (Margulis), Lynn (1967) 'On the Origin of Mitosing Cells', *Journal of Theoretical Biology* 14(3), pp. 225–274.
Shea, N. and Bayne, T. (2010) 'The Vegetative State and the Science of Consciousness', *British Journal for the Philosophy of Science* 61(3), pp. 459–484.
Shoemaker, S. (1970) 'Persons and their Pasts', *American Philosophical Quarterly* 7, pp. 269–285.
Simons, P. (2018) 'Processes and Precipitates', in: Nicholson, D. J. and Dupré, J. (Eds.), *Everything Flows. Towards a Processual Philosophy of Biology*, Oxford: Oxford University Press, pp. 49–60.
Skillings, D. (2016) 'Holobionts and the Ecology of Organisms: Multi-species Communities or Integrated Individuals?', *Biology and Philosophy* 31, pp. 875–892.
Snowdon, P. (2014) *Persons, Animals, Ourselves*, Oxford: Oxford University Press.
Snowdon, P. (this volume) 'The Nature of Persons and the Nature of Animals', in: Meincke, A. S. and Dupré, J. (Eds.), *Biological Identity. Perspectives from Metaphysics and the Philosophy of Biology*, (*History and Philosophy of Biology*), London: Routledge, pp. 233–250.
Swinburne, R. (1984) 'Personal Identity. The Dualist Theory', in: Shoemaker, S. and Swinburne, R., *Personal Identity*, (*Great Debates in Philosophy*), Oxford: Basil Blackwell, pp. 1–66.

Thompson, E. (2007) *Mind in Life: Biology, Phenomenology, and the Sciences of Mind*, Cambridge, MA: The Belknap Press of Harvard University Press.
Van Inwagen, P. (1990) *Material Beings*, Ithaca, NY, and London: Cornell University Press.
Varela, F. J. (1997) 'Patterns of Life: Intertwining Identity and Cognition', *Brain and Cognition* 34, pp. 72–87.
Varela, F. J., Thompson, E., and Rosch, E. (1991) *The Embodied Mind: Cognitive Science and Human Experience*, Cambridge, MA: The MIT Press.
Wheeler, W. M. (1911) 'The Ant-colony as an Organism', *Journal of Morphology* 22(2), pp. 307–325.
Wiggins, D. (1980) *Sameness and Substance, (Library of Philosophy and Logic)*, Oxford: Blackwell.
Wiggins, D. (2001) *Sameness and Substance Renewed*, Cambridge: Cambridge University Press.
Wiggins, D. (2016) 'Activity, Process, Continuant, Substance, Organism', *Philosophy* 91, pp. 269–280 (reprinted as Chapter 9 in this volume).
Williams, B. A. O. (1956/57) 'Personal Identity and Individuation', *Proceedings of the Aristotelian Society* 57, pp. 229–252.
Wilson, D. S. and Sober, E. (1989) 'Reviving the Superorganism', *Journal of Theoretical Biology* 136, pp. 337–356.
Wolpert, L. (1969) 'Positional Information', *Journal of Theoretical Biology* 25, 1–47.
Wolpert, L. (2011) 'Positional Information and Patterning Revisited', *Journal of Theoretical Biology* 269, pp. 359–365.
Zilber-Rosenberg, I. and Rosenberg, E. (2008) 'Role of Microorganisms in the Evolution of Animals and Plants: the Hologenome Theory of Evolution', *FEMS Microbiology Reviews* 32, pp. 723–735.

Index

Note: Page numbers followed by "n" denote endnotes.

activity 12, 59n8, 76f., 89f., 92–95, 97, 101, 103n8, 110–112, 114, 119–122, 135, 137, 142, 149–152f., 156f., 160, 167, 169, 174f., 175f.n7, 176n9, 180, 185f., 194f., 196n4, 208f., 217, 225, 253, 258, 260, 267, 272n17
  principle of, 12f., 121, 169, 174–176, 185f., 215; *see also* Wiggins, David
agency, agents, 49, 94–97, 100f., 102n8, 103n12, 220, 222–227, 273
action, 14, 44, 95, 97–99, 119, 174, 188, 220–223, 225f.
altruism, 45, 47
amoeba, 73f., 75, 79, 193, 255–257
animalism, xvi, 1f., 14f., 155, 162, 163n8, 200, 233–239, 247–249, 251–278
animals, 2, 11, 14, 24, 26, 28–31, 33f., 37, 43, 63–65, 77, 118, 128–148, 157, 161f., 173, 200, 210, 233–250, 251–255, 257, 261, 263, 268, 270n3n5, 271n7
Anjum, Rani L., 15n6, 221
Anscombe, Gertrude E. M., 200
ants, 26, 31, 45, 47, 51, 65, 154–156, 158, 163nn5–7, 262
Applebaum, Barbara, 227, 230
Aristotle, 2, 4, 6, 9, 11, 15n1, 23, 25, 42, 107, 109, 118–123, 123n7, 159, 161, 176n7, 215f., 221f., 228, 264f., 272n19; *see also* metaphysics – (neo-) Aristotelian
artefact, 1f., 68f., 184, 196n4, 229n1
asymmetric dependence, 190, 192, 194
autocatalysis, 88f.
autonomy, 11, 13, 45, 78, 87, 93–96, 99–101, 102f.n8, 110f., 118, 124n11, 140f., 152f., 157, 207f., 226f., 266f.
autopoiesis, autopoietic system, 10f., 89–99, 101, 102n4, 266, 272n24

atomism, 14, 225, 230, 264f., 272n19
atoms, 2, 64, 71f., 74–81, 83, 130, 151, 168, 175, 217, 220, 230, 265, 272n19n21, 274n36

bacteria, 23, 33, 47, 48, 64f., 77, 96f., 99, 103nn13–14, 154f., 163n7
Bechtel, William, 98, 114
bees, 22, 25f., 31, 45–47, 53, 262
Benhabib, Seyla, 226
bifurcation, 12, 128, 161, 267, 273n30
biofilms, 99f., 103n13, 103n15, 163n2
biogenesis, 10f., 88, 90, 92, 94, 101
biological identity
  Biological Continuity Criterion of, 14, 253–257, 268
  Brain Stem Criterion of, 272n18
  vs. biological individuality, 15n2, 271f.n15
  Life Criterion of, 14, 257–260, 263, 268
  metaphysical theory of, xvi, 4
  realism about, 4f.
birth, 13, 68, 161, 187, 193, 196n5, 203, 205, 209f.
blackberry plants, 12, 173f.
Bostrom, Nick, 160, 163n13
boundaries (spatio- and/or temporal), 3, 14, 22, 25, 64–70, 74, 78f., 81–83, 153f., 156, 177–179, 181–191, 193, 195, 196n7, 202, 204–207, 223–225, 267, 269
branching problem, 1, 14f., 252, 254, 257, 259, 267–269, 273n28; *see also* fission
Brogaard, Berit, 200–207, 209f., 212n17
budding, 9, 29, 35f., 189f., 257, 261
Burtt, Edwin A., 224
Butler, Judith, 214f., 223, 226f.

Campbell, Tim, 67, 238
capacities, 10f., 42, 51, 78, 86, 94f., 97–101, 102n5, 108–113, 115, 117, 119–122, 130, 132, 136, 139, 142, 158, 178, 181, 187, 230n12, 235, 246, 253f., 268; *see also* causal powers *and* potentiality
Cartwright, Nancy, 112f.
causal powers, 16n11, 37n, 112–117, 119–121
Cazzolla Gatti, Roberto, 96, 102n
change (over time), 1, 31, 74, 109, 112f., 119, 121, 128, 140, 150, 153, 161f., 167, 175n3, 178f., 181–186, 188, 194, 202f., 209, 216f., 241f., 255, 258, 265, 272nn20–21
chemistry, 88, 218
chimerism, chimera, 136f., 257
Chiu, Lynn, 3, 97, 262
Clarke, Ellen, 3, 23, 40, 44, 63f., 78, 82n6, 86, 177f., 194, 211n1, 260
classification, 75f., 194f.
  metaphysical, 25, 28
cnidarians, 26, 31, 33, 132f., 135f.; *see also* corals, siphonophores *and* Portuguese man-of-war
cohesiveness, 101, 171, 176n9, 216, 220
collaboration, 49, 99, 155, 210; *see also* cooperation
colonial organisms, 9, 25–37, 98–9, 173, 261, 262; *see also* cnidarians
colony, 25–37, 45, 47, 63–65, 69, 103n13, 154f., 156, 173, 261f.
commensalism, 48, 155, 192
complexity, 34, 90, 92, 97, 101, 102n1, 149, 155, 266
complexification, 11, 88, 92f.
composition (of material objects),2, 6f., 16n9, 23, 28, 37n7, 65f., 69–81, 82n3, 83n10n13, 87f., 91, 99, 102n3, 109, 111–114, 116f., 123n3, 134, 136, 173, 181, 236, 253, 256, 258, 260, 262, 265f., 271nn7–9nn11–12, 272n19n21; *see also* material constitution or constituents
  problem, 72, 74f.
  restricted, compositional restrictivism, 271n11
  unrestricted, compositional universalism, 10, 75–78, 83n14, 271n11
  compositional nihilism, 271n11
conceptual scheme, 40f., 215

conflict, 46f., 52f., 100, 208, 212n16, 261
conjoined twins; *see* twinning – conjoined
consciousness, 67, 220, 222, 225, 235, 251, 270n1; *see also* self-consciousness
constraints, 13, 91, 93, 180–184, 187f., 194, 197n12
  closure of, 13, 110f., 121f., 178, 180–188, 192, 194f., 196n3
continuant, 12, 14, 156, 163n10, 167–176, 185, 216f., 219f., 222f., 229nn3–4, 230n11, 242–244, 247, 272n25
continuity, 1, 13, 153, 160, 194, 252, 264, 267, 274n36
  biological, 14, 253–257, 259, 264, 267–269, 273n31
  causal, 13, 150, 184f., 194f.
  organisational, 13, 178f., 181, 183–188, 194, 196n6, 273n26
  physical, 34, 273n31
  psychological, 82n8, 252, 254, 264, 270n2, 273n31
  of soul stuff, 271n14
  spatiotemporal, 13, 79, 178f., 183, 185, 196n1, 270n2
  topological, 13, 209
  transtemporal, 209
cooperation, 3, 11, 46–48, 52f., 88, 91, 99f., 103n15, 155, 157, 163n9, 178, 261f.; *see also* collaboration
corals, 22, 26, 32, 46, 53, 135, 261
countability, 22, 37n1, 42, 63

dandelions, 35, 36, 40, 59n7
Darwin, Charles, 141, 157
Darwinian individuals, 57
Darwinism, 143
Dawkins, Richard, 155, 159
death, 68, 77, 97, 156, 159f., 181, 187, 196n5, 200, 206, 245, 265, 271n9
Democritus, 265
development, 9, 13, 34–36, 131f., 134–144, 154, 160, 177–182, 186f., 190f., 193–195, 200f., 254, 255–257, 273n8
Devitt, Michael, 129, 143
DiFrisco, James, 12f., 86, 178, 186, 194, 196n3, 273n26
division of labour, 49, 156, 158, 262
DNA, 22, 70f., 97, 102n5, 139, 190
downward causation, 115, 266
Dupré, John, xvi, 3f., 12, 15n5, 37n1, 49, 57, 71, 82n1, 83n17, 108, 123n3n8, 129, 142f., 149, 151f., 155, 162, 163nn1–3nn9–10, 167f., 173,

175, 210, 214–220, 227f., 229n2n4, 230n10n15n17, 231n23, 249n12, 261–263, 265
dynamical systems, 3, 135

earthworms, 59n8, 157
Eberl, Gerard, 97, 262
ecology, 15n4, 50–57, 58n1, 59nn8–9, 86, 99, 102n6, 190, 194f., 265
ecological interdependence, 218
ecological niche, 181
ecological systems or networks, ecosystems, 50, 64, 87, 93f., 103n14, 184f.
 individuals as, 10f., 41, 47–58, 63
emergence, 107–111, 115–118, 121f., 124nn20–21
enactivism, 268
encapsulation, 10, 88–92, 95
endosymbiosis, 48, 50, 103n16, 163n7, 261
 Endosymbiotic Theory, 33, 261
endurance, endurantism, 151, 230n17, 236, 241, 272n25
energeia, 119, 176n7, 185
energy, 3, 87–91, 94, 97f., 100, 109f., 122, 135, 142, 145n1, 149–151, 153, 180, 182, 265f.
entropy, 110, 152, 180
enzymes, 132, 151, 180f.
epigenetics, 3, 140, 256
essences, essential properties, 2, 11, 31, 129, 140, 150, 159f., 170, 175n7, 217, 242
essentialism, 5, 129, 143, 228, 242
eukaryotes, 33, 45, 48, 87, 128f., 181, 186
events, 12, 14, 37n1, 153, 167, 170, 175, 177–179, 186–190, 192f., 200, 206, 210, 216f., 220, 229n4, 255, 258–260, 263f., 272n22
evolution, 3, 10f., 32–34, 108, 128–145, 149f., 181f., 194, 208
 prebiotic, 10f., 87–96, 100
evolutionary biology or theory, 11, 15n4, 41, 44f., 123n1, 128f., 158
evolutionary transition, 45–47, 133, 135
 egalitarian vs. fraternal, 47f.
evolutionary synthesis; *see* modern (evolutionary) synthesis
extended evolutionary synthesis, 3, 15n4

family, 158
feminism, 161, 224

Ferner, Adam, 13f., 164n17, 219, 229n1n8, 230n9n18
fetus, 13, 59n5, 156, 161, 201f., 205, 208f., 211, 212n13n16n18, 247, 254, 262f., 267, 273n27n33
fission, 13, 71f., 177f., 187–194, 196nn9–10, 200, 238, 255–257, 264; *see also* branching problem
fitness, 44, 96, 100, 143, 177f., 208, 212n16
four-dimensionalism, 13f., 195, 221, 230n17, 241, 273n30; *see also* perdurance, perdurantism *and* temporal parts
function, functionality, 6, 9, 24–27, 28–32, 32–34, 36, 37nn1–2, 48–51, 59n6n8, 66, 86–88, 90f., 93–98, 102n5n8, 103n12, 107, 110, 114, 120f., 131f., 135f., 140, 142, 151–154, 156, 158, 160, 163n4, 178, 180–186, 188, 190, 192–194, 197n12, 205f., 208, 253, 256, 272n19
functional integration, 10, 13, 76–79, 94, 101, 197n12, 208, 267, 269
fusion, 13, 50, 59n5, 70, 130, 177f., 187f., 190–195, 197nn11–13, 200, 257, 264, 271n9

gastrula, 136, 254, 256
gastrulation, 133f., 136f., 200, 202f., 211n10
gene, 3, 43, 49f., 58n3n5, 63, 71, 78, 96, 99, 129, 131f., 135–137, 141–143, 159, 163n12, 262
 regulatory networks, 137, 141, 144, 256
 selfish, 155, 159
genidentity, 12f., 169f., 184, 194, 219, 230n16, 249n11
genome, 3, 10, 48, 50, 54, 91, 102n3, 128, 132, 135f., 142, 152, 163n3, 208, 262
genotype, 59n7, 70–75, 79, 99, 138, 173
Gilbert, Scott F., 3, 15n4, 47, 49, 87, 186, 262
Godfrey-Smith, Peter, 3, 15n7, 45f., 57, 63, 71, 82n4n6, 86, 196n7, 260
Goodnight, Charles, 44, 54
Gould, Stephen J., 27f., 30, 32–37
Griesemer, James R., 98, 189
growth, 9, 28, 34f., 71, 89, 91, 173, 177, 190, 210, 256
Guay, Alexandre, 3, 8, 15n3n7, 16n13, 167, 169–173, 175n3n5, 175f.n7,

176n8, 184f., 191, 194, 229n5, 230n16, 249n11
Guerrero, Ricardo, 94

Haack, Susan, 229n7, 230n12
Haber, Matt, 3, 15n7, 16n9, 40, 59n42, 62
Haeckel, Ernst, 27f., 34, 261
Hankinson, Robert J., 120
Haraway, Donna, 225f.
Haslanger, Sally, 214, 223–225, 230n20, 231n24
Hayles, N. Katherine, 223–226, 231n22
health, 141f., 150, 153, 160
Hennig, Willi, 43
Heraclitus, 168, 228
Herron, Matthew D., 40, 207
holism, 266, 272n19n21
holobiont, xvi, 3, 152f., 262
holozoan, 129f., 133, 135
homeostasis, 13, 149, 153, 160, 207f., 211n9
homeorhesis, 149, 153, 160
Hull, David L., 15n7, 44–49, 54, 57f., 83n19, 184f., 191f., 194f.
humans, 12, 14, 32, 35, 40, 46, 48, 53, 59n5, 141f., 153–162, 172f., 194, 200–203, 209, 211n4, 214, 219f., 224–229, 239, 251–257, 260, 263f., 268
Huneman, Philippe, 3, 9f., 15n4, 45, 50, 54f., 108, 116, 260f.
hylomorphism, 2, 11, 107–127, 272n19
hymenoptera; *see* ants

identity through time, diachronic identity, 1–3, 9, 12–15, 15n1, 42, 169, 177–199, 210, 251f., 263–269, 271f. n15, 273n26; *see also* persistence
immortality, 12, 159–161, 163n12
immune system, immunology, 11, 13, 50, 56, 64, 76f., 86, 96–99, 101, 103n11, 171, 209, 262, 274n35
indiscernibility of identicals, 172, 176n10, 237
individualism, 158, 162, 163n9
   promiscuous, 4, 155, 156, 158
individuation, xvi, 4, 14, 42, 54, 56, 86f., 90–94, 97f., 100, 103n10, 112, 117, 129, 137, 157f., 161, 168, 170, 177f., 181, 185f., 195, 196n4, 259f., 263–265, 269
indivisibility, 43, 175, 259
inheritance, 3, 11, 91–93, 100, 131, 140, 163n7
   statistical, 102n3

integration, 192–194, 208, 262
   bioelectrical, 138, 140, 144
   causal, 66
   functional; *see* functional integration
   metabolic, 208
   organisational, 10, 90

Janzen, Daniel, 40, 59n7, 83n10, 210
Johnson, William E., 167, 175n4
Johnston, Mark, 118, 123n4
Jonas, Hans, 110f.

Kant, Immanuel, 11, 111, 129, 144
Keller, Evelyn Fox, 92
Kim, Jaegwon, 108, 115–117, 124n21
Kingma, Elselijn, 12f., 161, 194, 201, 207, 209, 211nn5–6, 212n15n18, 263, 273n27

Ladyman, James, 6, 8f., 41
Leibniz, Gottfried W., 43, 237
Lennox, James G., 121
Lewin, Kurt, 184
Lewis, David K., 56, 83nn12–13n20
life; *see also* biological identity – Life Criterion of
   individuality or unity of, 2, 10, 14, 259–263, 269
   diachronic identity of, 266
   human, 156, 159, 161, 268
   process view of, 267
life cycle, xvi, 12f., 29, 50, 78, 149, 153f., 160, 171, 173, 177f., 181, 184, 186–188, 190, 192, 195, 208, 218, 262
life-form, 43, 222
life sciences, 63, 87, 233, 239
lineages, 48f., 87, 102n6, 128, 149, 155, 164n16
liquid tissues, 11, 130–144, 145n1
Locke, John, 68, 150, 235, 248n5, 251, 259f., 270n1
Lockeanism, 68f., 73, 82n8, 235–237, 248nn5–6, 270n2
Lovibond, Sabina, 226
Lowe, E. Jonathan, 2, 6, 8, 15f.n8, 22, 24, 41–44, 46f., 179, 185, 215

machines, 114, 159f., 225, 235
Macpherson, Crawford B., 225
Margulis (Sagan), Lynn, 3, 33, 48, 152, 261f
Marmodoro, Anna, 119–121, 123n4, 124n25

material constitution or constituents, xvf., 2, 4, 6f., 109f., 112, 122, 124n24, 187, 241, 265f.; *see also* composition
materialism, 11, 107–127, 196n3, 231n23
matter, 2f., 10f., 22, 66–69, 76, 78, 80, 87, 90, 94, 100, 107–112, 114, 117–122, 124n26, 131–133, 137, 142, 149, 153f., 168f., 180, 185, 194, 236, 241f., 245, 258, 265f., 272n19n21
Maturana, Humberto R., 89, 153, 266, 268, 273n32
McMahan, Jeff, 67, 238
McWhorter, Ladelle, 227
mechanistic explanation or modelling, 112–115, 151
medical science, medicine, 150, 159, 229, 234, 238, 267
Mei, Tsu-Lin, 224
Meincke, Anne Sophie, xvi, 4, 12, 14f., 108, 118, 153, 161f., 163n8n10nn14–15, 200, 210, 211n2, 212n19, 214, 227, 229n2n4, 230n17, 251f., 256, 263–265, 267f., 270n2, 271n14, 272n18n20n24, 273n25n32n34, 274n35
membranes, 88, 96, 137, 145n1, 180, 202, 205
memory, 29, 49, 59n6, 235, 248n5, 251; *see also* personal identity – memory criterion of
mereology, 6, 16n9, 70, 80, 118, 120, 202, 251
metabolism, 13, 57, 63, 76f., 88f., 91–97, 99f., 103n10, 110, 142, 149, 151–153, 155, 162, 171, 180, 182, 189, 191, 194, 205f., 208, 218, 253, 256f., 262, 265f., 271n12, 273n34
metamorphosis, 177f., 181f., 184–187, 194, 271n13
metaphysics, xvi, 1–21, 22–39, 40–62, 64, 69f., 72, 74f., 76, 80–82, 108, 112f., 115–117, 143, 152, 177, 214–232, 234, 247, 248n3, 264f., 267, 272n25, 273n26
    analytic, xvi, 1, 6, 16n8, 214, 227f., 251f.
    descriptive, 14, 40f., 43, 46, 52, 215, 217–220, 223f., 227f., 229n4, 230n9
    (neo-) Aristotelian, xvi, 2, 5f., 9–11, 14, 16n8, 22–27, 31, 37, 37f.n8, 108f., 118–123, 185, 202, 215f., 221–226, 229nn3–4, 230n10, 264f., 272n19
    of processes; *see* process ontology or metaphysics

revisionary, 52, 218–220, 229n4, 230n19
scholastic, 5, 170
scientific, scientifically informed or naturalised, 8, 41, 48, 54, 58, 218
of science, 6–9, 16nn11–12, 40f., 167, 218
of substances; *see* substance ontology or metaphysics
of things; *see* thing ontology or metaphysics
metazoan, 32, 37n7, 43, 45, 48, 50, 55, 57, 59n6, 129–136, 138, 140, 142, 144
microbes, 10, 152–154, 169, 261f.
microbiome, microbiota, 48, 50, 87, 210
Mill, John S., 113, 124n16
mind, 14, 67, 218–220, 222, 224
    too-many-minds problem, 14, 239
mitochondria, 33, 46, 48, 53
modern (evolutionary) synthesis, 108, 128f., 131, 143; *see also* evolution
monogenomic differentiated cell lineage, 152, 154f.; *see also* multicellularity
Montévil, Maël, 110, 122, 180
morals, 158, 211n5, 252
moral responsibility, 9, 13, 214–232
morphogenesis, 131–142, 196n2
morphology, 9, 33f., 36, 129–134, 136–144, 181, 185, 206
Moreno, Alvaro, 10, 12, 89, 91, 94, 102n1, 103n8n12, 108, 110, 132, 180, 261, 267f., 273n34
Moss, Lenny, 129, 135, 144
Mossio, Matteo, 12f., 108, 110, 122, 180f., 183, 196n3n6, 267, 273n26
multicellularity, 33, 43–45, 47f., 50, 58n2, 71f., 87, 128, 130, 135–139, 142–144, 152, 154, 185, 187, 189f., 192, 203, 208, 255f., 261f.
Mumford, Stephen, 8, 16n11, 221
mutualism, 47–51, 53, 155, 192, 262

natural kinds, 11, 129, 133, 140, 143f., 185, 233, 240
natural selection, 44f., 48f., 52–54, 57, 92f., 102n1, 134, 142f., 149, 152, 157, 178, 196n7; *see also* evolution
    community, 54
    group, 262
    holobiont, 3
    kin, 45, 47
    levels of, 45, 210
    multilevel, 3, 45, 47

superorganism, 3, 262
units of, 3, 44, 48f.
network, 10f., 55, 87–92, 94, 98, 100, 102n3n7, 158; *see also* gene – regulatory network
Newman, Stuart A., 11f., 33, 129–144
niche, 50f., 59n9, 143, 181, 203–208
niche construction, 3, 50, 53, 59n8, 94, 157
Nicholson, Daniel J., 4, 108, 114, 123n2n8, 124n9, 153, 163n1n10, 215, 217–219, 229n2, 230n15, 231n23, 265

Oderberg, David, 2, 5, 9f., 24f., 29, 118f., 123n4, 124n26, 200, 211n10, 212n13, 261
Olson, Eric T., 2, 10, 14, 67, 69, 82n7, 83n14, 200, 249n6n9, 251–258, 261, 263–265, 270nn3–5, 272nn17–18, 273n31n33
O'Malley, Maureen, 3, 49, 57, 82n1, 83n17, 155, 210, 261
organ, 23, 25f., 28–31, 33–36, 37n2, 131, 134, 136, 149, 151f., 203, 225
organisation, 7, 10–13, 25, 45, 87–101, 102n1n3n7, 102f.n8, 130, 177–199, 208, 262f., 267, 272n24, 273n26
levels of, xiii, 15n5
organismality, 13, 46, 57, 88, 111, 129, 131f., 142–144
ostension, 42, 241
Owen, Richard, 141

parasitism, 49f., 95f., 153, 155, 191f.
Parfit, Derek, 248n5, 252, 270n6
parthood, parts, 6f., 9–11, 13, 16n9, 22–26, 29–32, 34, 37nn2–3, 45, 48, 50, 54f., 64–66, 70–81, 82nn4–5, 83n10n12n14, 95, 98, 101, 110f., 113–118, 120–122, 124n27, 153–156, 161, 164n16, 167, 169f., 173f., 175n7, 176n9, 177, 179–184, 186–189, 191–194, 201–204, 206–210, 212nn17–18, 216f., 220–223, 228, 238, 243f., 246, 257, 266f., 271n12, 273n30; *see also* temporal parts
functional, 25, 91, 103n12, 184, 272n19; *see also* function
Pepper, John W., 40, 207
perdurance, perdurantism, 214, 230n17, 241, 272n25; *see also* four-dimensionalism *and* temporal parts
persistence, xiii, 1, 9, 12–14, 42f., 51, 58, 74f., 77, 81, 92, 109, 121, 142, 150, 156f., 159f., 167, 169, 175n2, 178f., 181, 184–186, 189–194, 196n1, 197n11, 202, 210, 215, 220, 223f., 230n17n19, 233, 236f., 242, 244–247, 248n5, 248f.n6, 251–253, 257–260, 265–267, 272f.n25; *see also* identity through time, diachronic identity
persistent vegetative state (PVS), 237, 254, 273n33
person, personhood, 1, 14f., 27f., 30, 32, 34f., 67–69, 73, 82n8, 155, 158, 160, 162, 200f., 214, 220, 223, 233–250, 264, 268
personal identity, xvf., 1f., 4, 9, 12, 14f., 67, 82n8, 162, 177, 200f., 216, 233, 236f., 249n9, 251–278
animalist or biological criteria of; *see* animalism *and* biological identity
bodily criterion of, 270n2
Complex View of, 252, 254
memory criterion of, 235, 248n5, 251
Psychological Theory or Psychological Continuity Criterion of, 248n5, 251, 254; *see also* continuity – psychological
Simple View of, 252, 259, 264
phenotype, 59n7, 93, 99, 136, 138, 140f.
physics, 12, 40f., 56, 110, 112, 129f., 137f., 144, 218
physiology, 13, 28f., 34, 49f., 57, 86, 99, 179, 181, 185, 194, 206–208, 211n4, 262, 274n35
placenta, 48, 59n5, 193–195, 200f., 204–206, 209, 211nn3–4, 212n13, 247, 273n28
plant cognition, 273n32
plasmid, 49, 87, 96, 99
pluralism, 4, 27, 36, 57f., 102n6, 173, 207, 217
Portuguese man-of-war, 27, 29, 46, 53, 58n2, 63, 69, 261; *see also* siphonophores
postmodernism, 223–226
potentiality, 2, 119f., 135; *see also* causal powers *and* capacities
Pradeu, Thomas, 3, 8, 15n3n7, 16n13, 48, 76, 79, 82n6, 83n17, 86, 97, 167, 169–173, 175n3n5, 175f.n7, 176n8, 184f., 191, 194, 209, 229n5, 230n16, 249n11, 262
pregnancy, 12f., 15, 159, 161f., 194, 200–213, 247, 254–256, 262f., 267f., 271n9, 273nn27–30, 274n35
principle of activity; *see* activity – principle of

processes
  biological, 3, 169, 178, 187, 195, 219, 253, 266
  as continuants, 163n10, 229nn3–4, 230n11, 272n25
  causal, 13, 186
  developmental, 9, 36, 135, 139, 178, 191, 193–195, 255
  hierarchy of, 149–151, 228
  living, 151, 161, 266
  metabolic, 88, 149, 151f., 162, 180, 257
  intertwined, 155, 157
  vs. events, 217, 229n4
  vs. (material) objects, substances or things, 150–152, 159, 167f., 215–217, 236, 239–247, 249n13, 265, 268
  organisms as, 149–166, 251–278
  reproductive, 36, 89, 178, 186, 190
process ontology or metaphysics, xvi, 4, 12, 14, 37n1, 108, 123n2, 150f., 159, 161f., 164n17, 167, 210, 217f., 221, 230n17, 265, 269, 272n23
prokaryote, 33, 87, 96f., 103n9, 103n9n16, 128f.
proto-species, 11, 94, 99, 102n6
psychological individuality, 67f.; *see also* personal identity – Psychological Theory or Psychological Continuity Criterion of
Putnam, Hilary, 124n14, 174

Queller, David C., 3, 46f., 52f., 178, 210, 261
Quine, William v. O., 41, 76, 83n16

regeneration, 43, 137f.
reidentification, 42f., 120f., 124n27, 158, 168, 172, 177
replication, 11, 95f., 102n1, 196n8
reproduction, 9, 13, 27–31, 35f., 45, 49, 63f., 71f., 78, 87, 89–92, 100, 102nn3–4, 103n10, 173, 177f., 181–195, 196n7n8, 197n11, 201, 208, 210
  asexual, 31, 36, 71, 89, 136, 174, 190, 196n9, 210, 257
  sexual, 31, 36, 136, 174, 192–194, 197n11n13, 200, 254f., 257
Rescher, Nicholas, 151
RNA, 92f., 95, 97
  world, 92, 95, 97, 102n5
Ross, Don, 6, 8f., 41

Sapp, Jan, 3, 87, 262
Segré, Daniel, 91, 102n3

Seibt, Johanna, 124n20, 186, 217, 219f., 227, 230n10n15
self-consciousness, 28, 68f., 235, 248n5; *see also* consciousness
selfishness, 45, 48, 158; *see also* gene – selfish
self-organisation, xiii, 108, 131, 135, 144
semaphoront, 43
Shirt-Ediss, Ben, 88
Shoemaker, Sydney, 82n9, 83n12, 161, 248n5, 254
Sidzinska, Maja, 161, 163n15
Simons, Peter, 6f., 152, 170, 265
siphonophores, 9, 12, 22–39, 49, 173f.; *see also* Portuguese man-of-war
slime moulds, 12, 47, 49, 173, 192
Smith, Adam, 156, 163n9
Smith, Barry, 200–207, 209f., 211n11, 212n14n17
Snowdon, Paul, 2, 12, 14, 163n8, 200, 248n4, 249n7n9n12, 251, 263, 271n10
sortalism, 172f., 186
sortals, 42, 47, 174, 179, 185f.
soul, 120, 159, 161, 251, 259, 264f., 270n1, 271n14, 272n19
space of (biological) individuality, 45–47, 49, 51–53, 56, 262
Special Composition Thesis, 258, 260, 271nn11–12; *see also* van Inwagen, Peter
species, 11, 22, 29, 34, 37n8, 42f., 49, 51, 59n9, 99f., 102n6, 129, 137, 139–141, 143, 164n16
  as individuals, 15n7, 16n9, 86
stability, stabilisation, 49f., 92, 99, 109, 138–140, 144, 149, 151–153, 156f., 160, 162, 167, 169, 266–269
  interactive self-stabilisation, 266, 268f., 274n35
Steward, Helen, 215–217, 229n4
Stout, Rowland, 216f., 229n4
Strassmann, Joan E., 3, 46f., 52f., 178, 210, 261
Strawson, Peter F., 40, 42, 179, 218f., 221, 224, 226f., 229n7, 230n12
Strevens, Michael, 113f., 124n18
substance, 2, 9, 11–14, 15n1, 23–25, 109, 112, 116–122, 123n7, 124n21n27, 142, 150, 152, 159, 161f., 164n16, 167, 169, 173–175, 175n1n3, 185f., 196n4, 202f., 209–211, 212n17, 214–226, 228, 229n3, 230n11, 240, 248n1, 253, 264f., 267, 269, 272n19n25

substance ontology or metaphysics, 4f., 9, 12–14, 179, 185, 201–207, 209f., 212n18, 215, 217, 221, 224, 229n4, 230n10n13, 253, 264f.; *see also* thing ontology or metaphysics
substantialism, 169f., 175f.n7, 184–186, 194f., 216f., 224
  anti-, 169, 230n16, 249n11
superorganism, 32, 34, 52, 65, 155, 156, 262
Swinburne, Richard, 259, 271n14
symbiosis, symbionts, 3, 10, 12, 23, 33, 47–49, 64, 77, 87, 96, 98–100, 103n11, 152–156, 169, 177f., 185, 187, 210, 261f.; *see also* endosymbiosis
systems biology, 3, 15n5, 265

Tauber, Alfred I., 3, 87, 209, 262
taxonomy, 27, 29, 37n8, 136, 139, 143, 178
temporal parts, 10, 13f., 74–78, 81, 83n12n14, 114, 186, 216, 220–223, 227, 229n2, 273n30; *see also* four-dimensionalism *and* perdurance, perdurantism
thermodynamics, 13, 110, 122, 153, 180, 266
  far from equilibrium or disequilibrium, 12, 109f., 153, 156, 180, 266
thing ontology or metaphysics, 14, 150, 156, 170, 230n10, 253, 265, 268f., 272n22, 272f.n25, 272n27, 274n36; *see also* substance ontology or metaphysics
Thompson, Michael, 215, 222
time scales, 149–151, 180, 262
transhumanism, 160
transition; *see* evolutionary transition
Trembley, Abraham, 43
twinning (monozygotic), 14, 161, 163n14, 172, 200, 211n2, 254–257, 259, 267f., 273nn29–30
  conjoined, 14, 67, 69, 238, 249n8, 260

van Inwagen, Peter, 2, 14, 70, 82n7, 83nn11–15n21, 124n19, 230n14, 251, 253–261, 263–266, 271nn7–9nn11–13, 272n19n22
Varela, Francisco J., 89, 153, 266, 268, 273n32
variability, 11, 13, 59n6, 138, 140, 178
variation, 129, 140, 195, 211n4
  epigenetic, 140f.
  functional, 182
  genetic, 129, 140
vesicles, 88, 90f., 131
virus, 48f., 59n5, 87, 95, 155, 184
vitalism, 107, 110, 123n2
von Bertalanffy, Ludwig, 109f.

Waddington, Conrad H., 109, 140, 149, 160
Walsh, Denis, 5, 11f., 15n4, 52, 108, 115, 123nn1–2, 124n22
weak individuality, 9f., 40–62
Wheeler, William M., 262
Wiebe, Kayla, 11f.
Wiggins, David, 2, 9, 12–14, 172, 175nn6–7, 176n11, 179, 185f., 194, 196n4, 214–222, 224, 226, 229nn3–4n6n8, 230n11n16n19, 243f., 249n11, 251, 254, 270n4
Williams, Bernard, 270n2
Wilson, David S., 3, 45, 65, 154, 262
Wilson, Edward O., 3, 27f., 32–34, 262
Wilson, Jack, 27, 49, 66f., 71, 76, 78f., 82n6, 83, 83n18, 173f.
Winther, Rasmus, 124n15n17

yeast, 189f., 195, 196n9
Young, John Z., 175f.n7, 217

zygote, 135f., 139, 152, 156, 172f., 193f., 200, 211n5, 216, 254f., 257, 259, 268, 271n9

For Product Safety Concerns and Information please contact our EU representative GPSR@taylorandfrancis.com
Taylor & Francis Verlag GmbH, Kaufingerstraße 24, 80331 München, Germany

www.ingramcontent.com/pod-product-compliance
Ingram Content Group UK Ltd.
Pitfield, Milton Keynes, MK11 3LW, UK
UKHW021443080625
459435UK00011B/353